Water in Himalayan Towns: Lessons for Adaptive Water Governance

In Focus – Special Book Series

Water in Himalayan Towns: Lessons for Adaptive Water Governance

Editors
Anjal Prakash and David Molden

IWA PUBLISHING

Published by

IWA Publishing
Alliance House
12 Caxton Street
London SW1H 0QS, UK
Telephone: +44 (0)20 7654 5500
Fax: +44 (0)20 7654 5555
Email: publications@iwap.co.uk
Web: www.iwapublishing.com

First published 2020

British Library Cataloguing in Publication Data
A CIP catalogue record for this book is available from the British Library

ISBN: 9781789061895 (paperback)
ISBN: 9781789061901 (eBook)
ISBN: 9781789061918 (ePUB)

Contents

Editorial
Mapping challenges for adaptive water management in Himalayan towns

Anjal Prakash[a,*] and David Molden[b]

[a]*Bharti Institute of Public Policy, Indian School of Business, Hyderabad, India*
**Corresponding author. E-mail: anjal_prakash@isb.edu*
[b]*International Center for Integrated Mountain Development, Kathmandu, Nepal*

1. Introduction

The Hindu Kush Himalaya (HKH) is characterized by complex topography, climate, hydrology, and hydrogeology. Each of these factors plays an important role in determining the availability of water for people living in the Himalayas (Scott *et al.*, 2019). These physical features are also a constraint to rapid urbanization (Mukherji *et al.*, 2018). Only 3% of the total HKH population live in larger cities and 8% in smaller towns (Basyal & Khanal, 2001). However, urbanization has increased in recent years with people from rural areas flocking to nearby urban centres in search of employment and other economic opportunities (Bajracharya *et al.*, 2019). As a result, the share of urban population is increasing in the region, while that of the rural population is declining. Projections show that by 2050, more than 50% of the population in HKH countries will live in cities (UNDESA, 2014). The unique geophysical situation of mountain cities and towns requires a special approach when conceptualizing water management. Moreover, rapid urbanization has led to new challenges that require a different perspective, a mountain perspective if they are to be solved.

Unplanned urbanization is causing significant changes in land use and land cover and reducing the recharge areas of springs (Jeelani *et al.*, 2017; Rani *et al.*, 2018; Scott *et al.*, 2019; Thakur *et al.*, 2019). Climate-induced changes in the physical environment have resulted in increased rainfall variability and heat stress. Extreme rainfall events are leading to more landslides, while an increase in average temperatures has caused glacier melt and subsequent changes in hydrological regimes in the region (Xu *et al.*, 2009). These critical stressors – climatic and non-climatic – are adversely affecting the socio-ecology of

doi: 10.2166/wp.2020.000

urban agglomerations in the HKH. The encroachment or degradation of natural water bodies and the disappearance of traditional water sources, such as springs, are evident (Sharma *et al.*, 2016). Most towns in the HKH meet their water needs from springs, streams, ponds, and lakes, which are largely interlinked systems, and their degradation and loss is leading to widespread water stress.

The demand for water has increased manifold with growing populations and increased consumption. At the same time, water availability in these towns has been adversely affected by the climatic and socio-economic change. Some towns are also major tourist destinations and host large floating populations during peak tourist season, challenging their carrying capacities (Shah & Kulkarni, 2015; Mondal & Roychowdhury, 2019; Ojha *et al.*, 2019). At such times, residents must cope with water scarcity as demand for water increases and water distribution through the public water supply systems becomes highly inequitable. Inefficient water utilities and governance challenges become much more critical as the sources of water are limited and the local geology hardly supports accessing groundwater unlike in the plains (Dame *et al.*, 2019). All these factors are resulting in increased water insecurity for the poor and marginalized in urban centres across the HKH.

This special issue looks at the challenges of water management in 12 towns from four corners of the Himalayan region. These include, from west to east, Murree and Havelian in Pakistan; Kathmandu, Bharatpur, Tansen, and Damauli in Nepal; Mussoorie, Devprayag, Singtam, Kalimpong, and Darjeeling in India; and Sylhet in Bangladesh (see Figure 1). All these cities face challenges of changing water budgets, increasing demand for water, and water scarcity, which is discussed in the papers. They also map future challenges that these towns will face in a 'business as usual' scenario. These cases draw from primary research and fill an important knowledge gap about the status of water resources and water supply in Himalayan towns. Authored by a multi-disciplinary team consisting of physical and social scientists, anthropologists, geographers, and planners, the papers highlight the concerns

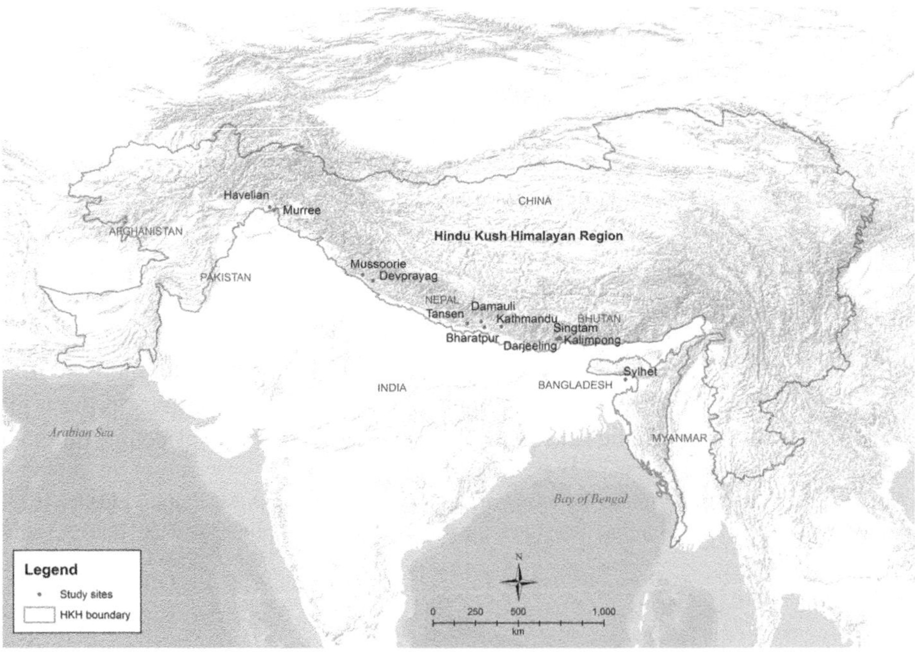

Fig. 1. Case study sites in the HKH region.

around unplanned and haphazard development in the region, which is leading to problems of inequity in water supply and unequal developmental outcomes. They also identify areas for future research and action on urban issues in the region.

2. Key messages from this volume

This special issue adds new findings and reaffirms some of the findings of recent studies on water trends in Himalayan towns. Below, we discuss some of the key concerns and messages emerging from these case studies.

2.1. Urban Himalayas running dry

This volume provides evidence that Himalayan towns are facing increased water insecurity. The review paper by Sreoshi Singh *et al.* (this volume) shows that the interlinkages of water resource availability, resources, water supply systems, rapid urbanization, and consequent increase in water demand (both every day and seasonal) is leading to increased water insecurity in towns in the HKH. The water insecurity is attributed to poor water governance, lack of urban planning, poor tourism management during peak season, and climate-related risks and challenges. Communities are coping via short-term strategies such as groundwater extraction, which is proving to be unsustainable. There is a lack of long-term strategies for water sustainability in urban centres, and this requires special attention by planners and local governments. In a study of two Himalayan towns in Pakistan's Indus River basin, such as Murree and Havelian, Virk *et al.* (this volume) show heavy dependence on groundwater as a primary source of water supply. Estimates of water availability, consumption, and water sufficiency ratio (WSR) reveal that available groundwater is insufficient to meet the requirements for consumption in Havelian. However, in the case of Murree, water availability is sufficient due to recent changes in the infrastructure. There is evidence that water is being mismanaged at the household level in both towns. Factors, such as rapid urbanization and population growth, are will increase the requirements of water in the future.

2.2. Failing governance of water towers

The papers in this volume show that water demand in urban locations is outstripping supply leading to a rise in the privatization of water supply, mostly through water tankers. These private tankers are fulfilling the demand by fetching water from the nearby rural locations, and this sector is largely unregulated. In their paper on understanding water resilience in Himalayan towns, Singh & Pandey (this volume) note that the world's urban population is expected to rise to 68% by 2050 with implications for Himalayan regions as well. Small towns will expand as this growth is expected to be greater in the developing countries. This review paper, based on the number of cases studies, shows that water governance in the HKH region remains a blind spot and challenges pertaining to urban water resilience are poorly understood. Himalayan towns depend on springs that are fed by numerous local aquifers and form the primary means of water supply. The springs also contribute significantly to base flows in Himalayan rivers. Even by gross, conservative estimates, there are more than a million perennial springs in the Indian Himalayan region alone (Shah & Kulkarni, 2015). These springs are under great threat from urban waste and sewage polluting the water bodies and catchments. This volume reaffirms the findings of other studies that have indicated water quality issues in spring sources. Kumar *et al.* (1997) assessed physico-chemical characteristics of water from 12 springs located within the municipal limits of Almora, a central Indian Himalayan town. It indicated a direct influence of unplanned sewage disposal on spring water quality. Bharti *et al.* (this volume) further reemphasize these points in their study of two other towns in the Indian Himalayan region – Mussoorie and Devprayag. They find that in the current scenario of a changing climate, natural springs, their main water resource, are drying up. Mussoorie experiences an acute shortage of water in summer, precisely when the town hosts numerous tourists. In Devprayag, religious tourism and in-migration from rural areas have contributed to rising demand. The reduced discharge in nearby streams has further widened the demand–supply gap.

2.3. Caste inequalities and water rights

The papers in this volume show that understanding issues of gender and caste are important in cities where individual and household water rights are mediated through the politics of water distribution. Addressing water inequities and insecurities demand research as well as development and policy responses, which means going beyond physical aspects of water management and delving into the social science of how water is distributed. Molden *et al.* (this volume) provide new insights on water security from their case of Kathmandu, the capital of Nepal. Their study of 47 household water managers highlights three dimensions of water security: experiences of water security vary greatly between households over the year; social connections and landownership play an important role in mediating these experiences; and coping with poor water supply places a burden on certain household members. Raina *et al.* (this volume) provide evidence of both horizontal and vertical inequity in Kathmandu's informal water market. This study shows that 20% of Kathmandu's poor households do not have access to the formal water supply system, and they end up paying more for accessing water as compared with the richer households who have access. The study also found that the poor spend a greater proportion of their income on meeting their household water needs which means a much greater financial burden. Rich households spend 38.2% less on the water than poor households. Poorer households stand to gain if the water economy shifts from informal to formal.

2.4. Decentralization and powers to local bodies

An important aspect of urban water management is decentralized governance and the devolution of power to local bodies and institutions managing water. A study of Tansen and Damauli, two small towns in Nepal, by Singh *et al.* (this volume) shows that water institutions have played a very significant role in water supply management at the local level and that proper management can help avert critical water shortages. The paper compares two towns that have different outcomes for water supply and access. In Tansen, infrastructural constraints and large-scale corruption in the systems' upkeep and maintenance have implications on water governance, while in Damauli, the systems have been well managed due to the involvement of local communities. Shah & Badiger (this volume) present the case of Darjeeling in the Indian Himalaya where the water crisis is the result of a conundrum due to the interlinked problems of political unwillingness, insufficient investments, failure of cooperation between the state and regional institutions, and inadequacies in local governance, including institutional capacity. Ghatani (2015) finds similar challenges in Darjeeling. The study reveals that water management can be rational only if the institutions responsible for such management are efficient. It shows how community participation in the enforcement of rules and regulations on illegal connections, and the misuse of water and optimization of the budget for public utility services has helped in maintaining the service levels although the challenges of access to water persist. Access is a larger issue of investment in water infrastructure and urban planning processes.

2.5. Looming climate risks

This volume reaffirms some of the studies that have indicated that climate risks, such as floods and landslides, are becoming more frequent, and therefore, they must be considered in the urban planning process. In the study of the Himalayan towns of Leh and Dharamshala in India, Sudhakar (2010) found clear linkages between climate change and water scarcity. Water supply in these towns comes from glacier melt. The author presents evidence of changes in the towns' local climate parameters, such as snowfall, rainfall, and temperature, and the hydrology of water bodies, which contribute to the water availability. These changes have caused an imbalance in the demand and supply of water, as the supply is further strained by changes in climatic conditions. The author notes that the response to this is largely supply-driven while adaptive measures are needed in the overall planning process to secure water for urban populations. Tiwari *et al.* (2018) note that expanding urban areas in the high mountains are now serving as centres of growth by creating opportunities for employment and access to services and infrastructure, thereby contributing to the development of their vast hinterlands through the trickle-down effect. However, rapid and unplanned urbanization have impacted the hydrological regimes of Himalayan watersheds – it has reduced groundwater recharge; decreased the availability of water for drinking, sanitation, and crop production; depleted forests and biodiversity; increased risks of natural hazards and disasters, both in urban as well as peri-urban areas; and increased the water, food, livelihood, and health insecurities of mountain communities. Moreover, climate change has stressed urban ecosystems by increasing the frequency, severity and intensity of extreme weather events. Briscoe *et al.* (2006) note that Pakistan's threatened wetlands were part of the drainage system and had averted floods in many cases. Especially in northern Pakistan, these wetlands were part of the flood control system, reducing the impacts of Indus floods while they accommodated water from Indus and torrential hill streams.

Pervin *et al.* (this volume) show how cities in South Asia are experiencing stormwater drainage problems due to a combination of urban sprawl, structural, hydrological, socio-economic, and climatic factors. The frequency of short duration, high-intensity rainfall is expected to increase in the future due to climate change. Given the limited capacity of drainage systems in South Asian cities, urban flooding and waterlogging are expected to intensify. The problems get worse when low-lying areas are filled up for infrastructure development and unplanned urban growth, reducing permeable areas and infiltration. In a study of Sylhet town in Bangladesh and Bharatpur in Nepal, they find that significant areas of both cities are at risk from flooding. With climatic changes, the threats to these areas will increase. The study suggests a number of measures to reduce long-term flooding risk in these cities.

3. Towards adaptive water management for Himalayan towns

From the case studies in this volume, we find that increasing urbanization and changing climate are two critical stressors that are adversely affecting the biophysical environment of urban areas in the HKH.

With development plans and policies focusing more on rural areas, the degradation of urban environments has not been a matter of concern. Across the region, the encroachment and degradation of natural water bodies (springs, ponds, lakes, canals, and rivers) and the disappearance of traditional water systems (such as stone spouts, wells, and local water tanks) are evident. The degradation and reclamation of water bodies affect wetland ecosystems and reduce retention capacities that prevent flooding. Consequently, urban drainage and flood management systems are impaired.

Rapid urbanization and climate change have short- and long-term implications for biophysical and socio-economic environments. Urban waste-water discharge and solid-waste disposal are polluting surface water bodies in urban and peri-urban areas. Waste-water re-use for agriculture, a common practice in peri-urban areas, often ignores potential health hazards and other adverse effects of polluted water. Groundwater is often contaminated with salts, chemicals, iron, and fluoride and characterized by hardness. The case studies point to declining groundwater reserves due to over-exploitation for domestic, industrial, commercial, and agricultural purposes. Groundwater recharge rates are reduced because of low rainfall and interventions in the recharge areas and processes such as sand mining.

The uniqueness in groundwater conditions in the HKH makes it a specific typology. Aquifers feed the springs which are the main water sources for urban and rural areas in the hills and mountains. Springs also contribute to base flows in the streams and rivers. However, there are challenges concerning contamination due to improper waste and sewage treatment and disposal and some aquifers that are recharged from greater distances are impacted by the sinking of wells. The recharge areas of these springs are not well protected because of the absence of such a component in the planning process. Increasing population pressure is also leading to competing claims over water from common aquifers.

The cases studies on water management in Himalayan towns highlight the need for course correction. Together, they indicate that the demand for water is outstripping supply in almost all the towns and the future looks bleak if the 'business as usual' scenario persists. Estimating and planning according to the seasonal carrying capacity of cities, waste management and springshed management are key to meeting future water demands. There is a range of water governance challenges and issues of equity that must be addressed to enhance water security for all.

We focus on five major points that can help in adapting to the issue of water scarcity and the looming crises induced by climate change in Himalayan towns and cities. *First*, we need to source water sustainably to bridge the gap between supply and demand. This could be done by reviving and protecting springs, increased water harvesting, and having multiple sources of water. More budgetary allocations could be made towards the above, so that water sources are protected and managed sustainably. We also found that in many towns, spring water is the only source and is insufficient to meet the needs across all seasons. Therefore, multiple sources of water could be part of the portfolio of the overall water supply for these towns. *Second*, we need to look at the governance and management of water beyond water utilities. Polycentric governance, meaning multiple governing bodies and institutions interacting with one another towards a coherent goal of providing access to water, could be a more suitable way to comprehend water governance in Himalayan towns and cities. *Third*, the equitable distribution of water needs more attention. We have found that the poor and marginalized are most affected when water supply dwindles. Many cities are faced with the challenge of providing access to safe water for the poor especially in the dry season when supply dwindles. *Fourth*, more appreciation is needed for women's multiple roles in water management and how they could be part of the planning and decision-making processes to provide solutions. *Fifth*, mountain cities need to be viewed in the broader context of mountain water, environment, and energy. Climate change impacts on these sectors are presenting new and growing challenges that Himalayan towns and cities have to grapple with. This volume discusses various choices and options – from demand management to supply enhancement, understanding ecological footprints of towns to managing water at a bioregional scale. In doing so, it is vital to address issues of equity and empower local institutions in managing water. Lastly, the focus for the future must be on building urban resilience by strengthening the adaptive capacities of affected communities while also understanding the limits to adaptation.

References

Bajracharya, R. M., Dahal, N., Neupane, K. R., Singh, V. & Habeeb, R. (2019). Urban water security challenges in the Nepal and Indian Himalaya in the context of climate change. *Resources and Environment* 9(1), 9–18. doi:10.5923/j.re.20190901.02.

Basyal, G. K. & Khanal, N. R. (2001). Process and characteristics of urbanization in Nepal. *Contribution to Nepalese Studies* 28(2), 187–225.

Briscoe, J., Qamar, U., Contijoch, M., Amir, P. & Blackmore, D. (2006). *Pakistan's Water Economy: Running Dry.* Oxford University Press, Karachi.

Dame, J., Schmidt, S., Müller, J. & Nüsser, M. (2019). Urbanisation and socio-ecological challenges in high mountain towns: insights from Leh (Ladakh), India. *Landscape and Urban Planning 189*, 189–199.

Ghatani, S. (2015). *Sustainable Urban Water Management in Darjeeling*. Doctoral Dissertation, Sikkim University. Available at: http://14.139.206.50:8080/jspui/bitstream/1/3112/1/Suvechha20Ghatani20(Geography).pdf (Accessed 23 November 2019).

Jeelani, G., Shah, R. A., Deshpande, R. D., Fryar, A. E., Perrin, J. & Mukherjee, A. (2017). Distinguishing and estimating recharge to karst springs in snow and glacier dominated mountainous basins of the western Himalaya, India. *Journal of Hydrology 550*, 239–252.

Kumar, K., Rawat, D. S. & Joshi, R. (1997). Chemistry of springwater in Almora, Central Himalaya, India. *Environmental Geology 31*(3–4), 150–156.

Mondal, T. K. & Roychowdhury, P. (2019). Water Scarcity in Himalayan Hill Town: A Study of Darjeeling Municipality, India. In: Ray B. & Shaw R. (eds) *Urban Drought. Disaster Risk Reduction (Methods, Approaches and Practices)*. Springer, Singapore, pp. 363–383.

Mukherji, A., Scott, C., Molden, D. & Maharjan, A. (2018). Megatrends in Hindu Kush Himalaya: Climate Change, Urbanisation and Migration and Their Implications for Water, Energy and Food. In: Biswas, A., Tortajada, C. & Rohner, P. (eds) *Assessing Global Water Megatrends. Water Resources Development and Management*. Springer, Singapore, pp. 125–146.

Ojha, H., Kovacs, E., Devkota, K., Neupane, K. R., Dahal, N. & Vira, B. (2019). Local experts as the champions of water security in the Nepalese town of Dhulikhel. *New Angle: Nepal Journal of Social Science and Public Policy* 5(1), 1–14.

Rani, M., Joshi, H., Kumar, K. & Tiwari, A. (2018). Recharge potential mapping in complex hydrological system of Kosi Basin. *Applications and Challenges of Geospatial Technology: Potential and Future Trends 9*.

Rani, M., Joshi, H., Kumar, K. & Tiwari, A. (2019). Recharge Potential Mapping in Complex Hydrological System of Kosi Basin in the Mid-Himalayan Region. In: *Applications and Challenges of Geospatial Technology*. Springer, Cham, pp. 9–23.

Scott, C. A., Zhang, F., Mukherji, A., Immerzeel, W., Mustafa, D. & Bharati, L. (2019). Water in the Hindu Kush Himalaya. In: Wester, P., Mishra, A., Mukherji, A. & Shrestha, A. (eds) *The Hindu Kush Himalaya Assessment*. Springer, Cham, pp. 257–299.

Shah, M. & Kulkarni, H. (2015). Urban water systems in India. *Economic and Political Weekly 50*(30), 57–69.

Sharma, B., Nepal, S., Gyawali, D., Pokharel, G. S., Wahid, S. M., Mukherji, A., Acharya, S. & Shrestha, A. B. (2016). *Springs, Storage Towers, and Water Conservation in the Midhills of Nepal*. Nepal Water Conservation Foundation and International Center for Mountain Development, ICIMOD Working Paper 2016/3, Kathmandu, Nepal.

Sudhakar, A. A. (2010). *Adaptation to Water Scarcity in Glacier-Dependent Towns of the Indian Himalayas: Impacts, Adaptive Responses, Barriers, and Solutions*. Doctoral Dissertation, Massachusetts Institute of Technology, USA.

Thakur, N., Rishi, M., Keesari, T., Sharma, D. A. & Sinha, U. K. (2019). Assessment of recharge source to springs in upper Beas basin of Kullu region, Himachal Pradesh, India using isotopic signatures. *Journal of Radioanalytical and Nuclear Chemistry* 1–9. https://doi.org/10.1007/s10967-019-06617-3.

Tiwari, P. C., Tiwari, A. & Joshi, B. (2018). Urban growth in Himalaya: understanding the process and options for sustainable development. *Journal of Urban and Regional Studies on Contemporary India 4*(2), 15–27.

UNDESA (2014). *World urbanization prospects, the 2011 revision*. Population Division, Department of Economic and Social Affairs, United Nations Secretariat.

Xu, J., Grumbine, R. E., Shrestha, A., Eriksson, M., Yang, X., Wang, Y. U. N. & Wilkes, A. (2009). The melting Himalayas: cascading effects of climate change on water, biodiversity, and livelihoods. *Conservation Biology 23*(3), 520–530.

Urbanisation and water insecurity in the Hindu Kush Himalaya: Insights from Bangladesh, India, Nepal and Pakistan

Sreoshi Singh[a], S. M. Tanvir Hassan[b,*], Masooma Hassan[c] and Neha Bharti[d]

[a]*International Centre for Integrated Mountain Development (ICIMOD), Kathmandu, Nepal*
[b]*Bangladesh Centre for Advanced Studies (BCAS), Dhaka, Bangladesh*
**Corresponding author. E-mail: tanvirhassan.bd@gmail.com*
[c]*Pakistan Agricultural Research Council (PARC), Islamabad, Pakistan*
[d]*The Energy and Resources Institute (TERI), Delhi, India*

Abstract

This paper reviews the interlinkages of critical state of water resources, supply systems, rapid urbanisation and demand regime, aggravated by tourism leading to increasing water insecurity in the Hindu Kush Himalaya (HKH). Urban centres in the HKH have been defined based on different criteria, but mountain-specific criteria are lacking. In the mountains, small settlements such as district headquarters perform a large number of functions, typical of an urban centre. However, they are not formally classified as urban centres because they do not meet the census-defined nationally set criteria of the respective country. Nonetheless, water insecurity is a reality, attributed to: (i) water governance issues; (ii) inappropriate urban planning, failing, and to some extent, unable to account for the floating population, such as tourists; and (iii) the scourge of climate change which could worsen the situation further. Short-term coping strategies to meeting water demands often involve unsustainable solutions, such as groundwater extraction, with long-term repercussions. However, long-term strategies for water sustainability by the governments have been beneficial while others are yet to show success. Initiatives by civil society and governments along similar lines in other countries could lead to a water-secure future for the fragile urban centres of the HKH region.

Keywords: Adaptation; Coping strategy; Hindu Kush Himalaya; Urbanisation; Water insecurity

doi: 10.2166/wp.2019.215

Introduction

More than half the world's population already lives in urban centres and this proportion is likely to increase to 66% by 2050. Therefore, urban areas need to be ready to serve another 2.5 billion people by mid-century, 90% of whom will be concentrated in Asia and Africa (UNDESA, 2014). Like the rest of the world, the Hindu Kush Himalayan (HKH) region is witnessing rapid urbanisation attributed to multiple bio-physical and socio-economic drivers leading to large-scale migration; therefore, water crises in the urban centres which are already recognised as one of the global high risk factors by WEF (2017) might be a potential issue of geopolitics. HKH region is regarded as remote and largely inaccessible, but has become the seat for several urban centres, propelled by infrastructural, economic and cultural changes, mostly tourism and religious pilgrimage. Others have been colonially strategic locations expanding into large settlements drawing tourists from across the borders. Further, urban growth has been fuelled by large-scale migration of the labour force for new opportunities, students from surrounding rural areas for better education and an immensely growing tourism services sector, leading to urbanisation becoming a reality in the Himalayan region despite its unique physiographic conditions. The absolute population in the mountains is lower than the plains, due to terrain constraints. However, the factors leading to the rapid urban growth in the HKH may be examined, although city-level case studies indicate that urbanisation trends are rather limited, while substantial literature focuses on environmental degradation mainly due to climate change.

The HKH system extends for about 3,500 km from Afghanistan in the west to Myanmar and China in the east, and runs through Pakistan, Nepal, India, Bangladesh and Bhutan. Since the protection of this ecosystem was stressed during the 'Sustainable Conference' in Johannesburg in 2002, it is important to identify the impact of urbanisation in this area. The paper highlights the trends and patterns of urbanisation in mountain states and provinces of Bangladesh, India, Nepal and Pakistan and factors that have led to the urban development. Table 1 shows the regions that are part of HKH in these four countries and

Table 1. Urban population in HKH regions of Bangladesh, India, Nepal and Pakistan.

Country	Names of states/provinces which are a part of HKH	Total population (in millions)	% of population in HKH to total population in the country	Data source
Bangladesh	Chittagong Hill Tracts (Khagrachari, Rangamati and Bandarban districts)	1.5	1.2	BBS (2014a, 2014b)
India	West – Jammu and Kashmir, Himachal Pradesh	19.4	1.6	Census of India (2011a, 2011c, 2011d)
	Central – Uttarakhand	10.1	0.8	
	East – West Bengal (only Darjeeling district), Sikkim, Nagaland, Manipur, Mizoram, Meghalaya, Tripura, Arunachal Pradesh and Assam (only Karbi Anglong and Cachar districts)	18.9	1.6	
Nepal	Mountain and hill districts	13.1	49.3	CBS (2014)
	Terai (plains)	13.3	50.7	
Pakistan	Khyber Pakhtunkhwa (KPK), Federally Administered Tribal Areas (FATA), Gilgit-Baltistan, and 12 districts of Baluchistan Province	24.5	18.6	PBS (1998)

their respective urban population. In most of these countries however, state reorganisation, boundary changes and changes in definition of what constitutes urban centres make it difficult to compare long-term trends of urbanisation in the region.

Mountain urban centres grow and prosper as a result of different physical and socio-economic drivers, e.g., large cities having grown by encroaching upon nearby hills, like Nainital in Uttarakhand State of India, which began expansion towards its hill sides in the 1970s (Tiwari & Joshi, 2014). Other instances have shown smaller towns growing through 'amenity migration' as urban centres offer new livelihood options to rural migrants, which has augmented the growth of a large resident population where additional demands for housing and other facilities have been created. The HKH region has varied geo-environmental conditions and resources available for development, and hence declared as ecologically sensitive by the Indian government (Kapoor *et al.*, 2009 in Kumar & Pushplata, 2013).

The HKH region produces one of the largest freshwater supplies on Earth with mainly Asia's largest six rivers, namely, Indus, Ganges, Brahmaputra, Mekong, Yangtze and Yellow rivers. The South Asian HKH region covers the major river basins of the Indus, Ganges and Brahmaputra in Pakistan, India, Nepal and Bangladesh. The HKH region is characterised by a variety of climatic conditions from tropical to alpine. The high mountains are a major barrier of the monsoon cloud flow from southwest to north-west (5,000 m), hence most of the precipitation occurs due to this barrier and this precipitation pattern dominates in the east rather than in the west. Mean annual precipitation ranges from 300 mm in Ladakh area in the west to 1,400 mm in Kathmandu and 4,000 mm in Pasighat in the Brahmaputra basin (ICIMOD, 2002). The duration of the rainy season increases from the west (two months) to the east (eight months) of Himalayas, therefore, the duration of high flow season increases from west to the east. Precipitation is also highly influenced by local orography, i.e., it increases from valleys to high mountains where the windward slope receives more precipitation than the leeward slope. Table 2 summarises the water resources of major river systems in South Asian HKH region with annual discharge, number of glaciers and lakes. The Indus river basin is the largest (1,165,000 km^2) while Brahmaputra is the longest river (3,848 km) with the highest mean annual discharge of 19,800 m^3/s (Wikipedia, 2017; Groot *et al.*, 2018). The numbers of glaciers and lakes are larger in Indus and Brahmaputra basins than in the Ganges.

Unfortunately, the unprecedented population growth has led to an overexploitation of water sources in the region pushing the inhabitants to a state of despair. Urban centres in the mountains largely depend on springs and rivers (supplied through piped networks). As these sources are snow- and glacier-fed, the impact of climate change may affect the quantity of water available from these sources leaving groundwater sources as more critical for these cities in different seasons. Several mountain urban centres

Table 2. Water resources of major river systems in South Asian HKH region.

Parameter	Indus	Ganges	Brahmaputra
Catchment area (km^2)	1,165,000	907,000	712,000
Total length (km)	3,200	2,500	3,848
Average annual discharge (m^3/s)	6,600	16,648	19,800
No. of glaciers	16,049	6,237	10,106
No. of lakes	411	87	474

Source: Adapted from Sharma *et al.* (2013a) and Wikipedia (2017).

are now augmenting their water supplies through water transfers from distant sources, as existing water sources are insufficient to meet rising water demands. However, due to the inherent fragility of mountain environments (Perlik, 2014; IHCAP, 2017) such water transfers may not always be feasible due to the high infrastructure and energy costs involved (Buytaert & De Bièvre, 2012).

Rising urban population leads to increasing demand for municipal water. Infrastructure development attempts to bring more people under the formal water supply systems, rather than increasing dependency on other sources (McDonald *et al.*, 2014). However, the burden of population increase leads to unplanned urban growth through increased pressure on water resources that are limited and unable to cope with increasing demands; and hence the need to access water from alternative sources, close or distant. When alternative water sources are tapped from long distances, it may help urban centres to escape water stress, but often are expensive projects. Alternative sources could be available within close proximity, but the terrain may be a constraint compared to that of the plains. Being located in developing countries, temporal solutions would be required for these urban centres through arranging either an interim water supply from sources close to them or adopting other strategies that are cost-effective.

Compared to most other ecosystems in the world, mountains are still relatively less explored and researched, therefore, it would be unwise to ignore this ecosystem from which half the global population benefits (Yi *et al.*, 2016). **Where do urban Himalayas stand in such a setting and what are their capacities to meet their water challenges?** This great research question is still not well documented in the literature, but an answer is necessary for the existence of urban Himalayas in the future. This review paper attempts to answer it through a substantial review of existing literature – journal articles, working papers, reports, online resources and the grey literature. It also attempts to highlight the key urbanisation trends in the four countries of the HKH region, using the latest available census data. Further, it enquires into the existing state of water sources that currently meet the water demands of growing populations; it evaluates individual and community coping and adaptation strategies regarding water scarcity. It also explains the history, geography and economics of urbanisation in the region, in order to identify the impact of this process on water demand in the future. Finally, it highlights the issue of data inconsistency in the region, and draws attention to further research questions for future research.

Urbanisation in Bangladesh, India, Nepal and Pakistan

Definitional conundrum: what is urban and should it be defined differently for the mountains?

In all these four countries, urban centres are broadly defined in terms of a population threshold and the nature of economic activities carried out by a majority of the population. Many of these countries have revised their definition of what constitutes urban centres over the last few decades, making it difficult to compare long-term trends of urbanisation in the mountains. In this section, a critical look is taken at what constitutes urban, and if urban so defined meets the special needs of the mountains as a region.

In Bangladesh, according to the censuses of 1981 and 2001, an urban area included city corporations, municipalities, upazila (sub-district) headquarters, growth centres, a cantonment, and urban agglomerations adjacent to large cities, i.e., city corporations termed as statistical metropolitan areas (SMA).

Table 3. Size class classification of urban centres in India.

Size class of urban areas	Population criteria
Class I	> 100,000
Class II	50,000–99,999
Class III	20,000–49,999
Class IV	10,000–19,999
Class V	5,000–9,999
Class VI	< 5,000

Source: Adapted from Census of India (2011a).

In 2011, the concept of SMA, growth centre, and some other urban areas was abandoned, thereby, urban areas are limited to only city corporations, paurashavas (municipality areas), upazila headquarters and cantonment areas (BBS, 2014b) with amenities such as metalled roads, improved communication, electricity, gas, water supply, sewerage, sanitation, and a high population density and people mostly engaged in non-agricultural occupations (Islam & Khan, 2013). One direct consequence of changing the definition was that the percentage of urban population of the country declined to 23.3% in 2011, which otherwise would have been around 28% according to the previous definition.

In India, a place is considered an urban centre, if it has a minimum population of 5,000 and at least 75% of the main workers[1] are engaged in non-agricultural pursuits and the density of population is at least 400 per km^2. Urban centres that meet the above criteria are designated as 'census towns' (see Table 3). Other places, such as municipality, corporation, cantonment board, notified town area committee, etc., are 'statutory towns' and mostly belong to size class VI (Census of India, 2011a). However, this definition poses difficulty for urban geography studies, since some settlements, especially in the Himalayan region, which truly are urban centres from a physical and economic standpoint, fail to meet the criteria laid down by the Census of India.

In Nepal, the term 'urban centre' has been defined differently across the census years from 1952, 1962, and 1992 (CBS, 2003). A final consensus only emerged as late as 1999. From merely counting people living in prominent settlements (with over 5,000 population), Nepal provided a specific definition of a *sahar* or urban centre for the first time in 1961. The 1962 Nagar Panchayat Act provided municipal status to urban centres with a minimum population of 10,000, which was again reduced to 9,000 in 1976, since several urban centres were excluded when the Village Development Committee (VDC) became the main unit of enumeration in the 1971 census. After the Local Self-Governance Act of 1999 came into force, these municipalities were categorised into three types, i.e., metropolitan city, sub-metropolitan, and municipality with different criteria for municipalities of the hills and the Terai (Table 4). These definitions are likely to be modified again, for under the new federal structure declared by the Constitution of 2015 of Nepal, existing districts and VDCs will be dissolved to form *Nagarpalikas* (urban areas) and *Gaopalikas* (rural areas). So far, Nepal is the

[1] As per the Census of India, workers are broadly classified into main workers and marginal workers. Those workers who had worked for the major part of the reference period (i.e., six months or more) are termed main workers. Main workers are again classified into nine categories. Conversely, workers who had worked for the minor part of the reference period (i.e., less than six months) are termed marginal workers. Both males and females are included in this classification.

Table 4. Post-1999 definition of urban centres in Nepal.

Category	Criteria
Metropolitan city (*Mahanagarpalika*)	Minimum population size of 300,000, annual revenue of at least NRs.400 million (US$3.6 million), facilities of electricity, drinking water, communication, paved main and subsidiary roads, provision of specialised health services, essential infrastructure for international sports events, adequate opportunities for higher education in different fields, at least one established university, adequate urban facilities, and an area that has already received the status of a sub-metropolitan city
Sub-metropolitan city (*Upamahanagarpalika*)	Minimum population size of 100,000, annual revenue of at least NRs.100 million (US$0.9 million), facilities of electricity, drinking water, communication, paved main roads, education and health services of a high standard, general infrastructure for national and international sports events, provision of public parks and a city hall and similar urban facilities, and an area that has already received the status of a municipality
Municipality (Terai/plains) (*Nagarpalika*)	Minimum population size of 20,000 and revenue of NRs.5 million (US$45 K). Facilities such as electricity, road, drinking water, communication, and other similar urban facilities should be present
Municipality (Hills) (*Nagarpalika*)	Minimum population size of 10,000 and revenue of NRs. 500,000 (US$4.5 K). Facilities such as electricity, road, drinking water, communication, and other similar urban facilities should be present

Source: Adapted from CBS (2003).

only country in the HKH which has a mountain-specific definition of an urban centre. It recognises that places with a lower population or economic activity threshold should qualify as an urban centre in the hills and mountain regions, given the limitations imposed by the physical geographies of the mountains.

In Pakistan, the census definition of an urban area was almost similar in the 1951, 1961 and 1972 censuses, where a place qualified as an urban centre if it had a minimum of 5,000 inhabitants. In 1981, the definition was changed by replacing the size-specific definition with an administrative criterion, whereby an urban area included a metropolitan corporation, municipal corporation, municipal committee or cantonment (Arif & Ibrahim, 1998). According to this criterion, 361 places identified as rural areas in the previous census but now showing more urban characteristics were added as urban centres in 1998 (Arif, 2003; Ali, 2013).

Definition of an urban area has been changing as the criterion for identifying a place as urban has been changing in different countries. These changes make it difficult to assess the actual population that should be considered for planning any developmental activity. If the changing definition of an urban area poses challenges for planning in any country, lack of separate definitions of an urban area for the mountains and plains may lead to neglecting several areas which need attention, especially that due to physical location. For instance, the inner and outer Terai region of Nepal has more urban centres vis-à-vis the higher mountains as they are points of contact for the hills and plains, sources of agricultural products and seats for industrial growth. However, in recent years, Nepal has seen the growth of several urban centres, in the hilly region, which have been given the status of municipality. The urbanisation process in the Himalayas in general has most often been boosted by tourism and development of road infrastructure, further leading to the emergence of smaller urban centres. The urban centres of touristic importance in the Himalayas receive seasonal population as tourists, as well as population migrating from surrounding rural areas. Nonetheless, Satterthwaite (2006) mentioned that there is a myth: 'smaller urban centres are growing faster than large cities, which are characterised by limited economic activities'. Satterthwaite (2006) further argues that the processes and nature of population growth varies between the small and large urban centres and within the small urban centres as well. It is, in fact, the unique character of locational advantage which predominantly plays a role in rapid growth of urban centres, both small and large. Some of these processes and dynamics have been examined in the next section.

As mentioned earlier, among the four countries of the HKH region, only Nepal has a mountain-specific definition of what constitutes an urban area by taking cognisance of the fact that relatively smaller conglomerations of population in a space restricted terrain, like the mountains, can function like an urban centre in terms of provision of amenities and services to a spatially dispersed rural hinterland. For other countries, a rather fixed definition of an urban area, irrespective of their physical location, could lead to excluding several large settlements of the mountain regions as being urban. Places like a small sub-district headquarter providing all urban services miss out on being called an urban centre, and hence are deprived of additional revenues and investments. Here, definitional changes by using a lower population threshold for the mountains can help ensure that unique characteristics of urbanisation in the mountains is recognised. This can also help in creation of suitable urban governance mechanisms for mountain cities, followed by the development of basic infrastructure for the large population.

Trends and drivers of urbanisation in the HKH

Urbanisation trends in the HKH display a unique character, with all four countries in the region showing that the urban population is increasing, while that of the rural population is declining. The literature also highlights that by 2050 more than 66% of the global population will live in urban centres which could add another 2.5 billion people to urban populations by 2050, with close to 90% of the increase being concentrated in Asia and Africa (UNDESA, 2014). Although rapidly urbanising, more than 61 million out of 200 million inhabitants living in the HKH region are below the poverty line (Hunzai *et al.*, 2011) and still suffer from low incomes, lack of employment opportunities, lack of infrastructure and lack of access to basic amenities. Migration to large cities for better economic opportunities is an obvious consequence and more random in nature as opposed to the past. Moreover, increased access to information, better connectivity and lower travel costs have enabled more people to move. As a result, most of the large cities and medium-sized towns in the HKH are witnessing their maximum growth; however, of course, continuing to be lower in size and population than the urban centres in the foothills and plains. One exception to this is the Chittagong Hill Tracts (CHT) in Bangladesh, where the urban population is 30% as opposed to 23% in the country as a whole (Table 5). This

Table 5. Percentage of urban population in Bangladesh, India, Nepal and Pakistan and in HKH parts of these countries.

Country	Urban population in the country (%)	Urban population in states and provinces of HKH region (in %)	Year	Sources
Bangladesh	23	30	2011	BBS (2014a, 2014b, 2012a, 2012b, 2012c)
India	31	26	2011	Census of India (2011a, 2011c, 2011d)
Nepal	17	5	2011	Central Bureau of Statistics (2014)
Pakistan	33	19	1998	PBS (1998)

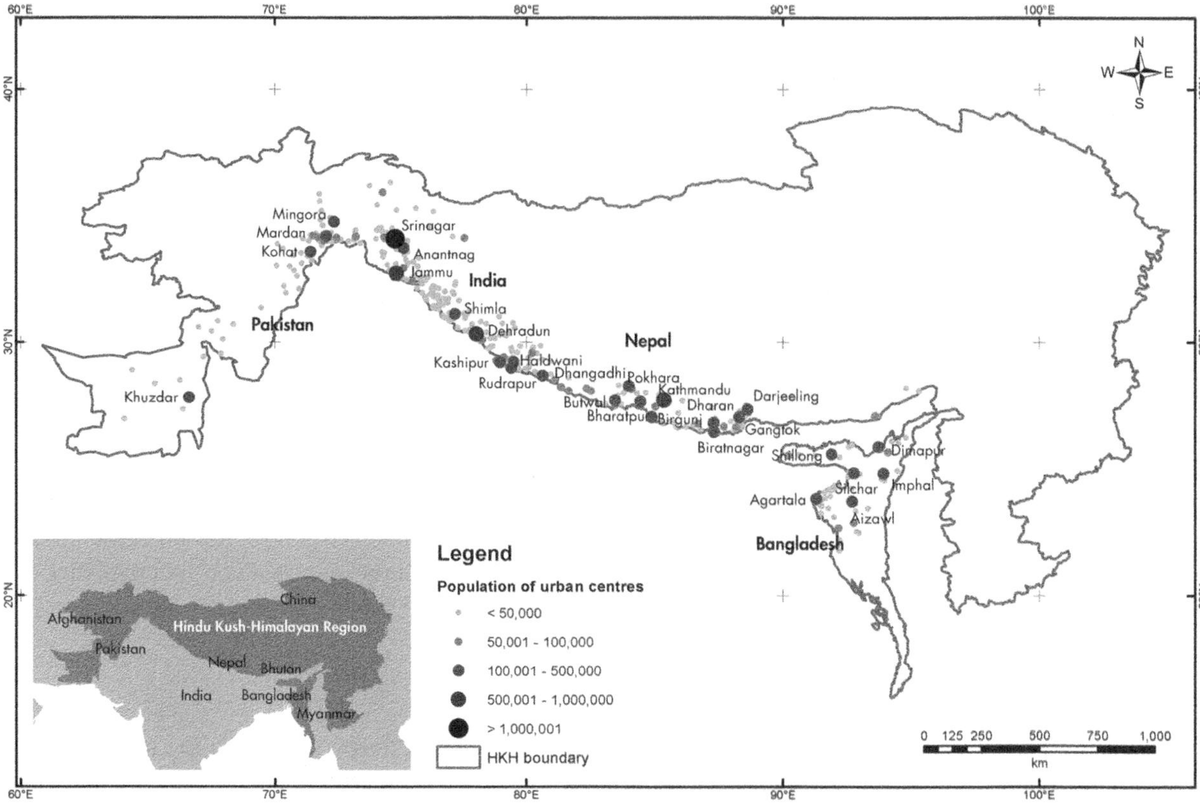

Fig. 1. Distribution of urban centres in HKH regions of Bangladesh, India, Nepal and Pakistan.

could be a consequence of the in-migration of the large Bengali population after mass out-migration of the local indigenous population from the region. Figure 1 shows the distribution of urban centres in the HKH regions of Bangladesh, India, Nepal and Pakistan.

Of the 200 million migrants worldwide, 15% (30 million) are from the HKH countries of Bangladesh, China, India and Pakistan, of which a significant proportion (although the exact figure is not known) are from the mountainous regions due mainly to: (i) rain-dependent agriculture leading to food insecurity (Sharma, 2011); (ii) lack of alternative sources of income due to strong feudal systems and internal conflicts (Hasan & Raza, 2009); and (iii) poor infrastructure, that has further accelerated the process of migration (Mamgain & Reddy, 2015). Entire family out-migration from the remote mountain regions in Uttarakhand in India to regions in the newly emerging urban areas within the state (between 2001 and 2011), like Dehradun, Kotdwara, Haldwani, Roorkee and Hardwar-Rishikesh have been witnessed (Census of India, 2011a). About 64% of all migrants in Pakistan have migrated to urban areas, while 25% of all migrants migrated to large cities like Karachi, Lahore and Rawalpindi where more job opportunities are available than in small towns (Hasan & Raza, 2009).

A recent study by Maharjan *et al.* (2018) on migration documented that 39% of rural communities have at least one migrant, of whom 80% are internal and the remaining 20% are international, while a quarter (10%) were reported as environmental displacement. Internal migration is mainly due to education purposes in mountainous areas and for employment opportunities in the plains. International migration is limited mainly to a south–south movement. Internal migration is mostly male-dominated (83%), mainly for the purposes of education and seeking employment. Female migration (17%) is mainly for higher education in urban destinations. Migration in mountainous areas is higher than in the plains. The major driver of the migration decision is economic, but environmental displacement was also prevalent in the study sites. However, Maharjan *et al.* (2018) evaluated the migration pattern of only households where some members are left behind in the rural areas. People who migrated with all household members were not considered. Moreover, how many people migrate from rural to urban areas was not directly documented.

Anbalagan (1993) pointed out that urbanisation has been very rapid in the Himalayas, but unplanned and unregulated. Ghosh (2007) states that greater accessibility of remote areas owing to improved road connectivity, publicity and marketing of new tourist locations and consequent growth of domestic and international tourism; development of horticulture; economic globalisation and gradual shift from primary to secondary and tertiary sectors; and the absence of urban land use policy have been responsible for the growth of urban centres, their growing numbers and complexity. However, the regional factors for migration do vary. For example, in the state of Jammu and Kashmir in the western section of the Indian Himalayas, tourism has contributed to in-migration into Srinagar while construction workers have migrated into Jammu city. The literature stresses the importance of religious power factors and long-distance trade for city formation in Ladakh region (Mathieu, 2003) where, in the last two decades, the population in the town of Leh has almost tripled. A study has shown that Leh has a substantially higher level of urbanisation (23%) than Kargil (9%). Some nomadic tribes have also migrated to Leh due to loss of grazing lands following the Sino–Indian war. The Changpa communities, for example, have also had to accommodate a large number of Tibetan refugees and their herds. However, introduction of subsidised food rations and improved road accessibility have reduced their need for arduous long-distance trade journeys. The town of Leh now has an entire colony where these nomadic tribes have settled, mostly in the working age group or within the age group of 10 to 19, the latter being unable to start schooling at such a late age and having to get into the workforce (Goodall, 2004). A study by Tiwari *et al.* (2018) revealed that 500 villages across the ten mountainous districts of Uttarakhand showed a decline in farm production and therefore reduced livelihood opportunities in the agricultural sector. This is mainly due to reduced rainfall, decreasing rainy days and resultant decline in irrigation potential, therefore increased frequent incidences of crop failure. These have resulted in increasing trends of rural out-migration contributing largely to urban growth in all parts of the Himalayas (Tiwari *et al.*, 2018).

Khawas (2005) points to the fact that nearly 28% of the urban population of the north-eastern Indian Himalayas is concentrated in about 4% of the cities. More than 73% of the small towns in the region collectively accommodate only just over 35% of the total urban population, as of 2001. Because of the hilly terrain, lateral expansion has not been possible in the north-east. However, several factors have played a role in the development of the urban centres and include reorganisation of boundaries after independence, attractive development projects from the state and central governments, contribution of Christian missionaries in terms of upgrading the quality of lives of tribal people, and heavy investment in the education and infrastructure sectors.

In Nepal, as many as 23 of 75 districts have seen negative population growth in 2001–2011 and they are all mountain districts, while the city of Kathmandu has seen a double-digit growth rate during the same period (Bakrania, 2015; Thapa, 2017). In Uttarakhand Himalayas (India), the plain districts of Dehradun, Udham Singh Nagar and Haldwani have seen a population growth of more than 30% between 2001 and 2011, while the mountain districts of Almora and Pauri have seen a negative population growth. The state has 1,053 uninhabited villages and another 405 villages with less than ten inhabitants – all in the five mountain districts of the state affecting the social composition of the population (Pathak *et al.*, 2017).

The variation in urban population concentration is also visible across the HKH, with Pakistan showing the highest growth rate of 5% per annum. This could be attributed mainly due to natural increase and internal migration, although reclassification of towns based on the changing definition of 'urban' may also have some role to play in this regard (Arif & Ibrahim, 1998). Within India, the urban population concentration is higher (41%) in the eastern Himalayan region than in the western (34%) and central Himalayas (25%) (Census of India, 2001, 2011a, 2011b, 2011c, 2011d), possibly because of lower average elevations in the eastern Himalayas. The growth rate is also higher in the eastern Himalayas compared to the western and central Himalayas (Table 6).

Much of the urban growth in some countries is restricted to the existing cities and towns (Table 7). For example, in Pakistan, not many new urban centres have emerged in the provinces of Khyber Pakhtunkhwa (KPK) and Baluchistan. In the Indian Himalayas, conversely, a number of rural areas have received the status of 'urban' from one census to another due to the change of definition of urban centres (for more information, please refer to the section 'Definitional conundrum: What is urban and should it be defined differently for the mountains?), a possible reason for drastic increase in the number of urban centres. For example, 173 new urban centres were added during 2001–2010, and most of these were in the size classes IV, V and VI (Census of India, 2011a, 2011b, 2001). In contrast, Nepal had no new urban centres in 2011, although this is likely to change with the newly created federal structure (Bakrania, 2015).

Table 6. Urban population concentration and growth rate across census years in HKH states and provinces of Bangladesh, India, Nepal and Pakistan.

Country	Province/State/District	Urban population (2001)	Urban population (2011)	Compound annual growth rate (CAGR) in %
Bangladesh	Chittagong Hill Tracts	433,989	728,904	5
India	Western Himalayan states	3,112,219	4,121,794	3
	Central Himalayan states	2,179,074	3,049,338	3
	Eastern Himalayan states	3,319,170	4,908,159	4
Nepal	Mountain and hill districts[a]			5.1 (AAGR)
	Terai (plains)[a]			3.2 (AAGR)
Pakistan[a]	HKH provinces	2,177,440	4,775,337	5

Sources: BBS (2014a); CBS (2012); Census of India (2011a, 2011c, 2011d); GOGB (2013); PBS (1998).
[a]*Note*: The data for Pakistan are based on the 1981 and 1998 censuses; urban population data (1981) exclude Gilgit-Baltistan, FATA and two districts of Baluchistan (data not available). Urban population across ecological divisions of Nepal not available. Growth rates have been shown from Bakrania (2015).

Table 7. Urban centres in HKH region of Bangladesh, India, Nepal and Pakistan (2001–2011).

Country/State/Province	Number of towns in 2001[a] (except Pakistan where data pertain to 1981)	Number of towns in 2011[a] (except Pakistan where data pertain to 1998)
CHT, Bangladesh	5	6
Western Himalayas, India	132	181
Central Himalayas, India	86	115
Eastern Himalayas, India	143	238
Hills, Nepal	27	27
Mountains, Nepal	2	2
Terai, Nepal	29	29
KPK, Pakistan	52	55
FATA, Pakistan	2	5
Balochistan, Pakistan	40	46
Gilgit-Baltistan, Pakistan	Data not available	5

Sources: BBS (2014a); Govt. of Gilgit-Baltistan (2013); Census of India (2001, 2011a, 2011b, 2011c, 2011d); CBS (2014); PBS (1998).

[a]*Note*: The data for Pakistan are based on the 1981 and 1998 censuses. Urban population data (1981) exclude Gilgit-Baltistan, FATA and two districts of Baluchistan (data not available).

Explaining urbanisation in the HKH: history, geography and economics

Local history, geography and economics of mountains impacted development activities and propelled urban growth. Historically, major trade routes ran parallel to the HKH, apart from the Ganga–Indus connection to the south and the Silk Road to the north that connected China to South and West Asia. Following this, some semi-urban and urban settlements emerged later in connection with Buddhism in the form of Lamaistic monasteries, which spread initially around Ladakh from the 12th century onwards (western Himalayas), then towards Sikkim (eastern Himalayas) by the 17th century (Mathieu, 2003). Consequently, many of these towns expanded and became gateways for pilgrims from Hindu and Buddhist communities. The towns of Joshimath in Uttarakhand and Dharamshala in Himachal Pradesh, for example, are gateways to the Badrinath Dham and the Dalai Lama monastery in McLeodgunj.

Urbanisation in the HKH also gained momentum during the British colonial period in the South Asian sub-continent. For instance, Shimla, Shillong and Darjeeling in India, and Muree in Pakistan, all served as summer capitals of British India, while the city of Dehradun grew during the early colonial period with the establishment of railroads, followed by the settling of ex-army personnel in the town during the early 20th century. Later, several national institutions were established in the city and, together with a growth in tourism, provided an optimum condition for several small industries to flourish, further attracting migrants from the rural hinterland (Sharma, 2001).

In the post-colonial era, events like border disputes and internal conflicts led to the initiation of development programmes by regional governments followed by growth of urban centres. For example, urbanisation in the Garhwal Himalayas in India was a consequence of the Indo–Sino war along the border, which led to the initiation of road construction and the emergence of urban corridors along highways (Maharjan, 2017). Rapid urbanisation in the north-eastern Indian Himalayas resulted from a government policy to facilitate rapid infrastructural development in the region (Pokhriyal, 2001). In Bangladesh, tribal insurgency in the CHT in the 1970s prompted the government to undertake multi-sectoral development activities in the region from the 1980s. Market development was one aspect, which coincided with the construction of roads under military supervision. Later, other forms of infrastructure development were undertaken through intervention of the Chittagong Hill Tract Development Board and several other agencies (Huq, 2001).

The unique geography of the HKH, its rich cultural diversity, favourable climate and picturesque landscape make several Himalayan towns like Rangamati, Bandarban and Khagrachhari (Bangladesh); Leh, Srinagar, Shimla, Haridwar, Rishikesh, Darjeeling, Gangtok, Shillong (India); Kathmandu, Pokhara, (Nepal); and Muree (Pakistan) be attractive tourist destinations. Table 8 estimates tourist arrivals in some of the major urban centres in the HKH based on various sources, including newspaper reports.

Tourism (specifically, religious tourism) has played an important role in creating new urban centres and expanding existing ones thanks to improved accessibility, drawing thousands of pilgrims each year (Sharma, 2001). Moreover, tourism for leisure is also gaining popularity with better road connectivity, thereby increasing economic opportunities in these urban locations.

Much of the urban growth in the mountains has also been due to economic development after the initial creation of market towns in the region, followed by further expansion when road networks were set up, boosting trade in the region. Additionally, Portnov *et al.* (2007) identified that indicators like road length, telephone lines, spatial association of urban centres, international boundaries, large rivers and large urban centres are major determinants for urban growth in Nepal. With a productive hinterland in the Terai, Nepal has promoted several industries like carpet weaving and food processing in the Kathmandu valley. Hence, the cities of Kathmandu, Lalitpur and Bhaktapur have almost become a conurbation (Muzzini & Aparicio, 2013) and the most urbanised region in Nepal (Thapa, 2017), attracting people from all over the country and abroad. In Pakistan, the HKH region, comprising the North-West Frontier Province (NWFP), is extremely remote with very low economic development that has led to migration to large cities like Karachi, where the migrants are preferred as labourers over those who live within the Sindh province. Of the total migrants across Pakistan, 12% of these migrants are from KPK, due to extreme vulnerability to crisis and conflicts (Hasan & Raza, 2009; Shahbaz *et al.*, 2012). In India, the city of Srinagar has become the largest city in the Kashmir Division

Table 8. Estimates of tourist influx in various urban centres in the HKH.

Urban centres/state/district	Total visitors	Source
Bandarban district (CHT), Bangladesh	0.65 to 0.70 million (yearly)	Hasan *et al.* (2013)
Jammu and Kashmir (J&K), India	About 1.3 million (yearly)	Correspondent (2017a)
Leh town, J&K, India	About 0.18 million (yearly)	LAHDC (2011)
Shimla, Himachal Pradesh, India	Over 2 million (peak season of April, May and June)	Das (2014)
Haridwar and Rishikesh towns, Uttarakhand, India	About 0.27 million (religious pilgrims only)	Correspondent (2013)
Darjeeling, West Bengal, India	0.55 million (April to June; September to mid-November)	Sarkar (2013)
Sikkim (entire state), India	0.61 million (yearly)	Govt. of Sikkim (2015)
All north-eastern states of India	Over 7 million (yearly)	Lyngdoh (2016)
Nepal	0.6 million (yearly)	Govt. of Nepal (2015)
HKH region of Pakistan	80 million (yearly)	Correspondent, (2017b)

(Bhat, 2008; Kuchay & Bhat, 2014a), attributed to migration, and accounting for over 73% of the urban population in the state (Census of India, 2011a).

From the above discussions, it may be summarised that the push and pull factors of migration across the HKH region vary from one country to another. In Pakistan (weak economic development and conflicts) and Nepal (remote), push factors play a more dominant role in migration as compared to the pull factors. Migrants from northern areas of Pakistan are mostly travelling to cities, like Karachi in the Sindh province, and not as many to urban centres within NWFP. In Nepal too, migrants are travelling mostly to Kathmandu valley or outside the country. In contrast, in the Indian Himalayas, pull factors seem to be stronger and migration mostly takes place to nearby towns within the state. This is due to the well-established tourism sector from Kashmir to the eastern states of Sikkim and West Bengal, which provides a large employment base for people migrating from rural areas, who have mostly settled here due to the access to social structures. This trend, of course, indicates very clearly the tremendous pressure that these urban centres are experiencing. Unfortunately, with population growing beyond capacity, the infrastructure and urban services may no longer remain adequate. Housing demands are increasing and this necessitates acquisition of agricultural lands and forests, which leads to environmental damage and affects other natural resources, particularly water. The next section will explore the change in the availability of water in the Himalayas due to urban expansion.

There are similarities in the processes of urbanisation between the plains and the mountains (migration-led growth) and also similarities in impacts of urbanisation (stress on water and resultant water insecurity). However, what sets apart mountain urban centres from their counterparts in the plains is that problems like water insecurity are much harder to solve, given the inherent natural fragility of these urban centres. The rivers of the Himalayan region get their water supply from snowmelt of the glaciers and the dwellers in urban centres get water pumped from the rivers for use. The flat terrain of the plains makes distribution of water easy compared to mountainous regions where slopes are dominant. The next section looks at issues of urban water stress in the HKH.

Status of water security in urban Himalayas

Urban growth in the HKH region has increased the demand for water in these urban centres. The spatial and temporal variability of precipitation is very high in the HKH region, with more than 80% of the annual rainfall occurring during the monsoon. While this often causes rivers to overflow their banks, bringing devastation to the people living in the floodplains, for the remaining months, acute water shortages due to seasonality in river flows are experienced (Molden *et al.*, 2014). Rural and urban areas bear the brunt of the changing physical conditions in the mountains, but the latter suffers more as water demands are very high. Over the years, when the urban centres of the mountains expanded, forests and agricultural lands have been replaced by concrete structures affecting infiltration, therefore, groundwater recharge decreases. Most of the urban centres are being provided water through a piped network; however, it is not sufficient considering the demand. The change in the availability of water in the urban areas of the HKH region is manifested in reduced discharge in the rivers, spring sources and overexploitation of groundwater resources to meet water demands. While catchment degradation was identified as the main cause for drying up of the springs in the last century, climate change is now emerging as the new threat in the 21st century (Tambe *et al.*, 2012).

Increasing demand and limited resources have culminated in the entry of private parties into the business of water. The new economic policy in 1991 gave a special boost to privatisation. In the water sector, privatisation takes place by treatment of water as a commodity (Bhattacharya *et al.*, 2013; Mukhopadhyay & Bhattacharya, 2013). Hence, big companies are in the business of water, sanitation, solid waste management, sewerage, bottled water, beverages and more in different forms of public–private partnership (Bhattacharya & Banerjee, 2015). The acute water crisis which Darjeeling faced in the late 1980s and early 1990s prompted the local government to supply water through tankers. Later in 1995, the Darjeeling Municipality requested the Darjeeling Gorkha Hill Council to form a committee to prepare a proposal for source augmentation options and to improve the supply infrastructure. The result was a proposal to lift water from the Balasun river into the two Senchal reservoirs (Boer *et al.*, 2011) at a cost of Indian rupees (INR) 400 million (approximately USD 6 million, as on 14 May 2018). While the World Bank had come forward to provide assistance, the project could not be accepted due to the irreconcilable differences on the project design, implementation and operation between the three parties involved. Later in 2004, this project was again considered as a necessary step towards improving water security and started, but the cost had increased to INR 560 million (USD 8.4 million) (Shah & Badiger, 2018). Community narratives collected during a study in the Darjeeling Himalayas by Rai *et al.* (2016a) indicate governance and convergence issues as primarily responsible for the inability to manage groundwater in the region. Additionally, despite clear mandates across departments being provided, the forest department has not been able to extend its mandate to include artificial recharge outside protected areas, which is another reason why rejuvenation of springs has not been possible across the region.

Spring sources which are the traditional water sources for the region have been largely affected during road construction and urban expansion (Sharma *et al.*, 2013b). While it still remains a source for some residents, the water quality is unreliable (Government of India, 2005), as experienced in Himachal Pradesh and Uttarakhand in the Indian Himalayas. However, in the eastern Indian Himalayas, monitoring of springs has shown that the quality of water from springs is excellent and suitable for drinking, but due to shrinking of the monsoon period as well as less snowfall in some areas, these springs are drying up (Sharma *et al.*, 2013b). Water quality deteriorates during the monsoon season as soil and mud is often washed out.

Water supplied by the municipal authorities also faces threat from degraded catchments and reduced recharge to existing storages. Further, encroachment of surface water bodies during urban expansion in several parts of the HKH (Ashraf, 2013; Kuchay & Bhat, 2014b) has been another factor leading to reduced 'sponge action' of land (Sharma *et al.*, 2013b). The network of pipelines and storage systems are old (IUCN, 2015), therefore, contamination from industrial pollution and domestic effluents (Annandale & Bailly, 2014) becomes inevitable. Several parts of these urban centres are also unserved due to lack of sufficient water in the sources. The crisis worsens during the dry seasons, particularly in large cities of touristic importance, as consumption levels soar, while recharge is limited. As local water sources have shrunk over the years, urban authorities have been compelled to augment supplies by developing planned water supply systems from other distant sources, like rivers, surface water bodies, etc. For instance, the city of Kathmandu is on the verge of importing 170 million litres per day (MLD) of water from the Melamchi river through a 26-km long tunnel. The Kathmandu valley on the whole faces a shortage of 210 MLD (Subedi, 2010). Most of the large cities of the HKH are facing shortages, as shown in Table 9 where the literature documented deficit of average daily gap of demand and supply is summarised. However, seasonal gap fluctuates, e.g., during summer demand increases as well as supply also increases due to an abundance of water during the monsoon.

Table 9. Gap between municipal demand and supply of water (in million litres per day (MLD)) in selected towns and cities in HKH regions in Bangladesh, India, Nepal and Pakistan.

Name of the urban centres and sources	Country	Population	Water demand (MLD)	Actual supply (MLD)	Deficit (MLD)
Rangamati (Chakma, 2008)	Bangladesh	84,000	9.1	6.4	2.7
Baramula (Baba, 2016)	India	71,434	22.7	7.9	14.8
Shimla (Correspondent, 2016a)		169,578	42	20.6	21.4
Mussoorie (Misra, 2016)		30,118	14.5	7.7	6.8
Darjeeling (Darjeeling Municipality, 2018)		118,805	7.0	1.9	5.1
Kohima (Yhokha & Resu, 2015)		99,039	10	1.2	8.8
Tansen (Khatri, 2016)	Nepal	29,095	2.5	0.5	2
KUKL service areas in Kathmandu valley (Udmale *et al.*, 2016)		1,346,518	362	184	178
Pokhara (Correspondent, 2016b)		255,465	65	45	20
Quetta (Balochistan) (Shahid, 2016)	Pakistan	489,046	170	106	64
Rawalpindi (KPK) (Ahmed, 2015)		1,410,000	79	20	59
Mansehra (KPK) (Ahmed & Kiran, 2011)		49,534	8.1	4.4	3.7
Karak (KPK) (Correspondent, 2016c)		27,893	1.3	0.6	0.7

Water shortage is an inevitable consequence of rapid urbanisation in the HKH, as indicated in the previous discussions. The advantage in the plains is that water can be sourced from long distances, while in the mountains this process may not always be feasible or can be very expensive. Investment in such projects is, therefore, usually limited, while old infrastructure disrupts the supply system and often causes water deficits. In this context, households struggle daily to meet their water needs, trying to find ways to respond to immediate shortages by coping methods, while a few long-term adaptation strategies can also be documented. The following sections identify these strategies by households in urban Himalayas.

Literature from across the HKH region indicates that most households in the urban Himalayas are shifting to distant spring sources when those within the close vicinity dry up; this increases their engagement time in fetching water. This has been experienced in the cities of Gilgit and Quetta (in Pakistan); Champawat, Gangolihat, Almora, Dwarahat, and Pithoragarh in Uttarakhand; Kurseong, Darjeeling and Kalimpong in West Bengal (PSI, 2003; Lepcha, 2013), Kohima (Correspondent, 2016d), and Mokukchung in north-eastern Himalayas. There are also instances of informal arrangements whereby temporary pipelines are organised for bringing water from distant springs. Another means through which households cope with water shortages in the urban centres of HKH, mostly in India, Nepal and Pakistan, is to buy water from tankers (ATREE, DGC, DLR Prerna & DNGON, undated; PSI, 2003; Shrestha & Shukla, 2014). This water is mostly sourced from distant springs or groundwater bore wells. While charges for water purchased from tankers may vary according to distance, it does have financial implications for poorer households, who often restrict the purchase of water, and therefore, the quantity used (Raina, 2016). Since the Bangladesh section of the HKH is characterised by low hills, additional water demands are met through community hand pumps (Khan, 2015). It may be understood that coping with water shortages is a household's individual decision.

The long-term strategies, better known as adaptation strategies, are undertaken by households as well as at the level of urban local governments towards managing chronic water shortages in HKH towns. Therefore, it would perhaps be a matter of greater relevance to say that households initially cope, and later as the problem escalates, they start adapting. In general, adaptation to water scarcity involves augmenting supplies from new sources and introducing conservation practices, which can improve water availability. Additionally, it may also help to reduce the damage from extreme climatic events which usually affect spring sources and increase surface runoff in the mountains. Most households in the HKH region, particularly in Nepal and India, have turned to groundwater sources, which they access through installing personal pumps. In Kathmandu, anyone who can afford it, digs a groundwater well (Raina, 2016). Kathmandu Upatyaka Khanepani Limited (KUKL), the government water utility in Kathmandu, also has invested in a large number of deep bore wells to supply water to the city (Dolma *et al.*, 2015). At this point, one can say that adaptation techniques of this kind to counter water scarcity by any household depend on the economic status. Further, rampant use of such technologies may also have an impact on the groundwater table and possibility of premature failure of wells, due to over-abstraction (Pandey *et al.*, 2016). As Shah (2016) correctly puts it, such pressures not only generate competition between springs and groundwater from 'common' aquifers, but water that was allocated to agriculture in the peri-urban pockets is also affected, when infrastructure interferes with 'natural recharge zones'. Conservation strategies have received much attention over the years, as they help to preserve the resources from being overexploited. Several households in parts of north-eastern Himalayas have pro-actively developed rainwater harvesting systems, a mission that was government supported through policies. Although knowledge about these systems is unknown to many people in other parts of the HKH, it has been considered a viable option which can be replicated by local governments. Such policies not only enable communities to adapt to water shortages, but also protect local surface and groundwater resources (Singh & Kandari, 2012).

Other planned adaptations by governments could secure water futures for urban residents in the HKH. For example, in Nepal, watershed management, afforestation programmes (by the UK's Department for International Development) and the construction of gabions and check dams (Government of Nepal, 2010) have helped improve the health of urban watersheds, improve groundwater recharge and reduce the frequency of flash floods. Although these are useful means of adaptation, they may not always be successful if not planned carefully. However, alongside strengthening local governments' capacities to prepare and implement integrated plans for urban areas (Nepal *et al.*, 2007), developing sustainable projects for augmenting water supplies can be an essential means to counter haphazard planning. The National Mission for Sustaining the Himalayan Ecosystem in India stresses the need to follow building codes and norms (SANDRP, 2013) while the Housing Policy of 2011 further strengthens this move. The National Action Plan for Climate Change (NAPCC) of India is also focused on sustainable urban development in mountain regions.

Pakistan has been successful in providing water and sanitation facilities in the rural areas of Gilgit-Baltistan through the Sanitation Extension Program by the Aga Khan Foundation, which could now be successfully replicated in the urban centres of Gilgit and Hunza. Augmentation of water supplies is an important aspect of adaptation by the national governments involving projects to draw water from alternative sources, although the scale varies. In Bangladesh, a BDT 400 million project (USD 4.9 million) has been developed for Rangamati to develop a water supply system by constructing a surface water treatment plant which uses Kaptai lake water. More recently, with Asian Development Bank funds, Local Government Engineering Department of Government of Bangladesh improved drainage networks (Asian Development Bank, 2015).

Only recently, Shimla has received funds from for the AMRUT project for integrated development of Shimla town, which includes most importantly, a water supply system under the State Annual Action

Plan (Government of Himachal Pradesh, 2015). These include progressive steps towards improvement and augmentation of the existing water supply schemes for Giri, Chair, Ashwini Khad, Gumma, Churat, Jagroti and Seog catchments, provision and rejuvenation of water supply lines, pumping machinery, leak detection machine, sluice valves, four water tankers and water meters (AMR), providing water ATMs in 25 wards of the city and filtration plants at natural springs to enhance the water quality and to provide citizens with good quality water along with rejuvenation of existing natural springs. Notwithstanding this, unfortunately, Shimla town recently saw the worst water crisis during the summer season in 2018, where supply from all the sources declined drastically from their capacities (Correspondent, 2018, May 26). Adaptation to urban water crisis requires a balance in the use of resources between the upstream and downstream users within and outside the urban watershed. In this context, suggestions indicate that those living in the immediate upstream of urban centres be encouraged to conserve natural ecosystems and be given financial assistance for doing so and that the concept of payment for ecosystem services (PES) be operationalised. Experiments, such as willingness to pay for water services, have been conducted in Nepal and India (Karn, 2008; Rai *et al.*, 2016b). However, so far, to the best of our knowledge, there are no urban centres in the entire HKH that incentivise and compensate their upstream neighbours for conserving springs and other water resources. In the fragile HKH region, adoption of a PES approach for proper management of water services may lead to a win–win situation for both water managers and users (Wunder, 2008), but is yet to be tested anywhere.

While several efforts are in place to improve the water security of people living in the mountains, Tortajada (2014) identifies that there is a need for constant accountability from users, donors, government and non-governmental organisations for their decisions and actions. With good information networks (Rouse, 2014), planning can be more precise and can yield better results.

Conclusions: towards a water-secure urbanisation process in HKH

This paper presents an outline of the current status of urbanisation as well as water insecurity in the HKH region of Bangladesh, India, Nepal and Pakistan. All these four countries are undergoing rapid urbanisation. This paper highlights three major points.

First, as the mountain sub-region is much smaller than the plains, globally, the urban population is also lower than in the plains, and attributed to differences in the terrain of the two regions, the former offers unfavourable conditions for human settlements to grow and prosper. Additionally, the uniform definition of urban centres applied with equal weight across plains and mountains, often tends to ignore important strategic locations as 'urban' in the mountains, although in terms of water security, the capacity of mountains is much lower than that of the plains. In the mountains, smaller settlements like district headquarters or market towns perform a number of functions typical of an urban centre. However, they are not formally classified as urban centres because they do not meet the nationally set criteria. This calls for a mountain-specific definition of urban areas, which takes into cognisance mountain specificities like fragility, limited water sources and remoteness. One exception is Nepal, which has a mountain-specific criterion for demarcating urban centres and which may also be followed in other countries.

Second, within each country and its mountain areas, the pace of urbanisation varies, with higher urban growth noted in lower elevations and valley bottoms, again implying the importance of geographical factors in creating urban centres. While historical and geographical factors explained urbanisation in the mountains in the past, in recent decades, the economics of migration and tourism have been the major drivers of urbanisation. While urbanisation in the mountains offers huge opportunities for people from rural hinterlands, the haphazard growth of these mountain urban centres coupled with their fragile geography makes them particularly vulnerable to water stress and insecurity. The majority of urban centres in the HKH cannot meet water demand from municipal sources. Some of these mountainous urban centres are of historical importance while some are popular tourist destinations. Therefore, there is seasonal flow of population in many urban centres. The present planning process has failed to provide alternative systems incorporating seasonal influx of population in these urban centres leading to acute water scarcity, congestion and pollution.

Third, there are many strategies undertaken by people in coping and adapting to water scarcity. This paper found that the most frequent practice is to tap groundwater, either from spring sources or through dug wells and bore wells. However, unless supplemented with adequate and well-planned recharge programmes, excessive reliance on groundwater will lead to further potentially deleterious consequences in the future, given that aquifers in mountainous regions are inherently fragile. Therefore, long-term strategies are needed, such as mountain-specific urban planning, that take into account the myriad fragilities of mountain ecosystems and ecological restoration of forested uplands that feed the urban water systems. Without long-term and sustainable urban planning and accountability of the stakeholders, many of these urban centres in the HKH are poised for a grim water future, which will only be exacerbated by climate change.

This paper has been able to identify critical issues around changing conduct to water demand of urban centres, seasonally, accounting for floating populations, water inequities arising from socio-economic factors, impact of governance practices and water budget of cities. Urbanisation provides multiple livelihood options and pointed research in the gap areas can improve urban and regional planning and help achieve the goal of building sustainable urban centres in the HKH region.

Acknowledgements

This work was carried out by the Himalayan Adaptation, Water and Resilience (HI-AWARE) consortium under the Collaborative Adaptation Research Initiative in Africa and Asia (CARIAA) with financial support from the UK Government's Department for International Development and the International Development Research Centre, Ottawa, Canada. This work was also partially supported by core funds of ICIMOD contributed by the governments of Afghanistan, Australia, Austria, Bangladesh, Bhutan, China, India, Myanmar, Nepal, Norway, Pakistan, Switzerland and the United Kingdom. The authors would like to thank Dr. Aditi Mukherjee who was part of the initial conceptualisation of the paper and has reviewed and commented on the initial drafts. The manuscript was reviewed by Dr. David Molden, Dr. Philippus Wester, Dr. Anjal Prakash and Mr. Nagraj Adve of ICIMOD for its consistency which improved its structure and arguments, therefore, overall quality of the manuscript, that is highly acknowledged. Finally, the manuscript benefitted from the comments and suggestions from two anonymous reviewers of the journal, whose efforts are greatly appreciated. The views expressed in this work are those of the creators and do not necessarily represent those of the UK Government's Department for International Development, the International Development Research Centre, Canada or its Board of Governors. In addition, they are not necessarily attributable to ICIMOD and do not imply the expression of any opinion by ICIMOD concerning the legal status of any country, territory, city or area of its authority, or concerning the delimitation of its frontiers or boundaries, or the endorsement of any product. No potential conflict of interest is reported by the authors.

References

Ahmed, S. I. (2015). *Real Estate Trumps Water in Pakistan's Capital: Water Storage is Sacrificed for Real Estate Development, Worsening Water Shortages in the Pakistan Capital of Islamabad and its Twin City Rawalpindi.* https://www.thethirdpole.net/2015/10/11/real-estate-trumps-water-in-pakistans-capital

Ahmed, S. & Kiran, F. (2011). *Mansehra City Profile.* UNHABITAT. http://urban.unhabitat.org.pk/Portals/0/Portal_Contents/KPK/Mansehra/Mansehra%20City%20Profile%20by%20NED.pdf

Ali, R. (2013). *Estimating Urbanization.* South Asia Institute Resources, Harvard University. http://southasiainstitute.harvard.edu/website/wp-content/uploads/2013/10/Estimating-Urbanization.pdf

Anbalagan, R. (1993). Environmental hazards of unplanned urbanization of mountainous terrains: a case study of a Himalayan town. *Quarterly Journal of Engineering Geology and Hydrogeology 26*(3), 179–184. doi: 10.1144/GSL.QJEGH.1993.026.003.03.

Annandale, D. D. & Bailly, H. (2014). *Strategic Environmental Assessment of the Master Plan for Gilgit City.* IUCN, Pakistan. http://api.commissiemer.nl/docs/mer/diversen/pos722-sea-masterplan-gilgitcity.pdf

Arif, G. M. (2003). *Urbanization in Pakistan: Trends, Growth and Evaluation of the 1998 Census, in UNFPA Population of Pakistan: An Analysis of 1998 Population and Housing Census.* Kemal, A. R., Irfan, M. & Mahmood, N. (eds) Pakistan Institute of Development Economics, Islamabad.

Arif, G. M. & Ibrahim, S. (1998). The process of urbanization in Pakistan, 1951–81. *The Pakistan Development Review 37*(4 Part II), 507–522.

Ashraf, A. (2013). Changing hydrology of the Himalayan watershed. In: *Current Perspectives in Contaminant Hydrology and Water Resources Sustainability.* Bradley, P. M. (ed.). InTech, Rijeka, pp. 221–241, doi: 10.5772/54492.

Asian Development Bank (2015). *BAN: Third Urban Governance and Infrastructure Improvement (Sector) Project – Rangamati Pourashava.* Asian Development Bank, Dhaka, Bangladesh. https://www.adb.org/sites/default/files/project-document/183295/39295-013-rpddr-02.pdf

ATREE, DGC, DLRPrerna, & DNGON (undated). Conserve water, reduce your waste. Ashoka Trust for Research in Ecology and the Environment (ATREE), Darjeeling Goodwill Centre (DGC), Darjeeling Ladenla Road Prerna (DLRPrerna), & Darjeeling NGO Network (DNGON). http://www.darjeelingprerna.org/Conserve%20Water%20Reduce%20Your%20Waste_large%20file.pdf

Baba, A. (2016). Baramulla residents forced to drink polluted water: No proper scheme for potable water. Greater Kashmir. http://www.greaterkashmir.com/news/kashmir/story/213513.html

Bakrania, S. (2015). *Urbanisation and Urban Growth in Nepal.* Governance. Social Development, Humanitarian Response and Conflict (GSDRC), Applied Knowledge Services of University of Birmingham, Birmingham, UK. http://www.gsdrc.org/wp-content/uploads/2015/11/HDQ1294.pdf

BBS (2012a). *Population Census, 2011: Community Report, Bandarban Zilla.* Bangladesh Bureau of Statistics. http://203.112.218.66/WebTestApplication/userfiles/Image/Census2011/Chittagong/Bandarban/Bandarban%20at%20a%20glance.pdf

BBS (2012b). *Population Census, 2011: Community Report, Khagrachhari Zilla.* Bangladesh Bureau of Statistics. www.khdcbd.org/khagracori/assets/uploads/files/Khagrachhari-at-a-glance.pdf

BBS (2012c). *Population Census, 2011: Community Report, Rangamati Zilla.* Bangladesh Bureau of Statistics. http://203.112.218.66/WebTestApplication/userfiles/Image/Census2011/Chittagong/Rangamati/Rangamati%20at%20a%20glance.pdf

BBS (2014a). *Population and Housing Census, 2011, National Volume 2: Union Statistics.* Bangladesh Bureau of Statistics. http://203.112.218.65/WebTestApplication/userfiles/Image/National%20Reports/Union%20Statistics.pdf

BBS (2014b). *Population and Housing Census, 2011, National Volume 3: Urban Area Report.* Bangladesh Bureau of Statistics. http://203.112.218.65/WebTestApplication/userfiles/Image/National%20Reports/PopulationHousingCensus2011.pdf

Bhat, M. S. (2008). *Urban System in Himalayas: A Study of Srinagar City Region.* Dilpreet Publishing House, New Delhi, India.

Bhattacharya, S. & Banerjee, A. (2015). Water privatization in developing countries: principles, implementations and socio-economic consequences. *World Scientific News 4*, 17–31.

Bhattacharya, S., Chattopadhyay, D. & Mukhopadhyay, A. (2013). Changing dimensions of food security in a globalized world: a review of the perspectives for environment, economy and health. *International Research Journal of Environmental Sciences 2*(3), 1–7.

Boer, L., Joshi, D., Rai, S. & Thomas, S. (2011). *The Perennial Springs of Darjeeling, A Survey to Community Based Conservation. Internship Report.* Wageningen University, The Netherlands.

Buytaert, W. & De Bièvre, B. (2012). Water for cities: the impact of climate change and demographic growth in the tropical Andes. *Water Resources Research 48*(8), 1–13. doi: 10.1029/2011WR011755.

CBS (2003). Population monograph of Nepal: Volume I. Central Bureau of Statistics, Government of Nepal. Retrieved from http://cbs.gov.np/nada/index.php/catalog/42/download/913

CBS (2012). National population and housing census 2011: National report. Central Bureau of Statistics, Government of Nepal. Retrieved from https://unstats.un.org/unsd/demographic-social/census/documents/Nepal/Nepal-Census-2011-Vol1.pdf

CBS (2014). Population monograph of Nepal: Volume I. Central Bureau of Statistics, Government of Nepal. Retrieved from http://cbs.gov.np/image/data/Population/Population%20Monograph%20of%20Nepal%202014/Population%20Monograph%20of%20Nepal%202014%20Volume%20I%20FinalPrintReady1.pdf

Census of India (2001). http://www.censusindia.gov.in/2011-common/censusdataonline.html

Census of India (2011a). *Primary Census Abstract.* Office of the Registrar General & Census Commissioner, India.

Census of India (2011b). *Administrative Atlas of India.* Office of the Registrar General & Census Commissioner, India.

Census of India (2011c). *District Census Handbook, Assam, Dima Hasao District. Series 19, Part XIIB.* Directorate of Census Operations, Assam.

Census of India (2011d). *District Census Handbook, West Bengal, Darjeeling District. Series 20, Part XIIB.* Directorate of Census Operations, West Bengal.

Chakma, S. (2008). Pollution takes serious turn at Kaptai lake. *The Daily Star.* http://www.thedailystar.net/news-detail-63043

Correspondent (2013). Uttarakhand towns ill-equipped to handle tourist surge, study says. *The Times of India.* http://timesofindia.indiatimes.com/india/Uttarakhand-towns-ill-equipped-to-handle-tourist-surge-study-says/articleshow/20735253.cms

Correspondent (2016a). In monsoon, Shimla suffers severe water scarcity. *The Times of India.* http://timesofindia.indiatimes.com/city/shimla/In-monsoon-Shimla-suffers-severe-water-scracity/articleshow/53590123.cms

Correspondent (2016b). JICA to help Rs. 5 billion for Pokhara water supply project, Recentfusion.com. http://www.recentfusion.com/2016/07/17/jaica-to-help-rs-65-billion-pokhara-water-project/

Correspondent (2016c). Karak people complain about shortage of water. *Dawn.* http://www.dawn.com/news/1260438

Correspondent (2016d). Nagaland's water challenges need immediate attention: WRC brings out draft Nagaland Water Policy and asks for feedbacks. *The Morung Express.* http://morungexpress.com/nagalands-water-challenges-need-immediate-attention

Correspondent (2017a). 80 m domestic tourists travel in Pakistan every year: PTDC. *Daily Times.* http://dailytimes.com.pk/business/16-Mar-17/80m-domestic-tourists-travel-in-pakistan-every-year-ptdc

Correspondent (2017b). Nearly 13 lakh tourists visited JK in 2016: Govt. *Kashmir Observer.* https://kashmirobserver.net/2017/city-news/nearly-13-lakh-tourists-visited-jk-2016-govt-14362

Correspondent (2018). Water Scarcity leaves Shimla High and Dry. *Tribune.* http://www.tribuneindia.com/news/himachal/water-scarcity-leaves-shimla-high-and-dry/594998.html

Darjeeling Municipality (2018). *Total Water Demand and Supply Calculation of Darjeeling Town.* Darjeeling Municipality. http://dm.gensoftindia.co.in/waterworks/Demand-Supply.aspx (accessed 27 April, 2018).

Das, P. K. (2014). Mass tourism & environ-infrastructural crises of Shimla city: a case study. *International Journal of Science and Research (IJSR) 3*(11), 1781–1790.

Dolma, K., Rishi, M. S. & Herojeet, R. (2015). Baseline study of drinking water quality – A case of Leh town, ladakh (J&K). *India. Hydrological Current Research 6*(1), 1–6. doi: 10.4172/2157-7587.1000198.

Ghosh, P. (2007). Urbanization – A potential threat to the fragile Himalayan environment. *Current Science 93*(2), 126–127.

Goodall, S. K. (2004). Rural-to-urban migration and urbanization in Leh, Ladakh. *Mountain Research and Development 24*(3), 220–227. doi: https://doi.org/10.1659/0276-4741(2004)024[0220:RMAUIL]2.0.CO;2.

Government of Gilgit-Baltistan, Pakistan (2013). *Gilgit-Baltistan at A Glance 2013.* Statistical Cell, Planning and Development Department, Gilgit Baltistan. http://www.gilgitbaltistan.gov.pk/DownloadFiles/GBFinancilCurve.pdf

Government of Himachal Pradesh, India (2015). *State Annual Action Plan – Himachal Pradesh*. http://amrut.gov.in/writereaddata/HPSaap.pdf

Government of India (2005). *Himachal Development Report*. Planning Commission, New Delhi, India.

Government of Nepal (2010). *National Adaptation Programme of Action (NAPA) to Climate Change of Nepal*. Ministry of Environment, Kathmandu. http://unfccc.int/resource/docs/napa/npl01.pdf

Government of Nepal (2015). *Nepal Tourism Statistics 2015*. Ministry of Culture, Tourist & Civil Aviation, Planning & Evaluation Division, Kathmandu. http://www.tourism.gov.np/downloadfile/Nepal_Tourism_Statistics_2015_forwebsite_edited_1486627947.pdf

Government of Sikkim, India (2015). *Statistics of Tourist Arrival in the State of Sikkim*. Tourism and Civil Aviation Department, Gangtok. http://scstsenvis.nic.in/WriteReadData/links/tourist%20arrival%20stats%20in%20sikkim-777822810.pdf

Groot, A., Singh, T., Pandey, A., Gioli, G., Ahmed, B., Ishaq, S., Raza, N., Hassan, M., Hassan, S. M. T., Syed, M. A., Namnun, N. & Eijrond, V. (2018). *Literature Review of Critical Climate-Stress Moments in the Hindu Kush Himalaya: A Resource kit*. ICIMOD. http://lib.icimod.org/record/33837

Hasan, A. & Raza, M. (2009). *Migration and Small Towns in Pakistan. International Institute for Environment and Development (IIED)*. http://pubs.iied.org/pdfs/10570IIED.pdf

Hasan, S. R., Islam, S. & Hasan, M. M. (2013). *Rapid Assessment of the Context and Current Status of Tourism Development and Management in Bandarban*. Dhaka University and International Centre for Integrated Mountain Development (ICIMOD). https://www.academia.edu/5367560/TOURISM_DEVELOPMENT_AND_MANAGEMENT_IN_BANDARBAN?auto = download

Hunzai, K., Gerlitz, J.-Y. & Hoermann, B. (2011). *Understanding Mountain Poverty in the Hindu Kush-Himalayas Regional Report for Afghanistan, Bangladesh, Bhutan, China, India, Myanmar, Nepal, and Pakistan*. International Centre for Integrated Mountain Development. http://lib.icimod.org/record/9376 (accessed 4 September 2018).

Huq, M. M. (2001). Markets, ethnicity and development in the Chittagong Hill Tracts of Bangladesh. In: *Market Towns in the Hindu Kush-Himalayas: Trends and Issues*, Sharma, P. (ed.), ICIMOD, Kathmandu, pp. 117–127.

ICIMOD (2002). *The Hindu Kush-Himalayan Hydrological Cycle Observing System (HKH-HYCOS): Establishment of A Flood Forecasting Information System in the Hindu Kush-Himalaya*. A regional component of World Hydrological Cycle Observing System (WHYCOS) of World Meteorological Organization, 60 pp. http://www.whycos.org/whycos/sites/default/files/public/pdf/projects/hkh/hkh-hycos.pdf

IHCAP (2017). *Urbanisation Challenges in the Himalayan Region in the Context of Climate Change Adaptation and Disaster Risk Mitigation*. http://ihcap.in/media/Urbanization.pdf

Islam, M. S. & Khan, M. N. (2013). Urban spatial dynamics. In: *Bangladesh Urban Dynamics*. Rahman, H. Z. (ed.). Power and Participation Research Centre, Dhaka, Chapter 2, pp. 11–50.

IUCN (2015). *Water Supply and Sanitation Situation in Selected Areas of Murree Hills*. http://waterinfo.net.pk/sites/default/files/knowledge/Water%20Supply%20and%20Sanitation%20Analysis%20In%20Selected%20Areas%20of%20Murree%20Hills%20-%20December%202014.pdf

Kapoor, M., Kohli, K. & Menon, M. (2009). *India's Notified Ecologically Sensitive Areas (ESAs): The Story So Far ...*. Kalpavriksh, Delhi & WWF-India, New Delhi. http://assets.wwfindia.org/downloads/indias_notified_ecologicallysensitive_areas.pdf

Karn, P. K. (2008). Making payment for environmental services (PES) work: a case study of Shivapuri national park, Nepal. In: *Shifting Paradigms in Protected Area Management*. Bajracharya, S. B. & Dahal, N. (eds). National Trust for Nature Conservation, Kathmandu, pp. 171–185.

Khan, Z. H. (2015). Study on water supply system at Rangamati municipal area, Chittagong Hill Tracts. *World Journal of Environmental Engineering 3*(1), 15–22.

Khatri, D. (2016). Tansen reeling under water shortage. *The Himalayan Times*. https://thehimalayantimes.com/nepal/tansen-reeling-water-shortage

Khawas, V. (2005). Urbanisation in the North-East: patterns, trends, and policy prongs. *Social Change 35*(2), 47–69. doi: 10.1177/004908570503500204.

Kuchay, N. A. & Bhat, M. S. (2014a). Analysis and simulation of urban expansion of Srinagar city. *Transactions of the Institute of Indian Geographers 36*(1), 110–121.

Kuchay, N. A. & Bhat, M. S. (2014b). Urban sprawl of Srinagar city and its impact on wetlands – A spatio-temporal analysis. *International Journal of Environment and Bioenergy 9*(2), 122–129.

Kumar, A. & Pushplata (2013). Building regulations for environmental protection in Indian hill towns. *International Journal of Sustainable Built Environment 2*, 224–231. http://dx.doi.org/10.1016/j.ijsbe.2014.04.003

LAHDC (2011). *Tourist Reception Centre, Leh*. Ladakh Autonomous Hill Development Council. http://leh.gov.in/depts/tour/def.html (accessed 18 February 2017).

Lepcha, N. K. C. (2013). Problems and prospects of water resource of Kurseong municipality, Darjeeling district. *West Bengal. Science Park Research Journal 1*(19), 1–11.

Lyngdoh, A. W. (2016). Northeast witnesses rise in tourists. *The Telegraph*. https://www.telegraphindia.com/1160724/jsp/northeast/story_98424.jsp

Maharjan, A. (2017). *Summary for Policy Makers: De-Populating Villages in the Kailash Sacred Landscape, India: Rethinking Policy Interventions*. International Centre for Integrated Mountain Development (ICIMOD). http://lib.icimod.org/record/32595/files/icimodSPM017.pdf

Maharjan, A., Hussain, A., Bhadwal, S., Ishaq, S., Saeed, B. A., Sachdeva, I., Ahmad, B., Hassan, S. M. T., Tuladhar, S. & Ferdous, J. (2018). *Migration in the Lives of Environmentally Vulnerable Populations in Four River Basins of the Hindu Kush Himalayan Region*. HI-AWARE Working Paper 20, Kathmandu. http://lib.icimod.org/record/34346

Mamgain, R. P. & Reddy, D. N. (2015). *Outmigration From Hill Region of Uttarakhand: Magnitude, Challenges and Policy Options*. http://www.nird.org.in/nird_docs/srsc/srscrr261016-3.pdf

Mathieu, J. (2003). *The Mountains in Urban Development: Lessons From A Comparative View*. Academy of Architecture, Mendrisio, Switzerland. http://www2.arc.usi.ch/index/aisa/ris_ist_labi_aisa_rivista.htm/2003_2_labi_rivista.pdf

McDonald, R. I., Weber, K., Padowski, J., Flörke, M., Schneider, C., Green, P. A., Gleeson, T., Eckman, S., Lehner, B., Balk, D., Boucher, T., Grill, G. & Montgomery, M. (2014). Water on an urban planet: urbanization and the reach of urban water infrastructure. *Global Environmental Change 27*, 96–105. http://dx.doi.org/10.1016/j.gloenvcha.2014.04.022.

Misra, P. R. (2016). Water worries back to haunt Mussoorie. *The Times of India*. http://timesofindia.indiatimes.com/city/dehradun/Water-worries-back-to-haunt-Mussoorie/articleshow/51565218.cms

Molden, D. J., Vaidya, R. A., Shrestha, A. B., Rasul, G. & Shrestha, M. S. (2014). Water infrastructure for the Hindu Kush Himalayas. *International Journal of Water Resources Development 30*(1), 60–77.

Mukhopadhyay, A. & Bhattacharya, S. (2013). Development and environmental issues vis-à-vis current perspectives of human rights. In: *Human Rights and the Third World: Issues and Discourses*. Bagchi, S. S. & Das, A. (eds). Lexington Books, Rowman and Littlefield Publishing Group, Lanham, New York, USA, pp. 59–80.

Muzzini, E. & Aparicio, G. (2013). *Urban growth and spatial transition in Nepal: An initial assessment*. The World Bank. Washington DC, USA. doi: 10.1596/978-0-8213-9659-9

Nepal, S., Bhavana, V. T. & Nirmana, V. (2007). *National Urban Policy, 2007*. Dept. of Urban Development and Building Construction, Ministry of Physical Planning and Works, Kathmandu, Nepal.

Pakistan Bureau of Statistics (1998). *Population Census 1998: Population by Administrative Units*. www.pbs.gov.pk/sites/default/files/population_census/Administrative%20Units.pdf

Pandey, V. P., Chapagain, S. K., Shrestha, D., Shrestha, S. & Kazama, F. (2016). *Groundwater Markets for Domestic Water use in Kathmandu Valley: An Analysis of its Characteristics, Impacts and Regulations*. Academia.edu, https://www.academia.edu/11198401/Groundwater_markets_for_domestic_water_use_in_Kathmandu_Valley_an_analysis_of_its_characteristics_impacts_and_regulations

Pathak, S., Pant, L. & Maharjan, A. (2017). *De-population Trends, Patterns and Effects in Uttarakhand, India – A Gateway to Kailash Mansarovar*. Working Paper 22, International Centre for Integrated Mountain Development, Kathmandu, Nepal, p. 46. http://lib.icimod.org/record/32787/ (accessed 5 March 2018).

Perlik, M. (2014). Mountain-specific urbanization or an urbanization in the mountains: Which role model for the future? In: *Paper Presented at the Second World Mountain Forum Conference*, Cusco, Peru. https://worldmountainforum.files.wordpress.com/2014/05/manfred-perlik.pdf

Pokhriyal, H. C. (2001). Characteristics and role of urban settlements in the Indian Himalayas. In: *Market Towns in the Hindu Kush-Himalayas: Trends and Issues*. Sharma, P. (ed.). ICIMOD, Kathmandu, pp. 57–73.

Portnov, B. A., Adhikari, M. & Schwartz, M. (2007). Urban growth in Nepal: does location matter? *Urban Studies 44*(5/6), 915–937.

PSI (2003). *Survival Lessons: Water Management Traditions in the Central-Western Himalayas*. Peoples' Science Institute, Dehradun. http://peoplesscienceinstitute.org/research/Water%20Management%20Traditions%20in%20W%20Himalayas.pdf

Rai, R. P., Shrestha, P., Sunar, D., Rao, P. & Tamang, L. (2016a). Water security in the Darjeeling Himalaya, unravelling the seen and unseen forces. Darjeeling Himalayan Initiative. In: *Paper Presented at the Sustainable Mountain Development Summit*, 20–22 September 2016, Ladakh.

Rai, R. K., Shyamsundar, P., Bhatta, L. D. & Nepal, M. (2016b). *Designing a payment for ecosystem services scheme for the Sardukhola watershed in Nepal*. SANDEE (South Asian Network for Development and Environmental Economics) Working Paper no. 108-16. ISSN 1893-1891. Kathmandu, Nepal. Retrieved from http://lib.icimod.org/record/32020

Raina, A. (2016). *Equity in Urban Water Service Delivery and the Role of Informal Water Vendors: The Case of Kathmandu Valley, Nepal*. Doctoral dissertation, National University of Singapore, Singapore.

Rouse, M. (2014). The worldwide urban water and wastewater infrastructure challenge. *International Journal of Water Resources Development 30*(1), 20–27. http://dx.doi.org/10.1080/07900627.2014.882203.

SANDRP (2013). *Uttarakhand and Climate Change: How Long can we Ignore This in the Himalayas?* South Asia Network on Dams, Rivers and People. https://sandrp.wordpress.com/2013/06/25/uttarakhand-and-climate-change-how-long-can-we-ignore-this-in-himalayas/

Sarkar, D. (2013). Darjeeling hills showing highly positive tourist inflow picture. *The Economic Times*. http://economictimes.indiatimes.com/industry/services/travel/darjeeling-hills-showing-highly-positive-tourist-inflow-picture/articleshow/20244004.cms

Satterthwaite, D. (2006). *Outside the Large Cities: The Demographic Importance of Small Urban Centres and Large Villages in Africa, Asia and Latin America*. Human Settlements Working Paper Series Urban Change No. 3, IIED, London. http://pubs.iied.org/pdfs/10537IIED.pdf

Shah, M. (2016). *Urban Water Systems in India: A way Forward*. Indian Council for Research on International Economic Relations Working paper 323. http://icrier.org/pdf/Working_Paper_323.pdf

Shah, R. & Badiger, S. (2018). Conundrum or paradox: deconstructing the spurious case of water scarcity in the Himalayan Region through an institutional economics narrative. *Water Policy*. doi: https://doi.org/10.2166/wp.2018.115.

Shahbaz, B., Shah, Q. A., Suleri, A. Q., Commins, S. & Malik, A. A. (2012). Livelihoods, basic services and social protection in north-western Pakistan. Retrieved from Secure Livelihoods Research Consortium. https://www.odi.org/sites/odi.org.uk/files/odi-assets/publications-opinion-files/7782.pdf

Shahid, S. (2016). International firms to be hired for Quetta water project. *Dawn*. https://epaper.dawn.com/DetailImage.php?StoryImage = 21_06_2016_005_009

Sharma, P. (2001). Market and small towns in HKH: perspective and issues. In: *Market Towns in the Hindu Kush-Himalayas: Trends and Issues*. Sharma, P. (ed.). ICIMOD, Kathmandu, pp. 1–16.

Sharma, J. R. (2011). Culture of migration in the middle hills in Nepal. In: *Labor Migration: Opportunities and Challenges for Mountain Livelihoods*, Sustainable Mountain Development Series Vol. 59, International Centre for Integrated Mountain Development (ICIMOD), pp. 18–20. Kathmandu, Nepal. Retrieved from http://lib.icimod.org/record/26975

Sharma, A. K., Singh, S. K., Kulkarni, A. V. & Ajai (2013a). Glacier inventory in Indus, Ganga and Brahmaputra basins of the Himalaya. *National Academy Science Letters 36*(5), 497–505. doi: 10.1007/s40009-013-0167-6.

Sharma, S., Kumar, K. & Singh, K. K. (2013b). *Water Security in the mid-Elevation Himalayan Watershed, East District with Focus in the State of Sikkim*. Paper presented at Kohima Workshop, September, 2013. http://www.indiawaterportal.org/sites/indiawaterportal.org/files/final_water_security_paper_santosh_sharma.pdf

Shrestha, D. & Shukla, A. (2014). Private water tanker operations in Kathmandu: analysis of water services and regulatory provisions. In: *Globalization of Water Governance in South Asia*. Narain, V., Goodrich, C. G., Chourey, J. & Prakash, A. (eds). Routledge, New Delhi, pp. 256–272.

Singh, T. & Kandari, L. S. (2012). Rainwater harvesting in the wake of climate change: a case study from Shimla city, Himachal Pradesh. *Universal Journal of Environmental Research and Technology 2*(4), 336–346.

Subedi, P. (2010). *Drinking Water Management in Newa State*. Centre for Constitutional Dialogue, Kathmandu, Nepal.

Tambe, S., Kharel, G., Arrawatia, M. L., Kulkarni, H., Mahamuni, K. & Ganeriwala, A. K. (2012). Reviving dying springs: Climate change adaptation experiments from the Sikkim Himalaya. *Mountain Research and Development 32*(1), 62–72. doi: 10.1659/MRD-JOURNAL-D-11-00079.1

Thapa, R. B. (2017). Kathmandu metropolitan area. In: *Urban Population in Asia and Africa: Geospatial Analysis of Metropolises*. Murayama, Y., Yamashita, A., Kamusoko, C. & Estoque, R. C. (eds). Springer Nature, Singapore, pp. 217–237.

Tiwari, P. C. & Joshi, B. (2014). Global change and mountains: consequence, response and opportunities. In: *Impact of Global Changes on Mountains: Responses and Adaptation*. Grover, V. I., Borsdorf, A., Breuste, J., Tiwari, P. C. & Frangetto, F. W. (eds). CRC Press, Boca Raton, FL, pp. 79–136.

Tiwari, P. C., Tiwari, A. & Joshi, B. (2018). Urban growth in Himalaya: understanding the process and options for sustainable development. *Journal of Urban and Regional Studies on Contemporary India 4*(2), 15–27.

Tortajada, C. (2014). Water infrastructure as an essential element for human development. *International Journal of Water Resources Development 30*(1), 8–19. http://dx.doi.org/10.1080/07900627.2014.888636

Udmale, P., Ishidaira, H., Thapa, B. & Shakya, N. (2016). The status of domestic water demand: supply deficit in the Kathmandu valley, Nepal. *Water 8*(5), 196.

UNDESA (2014). *World's Population Increasingly Urban with More Than Half Living in Urban Areas*. United Nations Department of Economic and Social Affairs. http://www.un.org/en/development/desa/news/population/world-urbanization-prospects-2014.html (accessed 25 March 2017).

WEF (2017). *The Global Risks Report 2017*. World Economic Forum, Geneva, Switzerland, p. 78.

Wikipedia (2017). List of rivers by discharge. Wikipedia. https://en.wikipedia.org/wiki/List_of_rivers_by_discharge

Wunder, S. (2008). How do we deal with leakage? In: *Moving Ahead with REDD: Issues, Options and Implications*, Angelsen, A. (ed.), Center for International Forestry Research (CIFOR), Indonesia, pp. 65–75.

Yhokha, V. & Resu, Z. (2015). Kohima water crisis: Short of 8.8 million liters: Nagaland state capital receives only 10% of the daily requirement of water. *The Morung Express*. http://morungexpress.com/kohima-water-crisis-short-of-8-8-million-liters/

Yi, Z., Yinfu, L. & Zenfeng, C. (2016) Research on resource environmental bear capacity of Yunnan mountain area – A case study of Yunnan Longchuan. *Yunnan Geographic Environment Research 28*, 29–34.

Received 30 October 2018; accepted in revised form 20 December 2018. Available online 12 February 2019

Urban water resilience in Hindu Kush Himalaya: issues, challenges and way forward

Vishal Singh* and Anvita Pandey

Centre for Ecology, Development and Research (CEDAR), Dehra Dun, India
**Corresponding author. E-mail: cedarhimalaya@gmail.com*

Abstract

The urban population is expected to rise up to 68% by 2050, adding 2.5 billion people to the urban areas of the world. The majority of the rise is expected to be in the low-income countries of Asia and Africa. Several cities/ towns in the Hindu Kush Himalaya (HKH) region are expanding at a rapid pace, putting additional pressure on water services and basic amenities for urban dwellers. Selected case studies undertaken by the authors suggest that the demand for water far exceeds municipal supply. Water governance in the HKH region remains a blind spot and challenges pertaining to urban water resilience are poorly understood. The paper is divided into three parts: the first outlines the development of towns and their water infrastructure through selected cases in the HKH, followed by key issues and challenges faced by urban systems and suggested measures to build urban resilience in order to deal with the projected rise in population, governance issues and anticipated changes in climate.

Keywords: Climate; Hindu Kush Himalaya (HKH); Population; Resilience; Urban; Water

Introduction

Water resources are gradually depleting and the issues related to water quality are taking an important space in the governance of the water sector (Srinivasan *et al.*, 2012). Currently, 55% of the population live in urban areas, which are expected to rise up to 68% by 2050, adding another 2.5 billion people to the urban areas (UNDESA, 2018). Ninety per cent of the growth is expected to be concentrated in the low-income countries (Shah & Kulkarni, 2015), posing serious challenges to water managers and city

doi: 10.2166/wp.2019.329

planners to build resilience against anticipated population and climate pressures. The term 'resilience' (Jalilov *et al.*, 2017) is defined as

> *'the capacity of urban water infrastructure to cope in the short term, and adapt and develop in the long term in the face of unforeseen changes such as major system failures, acute water infrastructure degradation, inadequate drinking water and wastewater infrastructure, and/or natural disasters.'*

In a broad sense, planning for sustainable water futures would require greater degrees of understanding around contemporary and future demand and supply (Cosgrove & Loucks, 2015). Development is the key concern of many low-income countries, which results in rapid and unprecedented urban growth (Stewart, 2017). This generates the need for larger supplies of resources, mainly water (Wakode *et al.*, 2018).

Approximately 1.3 billion people living in rural and urban areas across the Hindu Kush Himalaya (HKH) and downstream basin are affected by climate change (Eriksson *et al.*, 2009). The Himalayan region is warming at rates two to three times faster than the global average (Singh *et al.*, 2010). The towns and cities in the HKH are rapidly expanding and are subjected to extreme weather events as a result of climate change and unplanned urbanization (Sharma *et al.*, 2014).

Urban growth impacts several environmental components, i.e., land use change, hydrology, climate, biogeochemical cycle and biodiversity, which in turn, affects urban systems (Grimm *et al.*, 2008). The process of urbanization in the Himalayan region has largely been haphazard (Tiwari *et al.*, 2018), making it an important driver of environmental change (Walker, 2011). The fast rate of urbanization has impacted hydrological regimes and reduced groundwater recharge and decreased the availability of drinking water and sanitation (Tiwari *et al.*, 2018). The stress of climate change is manifested by an increase in extreme weather events in urban and rural areas of Himalaya (Balk *et al.*, 2009; Djebou & Singh, 2016).

In the HKH region, urbanization has largely remained outside the climate change debate, which has mostly been dominated by glacier melt and its consequences on river discharge (Hofer & Messerli, 2006). Water governance in mountains and hills remains a blind spot in the water governance literature. Compared to other geographies and water systems, mountain water and governance systems remain relatively unknown and largely unseen. In HKH states (Figure 1) and provinces of Bangladesh, India, Nepal and Pakistan, the urban population is growing between 3% and 5% (Singh *et al.*, 2019). This rapid pace of urbanization in the HKH raises important questions concerning the adaptive capacity of urban governance systems to changes in water availability and highlights the urgent need for crafting governance mechanisms for the management of a public good required by everyone. Floods and droughts, and the projected increase in extreme rainfall and storm events due to climate change in the region (Zhan *et al.*, 2018), pose serious challenges to urban water resilience.

This article will focus on cities/towns in the HKH region that are facing their own sets of challenges. In these towns, the influx of migrants, tourists, combined with the growth of the incumbent population, places considerable stress on the urban infrastructure to supply adequate clean water and sewage disposal, while the Himalayan towns are already grappling with issues and challenges for urban water resilience (Tiwari *et al.*, 2018), drawing on the limited information available and research and observations made by the authors. In order to build a holistic perspective, the paper is divided into three parts: the first part focuses on major urban centres and their water issues, followed by key issues and main challenges and issues facing urban water governance across selected case studies. We conclude with a number of recommendations concerning future strategies and actions that need to be taken to strengthen urban climate resilience in the region.

Fig. 1. Hindukush Himalayan region.

Development of urban centres and water crises in major cities of HKH

Several cities in the larger HKH region, from Afghanistan to Bhutan, are suffering from water crises and have put severe pressure on existing water resources (Table 1).

The Indian Himalayan Region (IHR) occupies 14% of the area of the Indian subcontinent, spreading across ten states (administratively) of Jammu and Kashmir, Himachal Pradesh, Uttarakhand, Sikkim, Arunachal Pradesh, Meghalaya, Nagaland, Manipur, Mizoram, Tripura, and the hill areas of the states of Assam and West Bengal.

India

In Jammu and Kashmir, Srinagar is the most important city as it is the state capital as well as an important tourist destination. Leh is a very high-altitude settlement of Jammu and Kashmir, with a sparse population, but has grown in importance as a tourist destination. Likewise, Shimla and Manali in the state of Himachal Pradesh and Mussoorie and Nainital in the state of Uttarakhand are the most popular tourist destinations of Northern India. These urban centres have witnessed a high rate of urbanization. In

Table 1. Demand and supply and major issues in selected cities of HKH Himalaya.

S. no.	City	Supply (Million Liters Per Day (MLD))	Demand (Million Liters Per Day (MLD))	Major issues
1.	Kabul (Afghanistan)	52.14	–	Limited distribution due to high energy cost. Only 20% of total population is supplied with water
2.	Quetta (Pakistan)	106	170	Rapid population growth, low rainfall, ground water depletion, water deficit of 50%
3.	Shimla (India)	54.5	64.7	Lack of new sources, high tourist influx, leakages in pipelines. 84.2% demand is met
4.	Mussoorie (India)	7.67	14.4	Poor infrastructure, tanker economy, high tourist influx, recharge zones degraded
5.	Nainital (India)	20.5	18.2	Depletion of lake level which is the main source of water, encroachment over recharge zones, increase in impermeable areas
6.	Kathmandu (Nepal)	86	280	Rapid influx of migrants, tourist influx, poor infrastructure, poor governance
7.	Xining (China)	–	–	Water shortage deterioration water quality
8.	Thimpu	–	9.9	Water contamination and shortage of water increase in impermeable surface causing flooding

addition, in Uttarakhand, a large number of pilgrims visit the temples of Kedarnath, Badrinath, Gangotri and Yamnotri (Char Dham) across the country. As per the figures released by Uttarakhand police, approximately 2.7 million pilgrims visited the Char Dhams of Uttarakhand in 2018. This has created multiple forms of opportunities in the large and small towns of the state, i.e., expansion of hotel industry, eateries along the route, taxi services and others. Further, in many cases, e.g., the government of the Indian state of Uttarakhand's failure to create opportunities and develop basic infrastructure in rural areas has also led to migration to cities. For example, since the formation of Uttarakhand state in 2000, the state has posted an impressive growth rate of GSDP between 2004–05 and 2015–16, with a focus on secondary and tertiary sectors of economy. The primary sector, agriculture, upon which approximately 70% of the population of the state depends, is not within this story of successful growth. Given the urban nature of secondary and tertiary sectors, the urban population, at roughly 26%, emerges as the major beneficiary of growth, and the lack of focus on agriculture and allied activities and industry has led to persistent and continuing urban–rural disparities (Singh & Pandey, 2018).

Nepal

It remains one of the least urbanized countries in South Asia and the world. The problems of definition in the study of Nepal's urbanization are considerable as the areas designated 'urban' have been defined and redefined over the years, and there is an evident lack of consistency in the definitions used (Devkota, 2018). Urbanization viewed cross three broad ecological regions – mountains, hills and Terai – shows enormous differences in urbanization volume and forms. In the hill region there is a stark difference between Kathmandu Valley, the traditional hub of Nepal's urbanization, and the rest of the hills. As a result of tourism, population pressure, poor infrastructure and governance, the demand and supply gap is approximately 70%. Kathmandu valley on the whole faces a shortage of 210 MLD. The mountain ecological region has only recently acquired a meagre share of the urban population, while the number of urban centres in the Terai has grown rapidly (Ishtiaque *et al.*, 2017). The hill region of Nepal is the meeting place of people coming from the north and the south of the country with a population density of 167.1 persons per square kilometre and comprises 44.3% of the total population of the country (Suwal, 2013). This zone has 39 districts with many of the large urban centres of the country, which have grown in importance as tourist and commercial places. Pokhara, for example, has grown at a very rapid rate due to tourism (Tiwari, 2012). Shortage of water, ineffective water management practices and rising population are key concerns. Pokhra's water supply is managed by Nepal Drinking Water Cooperation. The current supply is only intended for 8,000 households, and the households in 2014 were 34,000 (Pokharel, 2014). Baglung municipality is a banking and financial hub of the region, with the presence of branches of several national banks and regional offices. A large number of people from Baglung are employed in India and further afield and, therefore, the town in particular and the region in general receives the highest amount of remittances from foreign employment in Nepal (Sharma *et al.*, 2014). The town experienced a rapid expansion after the completion of the Pokhara–Baglung highway in the mid-1990s and a summer highway connecting Baglung to Chaura.

Xining in China, the capital of Qinghai Province suffers from the dual challenges of water shortage and deteriorating quality. Severe degradation of the catchment area and pollution in the nearby Huangshui River have been major concerns. To meet the growing demands for water in the area around the city of Xining, the Qinghai Xining Water Environment Project, worth US$150 million, is financed with an International Bank for Reconstruction and Development (IBRD) loan. The project aims to invest in storm and wastewater collection pipes, rejuvenate the environment along the banks of Beichuan River and strengthen the municipality's capacity in integrated water environment management.

Pakistan

A study by the World Bank that analysed the water situation in Pakistan declared Pakistan to be one of the world's most arid and water-stressed (soon to be water-scarce) countries, on a par with those of the Sahara Desert (Khan, 2009). Quetta, the provincial headquarters of Baluchistan, is plagued by a growing water security crisis. Located at an elevation of 1,680 m, it is experiencing rapid population growth due to in-migration. The gap between demand and supply is widening and becomes acute during the low rainfall period (Kakar *et al.*, 2018). Approximately three million people residing in Quetta require some 170 MLD of water per day to meet their demands; however, the supply is only 106 MLD. Groundwater is the only source to meet this demand which has receded down to 300 m due to overexploitation. The Water and Sanitation Agency (WASA) could only manage to produce 100 million per day through 400 tubewells, thus creating a deficit of 50%. The residents rely heavily on tankers operated through informal arrangements. This has resulted in the overexploitation of the groundwater through some 2,000 tubewells, and this unchecked installation of tubewells is dangerously depleting the water resources of Quetta. The proposed Mangi Dam in Quetta district is likely to provide an additional eight million gallons per day, however this scheme will take time to become operational.

Afghanistan

Kabul, the capital of Afghanistan, has a water crisis which is of major concern. The 4.5 million residents of Kabul (2010 data) suffer from water deficit on a daily basis. By 2050, the population of Kabul is expected to reach nine million. Data from 2013 indicate that, currently, Afghanistan Urban Water Supply and Sewerage Company (AUWSSC) is able to provide 15 Lpcd of water, which is only able to supply less than 20% of its population (Mukherjee *et al.*, 2006). Despite having over 7,900 Lpcd of water in the basin, the distribution is limited due to high energy costs (Integrity Watch Afghanistan). Water-borne diseases and infant mortality are concerns due to the lack of sewerage network (only 3%) (Himmelsbach, 2010). Open wells and hand pumps are the only source of water for the majority of the population. As in the case of Quetta, Pakistan, illegal borewells are in abundance in Kabul, and residents have to pay a high price (0.36 USD) to procure water from tankers (Stanikzai, 2014). These illegal borewells have contributed to the unsustainable depletion of groundwater, and the water table dropped by 15 m from 2008 to 2012 (Snyder, 2014). The United Nations Economic and Social Commission for Asia and the Pacific (UNESCAP, 2008) estimated per capita water consumption in Afghanistan to be less than half that of other Central and West Asian countries. With an improving standard of living and a projected population of nine million by 2057, an estimated six-fold water consumption increase in the Kabul Basin was simulated in the groundwater-flow model described by Mack *et al.* (2010).

Bhutan

Water is a means of subsistence for over 60% of the population (Dorji, 2016). The urban areas of Bhutan are facing drinking water and sanitation problems due to migration from rural areas. Cities, i.e., Thimou and Puentsholing, are finding it difficult to meet the demand and supply the gap (Bhujel, 2019). A limited supply of treated water, inadequate distribution systems, and flooding due to increase in non-permeable surfaces are primary issues faced by the two cities (Bhujel, 2019).

Key issues and challenges for urban water resilience in HKH

The HKH region is characterized by a unique hydrogeology, in which springs play a crucial role (Niti Aayog, 2018). Aquifers feed into springs which are the main water sources for urban and rural areas. Springs also contribute to base flows in the streams and rivers. In many parts of the HKH, springs fed during the monsoon by groundwater or underground aquifers are reported to be drying up (NITI Aayog, 2018), threatening a whole way of life for local communities. A recent assessment by the Advanced Centre for Water Resource Development and Management (ACWADAM) for the Springs Initiative – through its research in various mountain regions of India, including the Indian Himalaya states of Ladakh, Himachal Pradesh, Uttarakhand, Sikkim, Nagaland and a few pilot areas in Nepal – has estimated that there are nearly four million springs in the Himalaya and that more than 60% of the population living in the IHR is directly dependent on springs (NITI Aayog, 2018). Springs are a safe source of drinking water for rural and urban communities, and with increasing population and urbanization there is an ever-increasing demand for spring water in the region.

Many towns are now turning to borewells for their water supply (Wankhade *et al.*, 2014). This has led to depletion of groundwater and further drying up of springs. Recharge ponds and lakes also play an important role in feeding springs, lakes and streams (MoWR, 2016), but many of these are being filled in or encroached upon due to unplanned and unregulated construction. In response to water scarcity, many towns/cities are sourcing water from distant sources, even including inter-basin transfers.

The cases of Shimla and Almora have shown that there has been very little focus on conserving the existing water sources which have sustained these towns since their existence. For example, the town of Almora used to depend on springs (more than 300 springs at one point in time), but now lifts water from the Kosi River. The capacity of pumps and the water extracted has increased over time, but when the river dries up in summer the town is left with no options (Sharma *et al.*, 2014). In Mussoorie, an ambitious scheme has been proposed to lift water from the Yamuna River approximately 18 km from the town through a process of four-stage pumping. Nainital, conversely, has shifted from complete dependence on springs to 95% dependence on lake bank filtration (Dimri & Dash, 2012). Most springs in the catchment have either become seasonal or have completely dried up. Only one spring remains which provides the remaining 5% of water (Shah & Tewari, 2009).

Pumping water from far distances is likely to be disrupted in extreme weather events, due to blockage, damage to pipes, electricity failure or changes in the source from where water is extracted. Such experiences repeatedly demonstrate that engineering solutions, wherein water is lifted from distant sources at great costs, will not alone be able to sustain the growing urban clusters of the Himalaya in the decades to come. Extremely heavy capital investment, high running costs and depletion of even these distant water sources due to catchment degradation and climate change make this mode of water supply unsustainable in the long run unless augmented by some alternative solutions that reflect local preferences and extant water practices. There is huge scope for undertaking nature-based local level measures such as springshed treatment, protection and rejuvenation of recharge zones of springs and lakes and tapping into indigenous knowledge, but these practices are hardly in use by governments and municipalities even where the scope is high and robust research exists.

Climate change impacts have already been manifested in an increased occurrence of floods and droughts in the region, which affect urban centres more because of the haphazard urban growth that is underway. The Uttarakhand floods of 2013 resulted in havoc for the residents of the urban areas of Chamoli and Rudraprayag and further downstream, while incessant rains for over 24 hours caused severe water logging in Srinagar city in Kashmir in 2014 and 2015. Rapid expansion of the population in urban centres in Afghanistan, Pakistan and low-lying cities of the IHR of India have resulted in severe groundwater extraction and depletion (Rodell *et al.*, 2009). A modest body of work on mountain and hill irrigation exists, but an overarching perspective on water governance in mountain areas is lacking. This is especially true for urban water governance in the mountains and hills of the HKH region, where cities/towns are rapidly growing and facing increasingly serious water challenges. The provision of basic infrastructure such as roads, power, drainage, water, sanitation, recreation, communication and trading facilities demands strong government investment/role.

Challenges pertaining to urban water resilience are poorly understood in the HKH region, although in recent years some attention has been received from researchers and policy-makers. However, huge gaps exist at research and policy level. The HKH region generally evokes the image of idyllic environs untouched by rapid urbanization taking place in plain areas. Hence, the majority of water augmentation/security schemes is focused on large metro-political cities or rural areas. However, contrary to this, 27% of the population in the IHR is urban, which is just 4% below the national average (Sharma *et al.*, 2014). The majority of towns in the HKH region depends on spring water to varying extents. Towns such as Shimla and Mussoorie depend entirely on spring water. Climate change is projected to reduce the spring discharge by increasing evapotranspiration loss, runoffs due to increased rainfall intensity and an increase in impervious surface due to concretization. The current practice of lifting water from distant sources is the norm which is expensive and technically challenging.

Changing climatic conditions combined with population rise, tourism pressure and unplanned developmental activities have also been noticed in several Indian Himalayan urban settlements like Shimla, Manali, Palampur, Almora, Leh and Gangtok, as shown by Shah & Kulkarni (2015). In Gangtok, for example, migration from the surrounding rural areas is leading to the possibility of Gangtok reaching a million inhabitants by 2025. In Dharamshala, reduced rainfall is developing into a water crisis, which is further exacerbated by the tourism industry. The town receives only 2 hours of water supply daily and there are several mismanagement issues which are primarily responsible for the growing crisis.

Of direct relevance to urban water resilience in the region is the widely reported drying up of springs, depleting groundwater resources on which many urban water systems depend (Pandey, 2018). Some towns also depend on surface water bodies whose catchment areas are being degraded due to urbanization and increases in built-up areas. This unregulated growth prevents aquifer recharge into the ground due to the concretization of what used to be waterways, and soil compaction which also increases run-off that can result in flooding and soil erosion. For example, Phew Lake near the city of Pokhara in Nepal is subject to heavy siltation and dumping of wastes, which affects water availability and quality for residents. Similarly, Lake Nainital is experiencing consistent decline during the summer season as a result of mismanagement of its 'upstream' recharge zone (Dash *et al.*, 2008). There is currently no practice or protocol to include hydrogeology in urban planning, therefore no concept of 'protecting' recharge zones.[1] Recent initiatives have provided promising approaches to spring water management and 'springshed development' activity (Tambe *et al.*, 2012; Mahamuni & Upasani, 2011). Challenges and issues related to urban water are given in Figure 2.

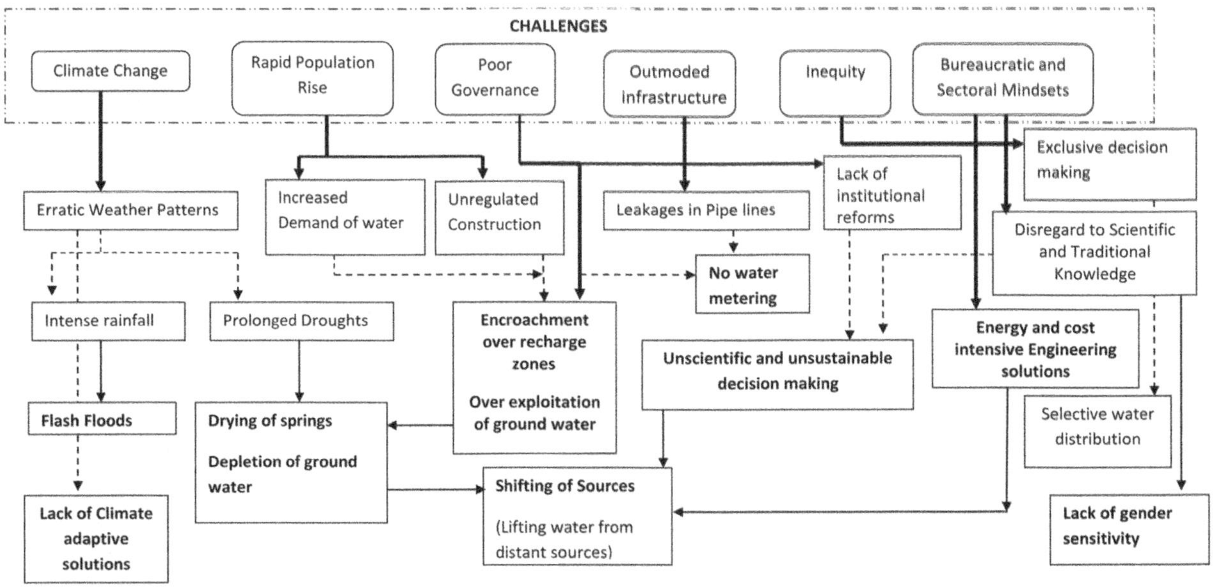

Fig. 2. Challenges and issues related to urban water management in HKH.

---

[1] Recharge zone is the area around an aquifer which replenishes water stored in an aquifer through precipitation or surface water.

Recommendations

Promoting urban water resilience will require an understanding of the vulnerability of urban water users and ecosystems to water-related hazards (Romero-Lankao & Gnatz, 2016), which needs to involve planning, policy and action, keeping in mind the mountain characteristics to align urban sustainability and sustainable urban livelihoods. The capacity of urban water users to perceive risks and develop effective responses needs to build.

In the context of the HKH region, climate change is an important factor that needs consideration in policy and practice through innovative approaches and tools that help create water-resilient cities and ensure quality water services. However, urban water systems cannot be looked at in isolation, hence, integrated multi-sectoral management will be required as industries, the environment and citizens pose competing demands within cities and catchments. These involve:

1. Mapping of water sources and important recharge, filtration or reservoir areas, understanding the demand and supply, keeping in mind the future population projections and changing climatic conditions.
2. Nature of the dependence of local communities on hydrological flows and developing a deeper understanding of recharge zones is also needed.
3. Research needs to be conducted in HKH towns to explore the links between urban and peri-urban water situations; to anticipate and understand possible future changes in water demand and supply with rising populations.
4. Document the links between climate-induced water issues at present and in the future, given the transforming pressures of urbanization in these towns; and to assess how far these towns have incorporated adaptive water governance in the context of changing climatic conditions and related extreme events.

In general, recharge areas of springs are also not well protected and Himalayan towns have generally not included spring protection and restoration in their planning process. Springs are further threatened by changing land use, ecological degradation, exploitation and climate change. Efforts to protect these vital water resources can help ensure water security for towns, but springs have not received adequate attention in the past and present. This has led to a significant data gap in terms of understanding spring water flow dynamics, watershed-related linkages, characteristics of local springs, and the factors leading to their decline in the HKH region.

The water supply of many cities in the HKH currently depends on or plans to source water from distant sources through emerging solutions which are cost and energy intensive and undesirable in the geologically and climate sensitive HKH.

5. Nature-based solutions such as rainwater harvesting, conservation of recharge zones and springshed management hold a great deal of promise; however, such practices have remained confined to municipal bye-laws.
6. Policies around conservation/protection of recharge zones, i.e., spring sheds, recharge areas of lakes and groundwater extraction and catchment areas of rivers catering to the urban populations need to be strengthened.

Governance structures across the urban centres in HKH seem weak and dominated by a certain section of society. Unregulated abstraction of groundwater has led to groundwater depletion in many urban centres of the region. The presence of an informal water market dominated by tanker mafia is seen across water-scarce cities of different countries in the HKH, due to which, water reserves have come under intense pressure. Ground water abstraction policies exist in municipal bye-laws in many cases, however they are seldom followed (Quetta, Kabul, Dehradun, Haldwani), and illegal borewells and overexploitation of groundwater resources remains a problem.

7. Ground water abstraction policies need to be developed/reassessed, keeping in mind the mountainous characteristics and emerging pressures. Groundwater extraction rules are seldom enforced due to the political clout of water-selling mafia to mint money.

Artificial low pricing of water has encouraged poor water management across countries in the HKH, promoting highly inefficient use in domestic, commercial, industrial and agricultural use.

8. Metering, differential pricing systems and incentivizing mechanisms need to be considered for promotion of the judicious use of water.

The cost at which water is supplied through tankers in most cases is extremely high, hence only the affluent members of the population are able to pay the premium. The poor and vulnerable are left at the mercy of government supply which is also not equitably distributed. Inequity in water distribution is a prominent issue across the cities and towns of the HKH. Socio-economically weaker areas within cities experience a 'zero day' at regular intervals, even in cases where water scarcity is not high. Gender sensitivity in terms of planning and community needs is limited to symbolism.

9. Equitable distribution of water and gender sensitive water management practices to benefit the population at the base of the pyramid.

Institutional capacities and urban governance mechanisms to manage water are generally low and poor. Moreover, multiple institutions/departments are involved in supply, management and maintenance of infrastructure related to water, which work in silos, and the lack of engagement within institutions/departments leads to unconnected decisions and actions, only aggravating the problem.

10. Strengthening institutional capacities and improved coordination between departments.

Strong cultural links and the peri-urban nature of many mountain towns helps citizens maintain stronger links with their natural environment than can be seen in larger cities. This has led to successful efforts led by citizens, for example, in Nainital, a group 'citizens for Nainital' have been constantly influencing decision-makers to make sustainable choices for lake conservation. Such efforts can be sustained for long periods. An aware and interested citizenry also lends itself well to citizen science efforts. People in towns like Mussoorie, Haldwani and Dehradun have shown an interest in long-term monitoring, collecting data and compiling long-term records which will reflect trends and changes. Citizen science could play a key role in data gathering; required are productive linkages between scientists, citizen groups and policy-makers. Although a significant amount of awareness exists among scientists and policy-makers, urban water management from the climate change perspective has not really received enough attention in policy documents, i.e., state action plans for climate change.

11. Strengthening links between scientists, policy-makers and citizens.

Behavioural change through awareness programmes, knowledge empowerment and citizen/stakeholder engagement in data collection and inclusiveness in decision-making is necessary to build robust and equitable policies around water management in urban centres. Several projects have also focused on sensitization at different levels, building awareness, generating a substantive database in different countries at different levels, involving communities as well as water utilities and municipal governments. Data deficiency is another issue; basic data on rainfall, temperature is not available even for established urban centres, and researchers have to rely on regional data which seldom provide accurate results, given the heterogeneity of HKH and the mountains and microclimatic conditions of the towns.

12. Knowledge empowerment and involving citizens in data collection and decision-making.

Today, people have access to better tools to connect, and the translated information may result not only in sharing of information but also experiencing citizens' knowledge. Networking of citizens may create an active role for them to participate in science and decision-making. The internet has opened up space for citizens to seek scientific empowerment through the web. Large amounts of technical data exist within institutions; however, it is seldom translated to a wider audience. Using apps and the internet is a possible way to disseminate information to achieve smart water management practices, and smart citizens.

Acknowledgements

We are grateful to International Development Research Centre (IDRC), Canada and HI-AWARE ICIMOD, Kathmandu for providing research funding and facilities. This work was carried out by the Himalayan Adaptation, Water and Resilience (HI-AWARE) consortium under the Collaborative Adaptation Research Initiative in Africa and Asia (CARIAA) with financial support from the United Kingdom's Department for International Development (DFID) and the International Development Research Centre (IDRC), Ottawa, Canada.

References

Balk, D., Montgomery, M. R., McGranahan, G., Kim, D., Mara, V., Todd, M. & Dorélien, A. (2009). Mapping urban settlements and the risks of climate change in Africa, Asia and South America. *Population Dynamics and Climate Change 80*, 103.

Bhujel, G. D. (2019). *Water Woes: Investments, Infrastructures, Policies and Advocacies, Kuensel*. Available at: http://www.kuenselonline.com/water-woes-investments-infrastructures-policies-and-advocacies/.

Cosgrove, W. J. & Loucks, D. P. (2015). Water management: current and future challenges and research directions. *Water Resource Research 51*, 4823–4839. https://doi.org/10.1002/2014WR016869.

Dash, R. R., Mehrotra, I., Kumar, P. & Grischek, T. (2008). Lake bank filtration at Nainital, India: water quality evaluation. *Hydrogeology Journal 16*, 1089–1099. doi 10.1007/s10040-008-0295-0.

Devkota, K. (2018). Challenges of inclusive urbanization in the face of political transition in Nepal. In: *Handbook of Research on Urban Governance and Management in the Developing World*. Makerere, J. M. & Katusiimeh, M. W. (eds). IGI Global, Hershey, PA.

Dimri, A. P. & Dash, S. K. (2012). Wintertime climatic trends in the western Himalayas. *Climatic Change 111*(3), 775–800.

Djebou, D. C. S. & Singh, V. P. (2016). Impact of climate change on precipitation patterns: a comparative approach. *International Journal of Climatology 36*(10), 3588–3606.

Dorji, Y. (2016). *Water Securing Bhutan's Future*. Asian Development Bank/National Environment Commission. The Asian Development Bank/National Environment Commission Royal Government of Bhutan Thimphu, Bhutan. www.adb.org/ www.nec.gov.bt.

Eriksson, M., Xu, J., Shrestha, A. B., Vaidya, R. A., Santosh, N. & Sandström, K. (2009). *The Changing Himalayas: Impact of Climate Change on Water Resources and Livelihoods in the Greater Himalayas*. International Centre for Integrated Mountain Development (ICIMOD), Lalitpur, Nepal.

Grimm, N. B., Faeth, S. H., Golubiewski, N. E., Redman, C. L., Wu, J., Bai, X. & Briggs, J. M. (2008). Global change and the ecology of cities. *Science 319*(5864), 756–760.

Himmelsbach, T. (2010). Groundwater Resources at Risk – Kabul Basin, Afghanistan – Assessment of the present situation and consequences for the future. In: *International Conference, Integrated Water Resources Management*, Karlsruhe, Germany, pp. 24–25.

Hofer, T. & Messerli, B. (2006). Floods in Bangladesh: history, dynamics and rethinking the role of the Himalayas. *Ecology 29*, 254–283.

Ishtiaque, A., Shrestha, M. & Chhetri, N. (2017). Rapid urban growth in the Kathmandu Valley, Nepal: monitoring land use land cover dynamics of a Himalayan city with Landsat imageries. *Environments 4*, 72.

Jalilov, S-M., Masago, Y., Nishikawa, J., Bai, X., Elmqvist, T., Takeuchi, K. & Fukushi, K. (2017). Enhancing Urban Water Resilience: Science-based Approaches and Strategies for Asian Megacities. Policy Brief (No. 9, 2017).

Kakar, Z., Shah, S. & Khan, M. (2018). Scarcity of water resources in rural area of Quetta District; challenges and preparedness. In: *IOP Conference Series: Materials Science and Engineering*, Quetta, Pakistan, p. 414. doi:10.1088/1757-899x/414/1/012013.

Khan, F. (2009). *Water Governance and Corruption in Pakistan. Running on Empty: Pakistan's Water Crisis*. Woodrow Wilson International Centre for Scholars, Asia Program, Washington, DC.

Mack, T. J., Akbari, M. A., Ashoor, M. H., Chornack, M. P., Coplen, T. B., Emerson, D. G., Hubbard, B. E., Litke, D. W., Michel, R. L., Plummer, L. N., Rezai, M. T., Senay, G. B., Verdin, J. P. & Verstraeten, I. M. (2010). *Conceptual Model of Water Resources in the Kabul Basin, Afghanistan*. U.S. Geological Survey scientific investigations report 2009–5262, 240 pp. Available at: http://pubs.usgs.gov/sir/2009/5262.

Mahamuni, K. & Upasani, D. (2011). *Springs: A Common Source of A Common Resource, IASC-2011, Digital Library of The Commons*. Indiana University, Bloomington, IN.

MoWR (2016). Repair, Renovation and Restoration of Water Bodies – Encroachment on Water Bodies and Steps Required to Remove the Encroachment and Restore the Water Bodies. In: *10th Report, Sixteenth Lok Sabha Standing Committee on Water Resources, Ministry of Water Resources, River Development and Ganga Rejuvenation*, New Delhi, India.

Mukherjee, A., Sengupta, M., Hossain, M., Ahamed, S., Das, B., Nayak, B., Lodh, D., Rahman, M. M. & Chakraborti, D. (2006). Arsenic contamination in groundwater: a global perspective with emphasis on the asian scenario. *Journal of Health, Population and Nutrition 24*(2), 142–163. Available at: http://www.jstor.org/stable/23499353.

NITI Aayog (2018). *Report of Working Group I Inventory and Revival of Springs in the Himalayas for Water Security*. Available at: http://niti.gov.in/writereaddata/files/document_publication/doc1.pdf

Pandey, K. (2018). Crisis in the Himalayas: nearly 50% perennial springs in the region have dried up. *Down to Earth*. Available at: https://www.downtoearth.org.in/news/india/crisis-in-the-himalayas-nearly-50-perennial-springs-in-the-region-have-dried-up-61482.

Pokharel, S. (2014). Pokhara making do with three decades old water supply system. *Republica*. Nepal Republic Media, 25 May 2014. Web. 16 August 2014.

Rodell, M., Velicogna, I. & Famiglietti, J. S. (2009). Satellite based estimates of groundwater depletion in India. *Nature 460*, 999–1002.

Romero-Lankao, P. & Gnatz, D. M. (2016). Conceptualizing urban water security in an urbanizing world. *Current Opinion in Environmental Sustainability 21*, 45–51.

Shah, M. & Kulkarni, H. (2015). Urban water systems in India typologies and hypotheses. *Economic and Political Weekly 50*, 57–69.

Shah, S. & Tewari, A. (2009). Impact of human disturbance on forest vegetation and water resources of Nainital Catchment. *Nature and Science 2009*, 74–78. Available at: http://www.sciencepub.net/nature.

Sharma, D., Singh, R. & Singh, R. (2014). Building urban climate resilience: learning from the ACCCRN experience in India. *International Journal of Urban Sustainable Development 6*(2), 133–153.

Singh, V. & Pandey, A. (2018). Climate and development; a diagnostic assessment since the formation of uttarakhand. In: *Development Dynamics of Himalayan State*, Vol. 1. Nutiyal, R. R. & Datta, R. (eds). Kalpraz Publications, New Delhi, India, pp. 157–179.

Singh, S. P., Singh, V. & Skutsch, M. (2010). Rapid warming in the Himalayas: ecosystem responses and development options. *Climate and Development 2*(3), 221–232.

Singh, S., Hassan, S. T. M., Hassan, M. & Bharti, N. (2019). Urbanisation and water insecurity in the Hindu Kush Himalaya: insights from Bangladesh, India, Nepal and Pakistan. *Water Policy* wp2019215. https://doi.org/10.2166/wp.2019.215.

Snyder, A. (2014). *Shortage in the Mountains of Plenty: Water Supply in Mountain and Hill Cities Throughout the Hindu-Kush Himalayan Region*, ICIMOD and The World Food Prize. Available at: https://www.worldfoodprize.org/documents/filelibrary/images/youth_programs/2014_interns/2014_br_research_papers/SnyderAbigail_LONGReport_56ED38F157B76.pdf

Srinivasan, V., Lambin, E. F., Gorelick, S. M., Thompson, B. H. & Rozelle, S. (2012). The nature and causes of the global water crisis: syndromes from a meta-analysis of coupled human-water studies. *Water Resources Research 48*(10), 1–16.

Stanikzai, J. (2014). Illegal wells hurting Kabul's water supply. *Tolo News* 14 August 2013.

Stewart, J. (2017). *Growth for Low-Income Countries?* United Nations University. Available at: https://unu.edu/publications/articles/growth-for-low-income-countries.html.

Suwal, B. R. (2013). *Internal Migration in Nepal. UNESCO Gender and Youth Migration*. A UNESCO online initiative on migration. Available at: http://www.unescogym.org/.

Tambe, S., Kharel, G., Arrawatia, M., Kulkarni, H., Mahamuni, K. & Ganeriwala, A. (2012). Reviving dying springs: climate change adaptation experiments from the Sikkim Himalaya. *Mountain Research and Development 32*(1), 62–72.

Tiwari, A. (2012). Water quality and quantity analysis in Sikkim, North Eastern Himalaya. *Current Science 103*, 41–45.

Tiwari, P. C., Tiwari, A. & Joshi, B. (2018). Urban growth in Himalaya: understanding the process and options for sustainable development. *Journal of Urban and Regional Studies on Contemporary India 4*(2), 15–27.

UNDESA (2018). *Revision of World Urbanization Prospects*. UN Department of Economic and Social Affairs, New York, p. 16.

United Nations Economic and Social Commission for Asia and the Pacific (UNESCAP) (2008). *Statistical Yearbook for Asia and the Pacific 2007*. UNESCAP. Available at: http://www.unescap.org/stat/data/syb2007/ESCAP-SYB2007.pdf (Accessed 5 September 2012).

Wakode, H. B., Baier, K., Jha, R. & Azzam, R. (2018). Impact of urbanization on groundwater recharge and urban water balance for the city of Hyderabad. *India. International Soil and Water Conservation Research 6*(1), 51–62.

Walker, B. (2011). *Urban Peaks in the Himalayas. China Dialogue*. Available at: https://www.chinadialogue.net/article/show/single/en/4306Urban-peaks-in-the-Himalayas.

Wankhade, K., Krishnachandran, B. & Vishnu, M. J. (2014). *Sustaining Policy Momentum, Urban Water Supply & Sanitation in India*. IIHS RF Paper on Water Supply and Sanitation, IIHS, Bangalore.

Zhan, M., Li, X., Sun, H., Zhai, J., Jiang, T. & Wang, Y. (2018). Changes in extreme maximum temperature events and population exposure in China under global warming scenarios of 1.5 and 2.0 °C: analysis using the Regional Climate Model COSMO-CLM. *Journal of Meteorological Research 32*(1), 99–112.

Received 9 July 2019; accepted in revised form 17 September 2019. Available online 22 October 2019

Water availability, consumption and sufficiency in Himalayan towns: a case of Murree and Havellian towns from Indus River Basin, Pakistan

Zeeshan Tahir Virk[a,b,*], Bilal Khalid[b], Abid Hussain[c], Bashir Ahmad[a], Salaar Saeed Dogar[a], Nelufar Raza[a] and Bilal Iqbal[a]

[a]*Pakistan Agricultural Research Council (PARC), Islamabad, Pakistan*

[b]*Leadership for Environment and Development (LEAD), Islamabad, Pakistan*

**Corresponding author. E-mail: zvirk@lead.org.pk*

[c]*International Centre for Integrated Mountain Development (ICIMOD), Kathmandu, Nepal*

Abstract

This study attempts to analyse the status of water availability, consumption and sufficiency in two Himalayan towns – Murree and Havellian from Pakistan's Indus Basin, using the primary data collected from 350 households, 26 town level focus groups and key informants. Findings revealed that groundwater is the main source of water on which around 85% of households are heavily dependent. Estimates of water availability, consumption and water sufficiency ratio (WSR) show that available groundwater is not sufficient (0.89) to meet the requirements for consumption in Havellian. However, in the case of Murree, available water is sufficient (1.92). Taking into account the national standards of water consumption, WSR estimates show that both towns have insufficient water availability (Murree: 0.68, Havellian: 0.50). There is evidence that in both towns, water is being mismanaged at household level. In addition, one-fifth of households reported that climate change has also affected the water availability in the towns over time. Factors such as rapid urbanization and population growth are likely to result in increased requirements of water in the future. Based on the findings, the study has suggested policy actions on protection, efficient use, diversification and governance of groundwater resources.

Keywords: Climate change; Governance; Groundwater; Himalayan towns; Indus Basin; Sufficiency ratio; Urbanization; Water availability; Water consumption

doi: 10.2166/wp.2019.012

Introduction

The Hindu-Kush Himalayan (HKH) region extends 3,500 km across eight countries – Afghanistan, Bangladesh, Bhutan, China, India, Nepal, Myanmar and Pakistan – covering an area of almost 3.5 million square kilometres, and accommodating nearly 240 million people (Sharma *et al.*, 2018). Almost one-fifth of the population of the HKH region is accommodated in the mountain areas of the Indus River Basin that lies in Pakistan (Sharma *et al.*, 2018). Over the past three decades, the HKH region has observed a rapid growth in urban population (Tiwari *et al.*, 2018). In addition to the natural increase in population of the region, the in-migration from the adjacent peri-urban and rural areas to the urban centres, for accessing better livelihood options, is also contributing significantly to the population rise in Himalayan towns (Tiwari *et al.*, 2018). It has led to an increased pressure on water resources in the Himalayan towns.

Another important factor which is causing a significant impact in the outlook of Himalayan towns in the Indus Basin is climate change (Bolch *et al.*, 2012; Immerzeel *et al.*, 2013; Palazzi *et al.*, 2013; Lutz *et al.*, 2014; Shrestha *et al.*, 2015a). A recent assessment report (Wester *et al.*, 2019) has revealed that 36% of the volume of the glaciers in the HKH region will melt by 2100 even if the world manages to keep warming below 1.5 °C. If the temperature increases by 2 °C, around half of the volume of these glaciers is likely to melt. It will be more alarming for the Indus Basin, where snowmelt and glacier melt contribute nearly 80% of its water flow – compared to the Ganga and Brahmaputra, which are mostly dependent on rainfall. Due to a high rate of glaciers melting, the Indus Basin will have more water flowing in, but in an increasingly unpredictable manner. In the HKH region, winters are seeing more warming than the summers with high mountain areas being the most affected. Precipitation across the Himalayas of the Indus Basin is likely to increase by 5% on average by 2050 (Shrestha *et al.*, 2015a). In addition to losses in glacial mass, temperature rise, increased liquid precipitation, there are high chances of a decline in solid precipitation (Wiltshire, 2014). A growing mismatch may occur between the timing of peak flows and the requirements for water in achieving food security in the basin (Lutz *et al.*, 2014). Increased temperatures and precipitation in the Indus Basin are causing erratic changes in rainfall and runoff patterns, which can cause serious implications for the residents of the mountain towns of Himalayas in the Indus Basin (Lutz *et al.*, 2016). These impacts range from reduced water availability and food production from mountain agriculture (Rasul *et al.*, 2019), to severe hydro-meteorological disasters such as flash floods, droughts and landslides. These impacts are significantly precarious for Himalayan towns as the climate risk and vulnerability associated with mountain communities is generally greater as compared to that of the plains (Tshe-ring *et al.*, 2010; Hunzai *et al.*, 2011; Gerlitz *et al.*, 2015). Therefore, increased uncertainty in the availability of natural resources caused by climate change will further increase the vulnerability of residents of Himalayan towns.

Himalayan towns in the Indus Basin have limited water resources available for domestic use. The terrain of these towns limits harvesting of domestic water from rivers and streams (Smadja *et al.*, 2015). The biggest challenge is that water storage projects involve high initial capital costs due to the rugged topography in the mountains (Hussain *et al.*, 2019). These limitations place the water availability scenario of these towns under inherent check. Since surface water sources are marginal and water demand has started exceeding the supply, these towns are becoming water stressed. Many of these towns have per capita water availability below the national and international standards (Laghari *et al.*, 2012). The reduced water availability for domestic use has been compensated through overexploitation of groundwater as an adaptive strategy both at household and town level. Groundwater appears to be a substantial source of urban water security as it is the major source of domestic water in Himalayas of the Indus Basin (Qureshi, 2011). Most of the groundwater in Himalayan towns is accessed in the form of fresh water springs or extracted via tube wells. Groundwater is a vital source of domestic water and thus water security for urban areas in the Indus Basin. Increasing urbanization in Himalayan towns is also threatening the recharge zones of groundwater aquifers. At the current rates of abstraction, aquifer depletion seems imminent. Despite its significance for water security groundwater resources are often unexplored and understudied in the Himalayas (Miller *et al.*, 2012; Smadja *et al.*, 2015). Further, the knowledge base regarding utilization of groundwater in Himalayan towns of the Indus Basin is very limited (Andermann *et al.*, 2012).

In the Himalayan towns of the Indus Basin, potential impacts of increasing urbanization, unsustainable urban development and climate change are set to exacerbate water stress. As a result, groundwater is increasingly supplanting the dwindling surface water supplies. This study attempts to analyse the current situation of groundwater availability, consumption and sufficiency in two Himalayan towns, i.e., Murree and Havellian, in the Indus Basin, using the primary data collected at both household and town levels. Moreover, the study also comprehensively investigates the key factors affecting water availability and consumption in the towns. To the best of our knowledge, this is the first study on comprehensive water availability, consumption and sufficiency analysis in the Himalayan towns in Pakistan's part of the Indus Basin. It is hoped that the findings of this study will significantly contribute to strategies and plans for urban centres' development in the Indus River Basin.

Study area and methodology

Study area

In the Indus River Basin, Himalayan towns are rapidly urbanizing, adding more pressure on groundwater resources. For this study, two highly urbanizing towns, i.e., Murree and Havellian, have been selected (Figure 1). These towns may present strong cases regarding the situation of urban groundwater sufficiency in Himalayas and the key factors influencing it. Murree is located in Rawalpindi District of Punjab Province in Pakistan. It is the most popular hill station of the country, lying at 2,300 m above sea level (masl), about 50 km away in the north east from the capital city Islamabad. It is situated in the sub-Himalayan foothills. Havellian is located on the famous Karakorum Highway, at about 833 masl, on the banks of the Dor River, and is the second largest municipality in Abbottabad District of Khyber Pakhtunkhwa Province in Pakistan.

In terms of climate, Murree and Havellian are different from each other. Murree is situated in a subtropical highland climate zone with mean annual precipitation of nearly 1,800 mm (NOAA, 1995) and mean annual temperature of 12.7 °C (Rasul, 2012). Conversely, Havellian is warmer with a mean annual temperature of 20 °C, and receives mean annual precipitation of nearly 1,200 mm (Climate-Data, 2018). Significant characteristics of both the towns are presented in Table 1.

An overview of the historical precipitation and temperature data shows a significant variability over time for both towns (Figures 2 and 3). In Murree, mean annual rainfall trend shows a decline. Mean annual maximum temperature shows a gradual increase, and mean annual minimum temperature shows a decline with a few exceptions of rise from 2004 to 2008. In Havellian, annual maximum temperature has shown a gradual rise (Figure 3). These changes in rainfall and temperature trends in both

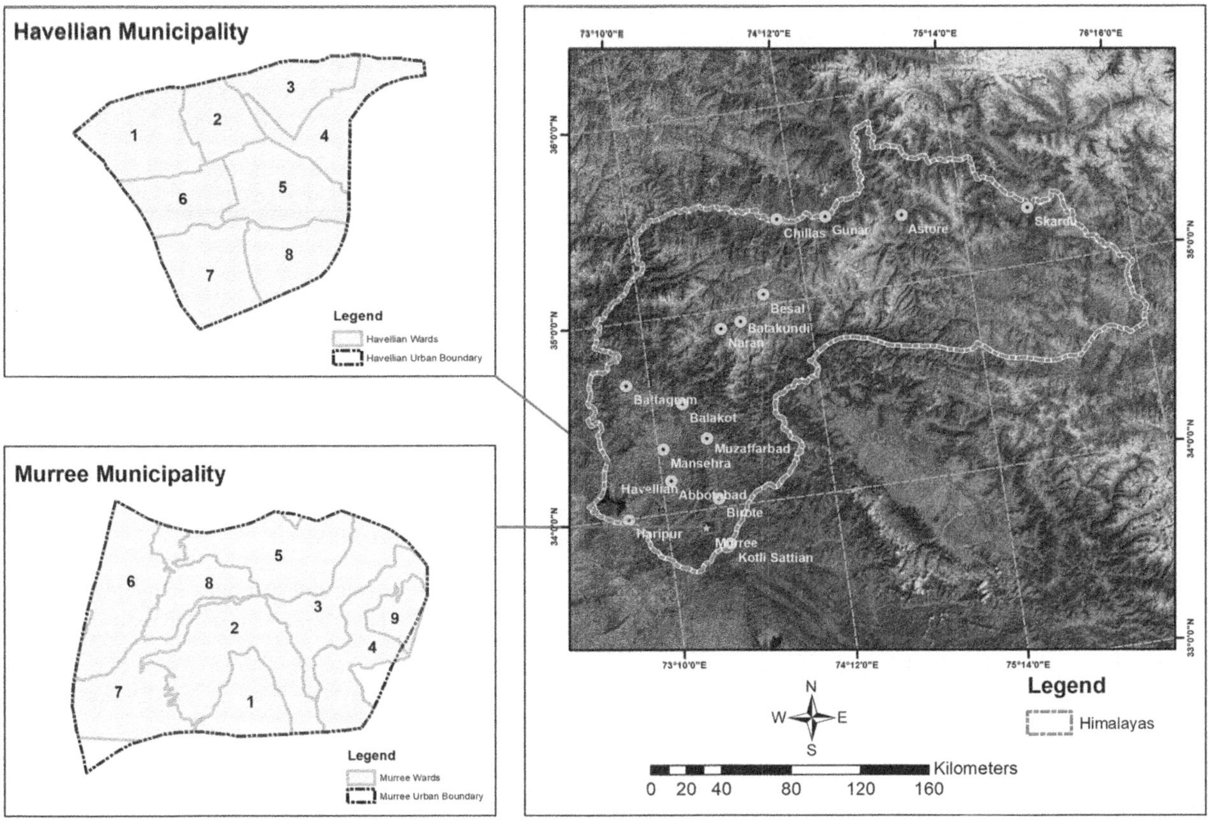

Fig. 1. Study towns in the Himalayas of the Indus River Basin.

Table 1. Study towns' salient features and characteristics.

Features	Murree	Havellian
Geolocation	33.55′ N and 73.23′ E	34.05′ N and 73.14′ E
Altitude	2,300 m (7,517 ft)	833 m (2,733 ft)
Municipal area	29.5219 km^2	0.89763 km^2
Administrative unit	Tehsil Murree, District Rawalpindi	Tehsil Havellian, District Abbottabad
Administrative unit area	434 km^2	406 km^2
Mean annual rainfall	1,789 mm/annum	1,194 mm/annum
Mean annual temperature	12.7 °C	20 °C
Climate zone*	Cwb (Sub-tropical, highland climate)	Cwa (Sub-tropical, humid)
Main water source	Groundwater (springs)	Groundwater (tube wells)
Sub-basin	Harro and Korang River Basin	Dorr River Basin

Source: Author's own calculations and District Profiles, Pakistan Bureau of Statistics.
*Koppen-Gieger Climate Classification (Kottek *et al.*, 2006).

towns have a direct implication for groundwater availability, as at present, both towns are relying exclusively on groundwater for domestic purposes.

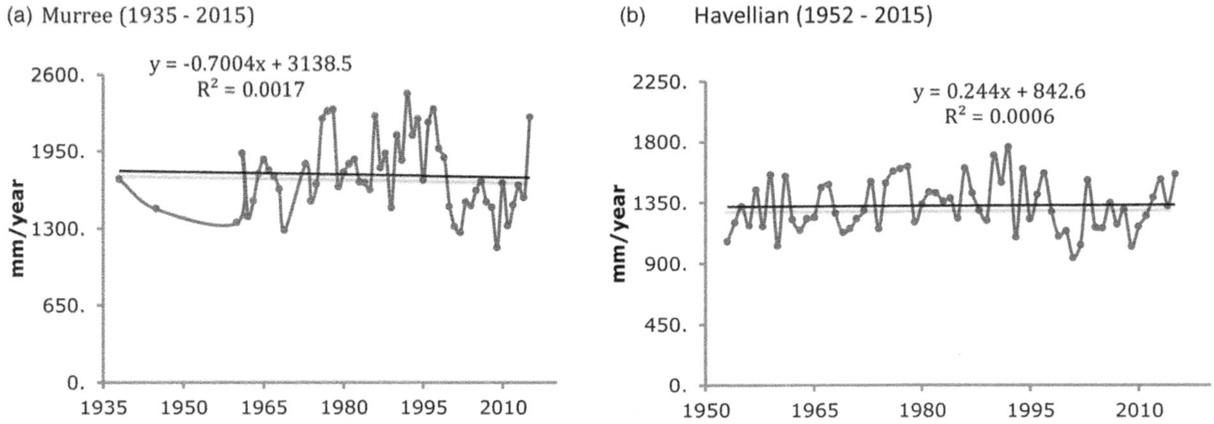

Fig. 2. Trends in mean annual rainfall. *Source:* Original data procured from Pakistan Meteorology Department.

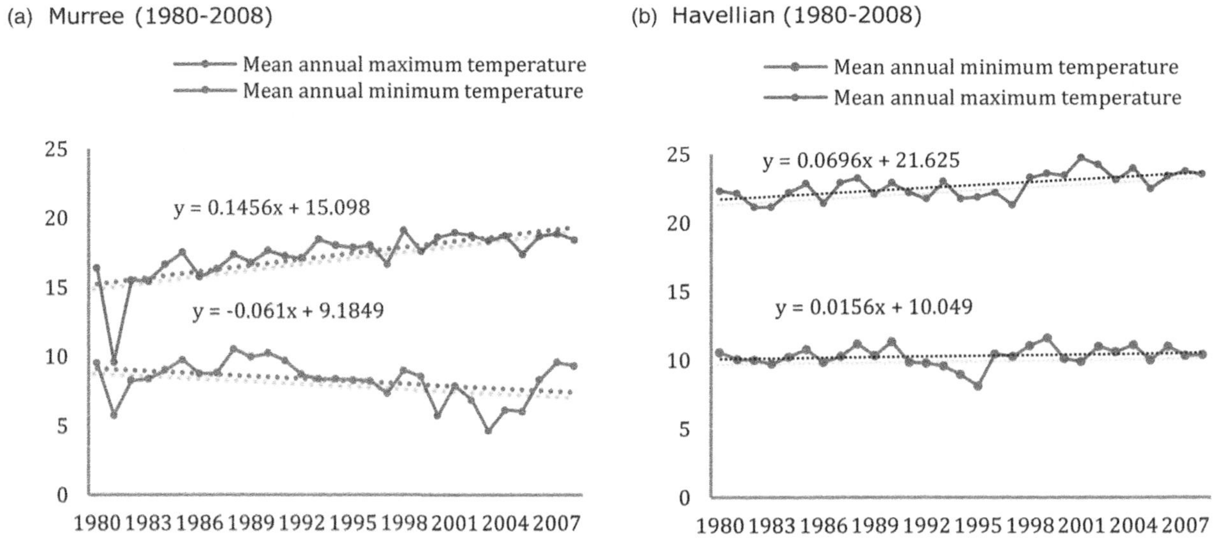

Fig. 3. Trends in mean annual maximum and minimum temperatures. *Source:* Original data procured from Pakistan Meteorology Department.

Study design and data collection

This study is mainly based on small-scale survey design. However, it also relies on some secondary data sources and qualitative information collected through focus group discussions and key informant interviews. For survey, Cochran's formula (Cochran, 1977) (Equation (1)) was used to estimate the sample size:

$$n = D \times \left[\frac{Z^2 \times (p) \times (1-p)}{e^2}\right] \tag{1}$$

In Equation (1), n denotes the sample size, and p is the percentage of households picking a choice (expressed as decimal = 0.5) and $(p) \times (1-p)$ expresses an estimate of variance. Z denotes Z-value (1.96 for 95% confidence interval), and e denotes the margin of error (± 0.065). It was planned to distribute the sample size across two towns. To compensate for any loss of statistical robustness in the sample distribution, a design effect 'D' (1.50) is considered in the formula. Using Equation (1), a sample size of 341 households was determined. However, in total, 350 households were actually surveyed. The sample was equally distributed among the two towns as the current population of both Murree and Havellian towns was approximately equal in 2017. Sub-samples of each town were distributed across streets/wards using the probability proportional to size (PPS) method. Within wards/streets, households were selected randomly using random route method. The household data were collected in 2017 using a standardized questionnaire that was digitized on an android-based application called AKVO-FLOW.

Secondary data sources for the study included: population data from Pakistan Bureau of Statistics (PBS) acquired through district census reports of districts Abbottabad and Rawalpindi (Pakistan Bureau of Statistics 1981, 1998, 2017); time series climate data for the past 30 years from Pakistan Meteorological Department (PMD) for daily temperature and precipitation of the towns of Havellian and Murree; and open source satellite images from Landsat and Sentinel for the years 1998 and 2017. Additional data were collected through focus group discussions (FGDs) and key informant interviews (KIIs). In total, 10 FGDs and 16 KIIs were conducted in the two towns, respectively, with community groups and the representatives of related government departments and NGOs. Some qualitative information was also collected on the areas of water sources, distribution mechanism, pricing, consumption, and governance issues in the towns.

Analysis tools

The study has mainly used descriptive statistics to analyse the data. In the questionnaire survey, data on household perception of the water resources situation, water sufficiency and climate change impacts on water resources were collected. However, the study has not entirely depended on this perception data. Additionally, it has quantitatively estimated water availability, consumption and sufficiency ratio, using both household and town level data. Quantitative estimations have helped in validating the household perception data. To assess the water sufficiency in towns, the study used the following equation (Equation (2)):

$$Si = \frac{W_a}{W_c} \quad (2)$$

where W_a is available groundwater, W_c is the consumed groundwater and S_i is the water sufficiency ratio (WSR). The value of WSR reveals the following outputs: $S_i < 1$, insufficient; $S_i > 1$, sufficient; $S_i = 1$, balanced or equilibrium state.

Water availability (Equation (2)) is based on two assumptions. First, at the town level, groundwater availability is equal to water extracted from ground sources for supply. Second, at household level, groundwater availability is equal to water stored in the house.

At household level, water consumption was estimated based on the quantitative data collected about the different uses of water in the household.

Statistical Package for the Social Sciences (SPSS-22) was used to conduct quantitative analysis of survey data and collected secondary data (i.e., population and climate data). Arc GIS 10.1 software was used to analyse the satellite images of Murree and Havellian to determine the increase in built-up area.

Results

Household characteristics

In both surveyed towns, the average household size is equal (seven members per household). However, in terms of type of house, there is a significant difference across towns. In Murree, more than 90% households have bricked or cemented houses. Conversely, in Havellian, only 56% households have bricked and cemented houses (Table 2). Moreover, in Murree, more than 90% of households own their houses. Conversely, in Havellian, nearly 30% of households live in rented houses. In Murree, the percentage of immigrating households is also higher compared to that of Havellian. In terms of income levels, in both towns almost 40% of households reported that their monthly income ranges from 151USD to 300USD per month (Table 2).

Table 2. Household socioeconomic characteristics.

Variable	Response	Towns	
		Murree	Havellian
Average household size (number of persons)	–	7	7
Average male household members		4	4
Average female household members		3	3
Respondent's education status (%)	Illiterate	12.0	12.6
	Primary	18.3	6.9
	Middle school	13.7	16.0
	High school	32.0	23.4
	Higher secondary	12.6	19.4
	Graduation	8	18.9
	Post-graduation and above	3.4	2.8
House ownership status (% households)	Rented	7.3	25.9
	Owned	92.7	74.1
House construction type (% households)	Brick/Cemented	89.8	56.4
	Mud/Thatch/Wooden	10.2	43.6
Migration status (% households) (immigrants)	Migrant	16.0	9.7
	Non-migrant	84.0	90.3
Toilet facility (% households)	Private	99.4	100
	Community	0.6	0
Household monthly income levels (income classes in USD) (% households)	<100	8.6	7.5
	100–150	25.7	12.6
	151–300	37.1	37.4
	301–500	15.4	25.9
	501–1,000	6.3	12.6
	>1,000	6.9	4.0

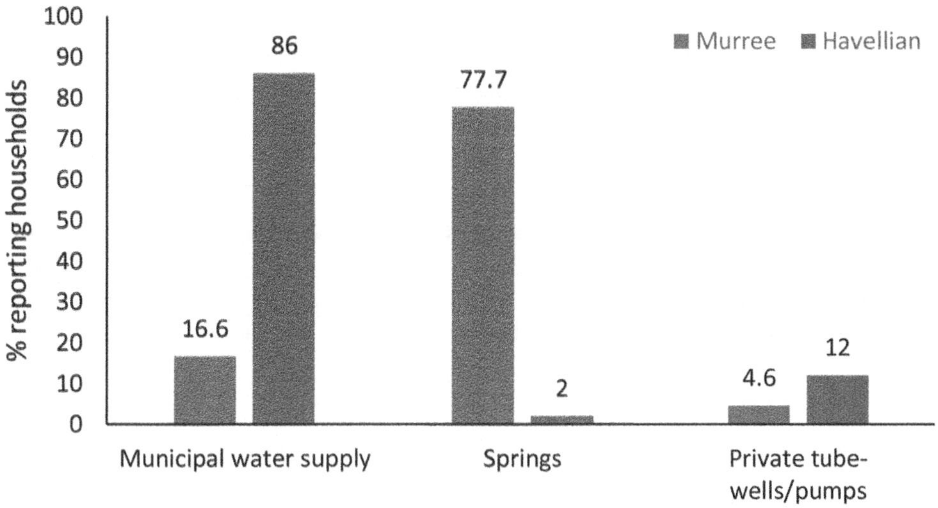

Fig. 4. Household water sources in selected Himalayan towns.

Main water sources

Survey data analysis reveals that in both towns, households are heavily dependent on groundwater sources, e.g., municipal water supply (mainly pumped by tube wells), springs and private tube wells/pumps. In Murree, almost 80% of households depend on spring water for domestic purposes (Figure 4), and in Havellian, 86% of households are dependent on municipal water supply. In Havellian, almost 5% and in Murree 12% of households depend on groundwater that is directly withdrawn at community level by using private tube wells/pumps.

In Murree, nearly 65% of households reported that the quantity of water available from springs is always sufficient, and 20% reported that water quantity is occasionally sufficient (Table 3). In Havellian,

Table 3. Household perception of water sufficiency.

Indicators	Response	Murree	Havellian
Sufficiency of spring water (% households)	Always sufficient	64	NA
	Occasionally sufficient	20	NA
	Never sufficient	4.6	NA
	Not applicable	11.4	NA
Sufficiency of municipal water to meet household water requirements (% households)	Always sufficient	9.14	57.7
	Occasionally sufficient	7.43	29.2
	Never sufficient	4	4.6
	Not applicable	79.42	8.5
Frequency of municipal water supply	Everyday	9.7	89.4
	Once in 2 days	5.7	8
	Once in 3 days	1.7	1
	Not applicable	82.9	1.6
% households reporting a change in water availability in last 20 years	70	43	70

NA, not available (data were not collected).

nearly 60% of households reported that water from municipal supply is sufficient to meet their water requirements, and almost 30% of households reported that supply is occasionally sufficient. Around 5% of households reported that they never get sufficient water (Table 3). Overall, 35% of households in towns reported that available groundwater is not sufficient to fulfil their domestic requirements. Around 90% of households receive municipal water every day, 8% receive water once in 2 days and just under 1% receive municipal water once in 3 days. More than 70% of households in Murree reported a change in water availability in the past 20 years. In Havellian, 43% of households reported a change in the water availability over the past 20 years (Table 3). In discussions, households in both towns reported seasonal fluctuations in water availability. In Havellian, less water is available in summer corresponding to increased overall demand, and in Murree, limited water is available in winter due to reduced discharge from the springs.

Water availability

Estimates of water availability at town level are based on the quantity of available water reported during the FGDs and KIIs conducted during the qualitative survey with officials of Town Municipal Authorities. In Murree, gross water availability (supply) is 7.02 MLD. However, after 25% conveyance losses, it reduces to net availability of 5.3 MLD (Table 4). The net available water is further divided across cantonment and civilian areas. Civilian users, including residential and commercial users, receive 2.9 MLD of water. The residential areas receive 60% of allocated water to civilian areas (1.74 MLD).

In Havellian, gross water availability (supply) is 1.7 MLD. However, after 10% conveyance losses, it reduces to net availability of 1.6 MLD (Table 5). The net available water is further divided across residential and commercial users. The residential areas receive 80% of net available water (1.252 MLD). Overall, there is less water available in Havellian (Table 5), compared to Murree (Table 4).

Conversion of town level water availability into litres per capita per day reveals that nearly 102 litres and 74 litres per capita per day water is available, respectively, in Murree and Havellian (Table 6). However, at household level, around 78 litres and 52 litres per capita per day water is stored. It implies that there is a substantial gap between town level water availability and household level water storage (Table 6). This is due to household level water losses, and direct use of supplied water without storing it. This indicates mismanagement issues of water at household level.

Water consumption and water sufficiency ratio

In Murree and Havellian, estimated per capita water consumption is 53 litres and 83 litres per capita per day, respectively (Table 6), which is significantly lower than the international standard of 100 litres per capita per day (WHO, 2003) and national standards of 150 litres per capita per day water required for consumption (Government of Pakistan, M.o.H.W., 1993). Water consumption in Murree is significantly lower than that of Havellian. This can be attributed to the weather differences in the towns. Murree is comparatively cooler (see sub-section *Study area*) and the people may consume less water for bathing and washing purposes. Estimate of water sufficiency ratio (WSR) reveals that available water is still sufficient to meet consumption requirements in Murree ($S_1 = 1.92$, $S_2 = 1.47$) (Table 6). However, In Havellian, available water is not sufficient (($S_1 = 0.89$, $S_2 = 0.63$) to meet the consumption requirements (Table 6). Taking into account the national standards of water consumption

Table 4. Water availability in Murree.

			Million litres per day (MLD)					
Access point	Water source	Date of commission	Daily supply (G)	After deduction of 25% conveyance losses (N)	Supply to cantonment areas (45% of N)	Supply for residential and commercial uses (RCU) (55% of N)	Commercial (40% of RCU)	Residential (60% of RCU) (water availability at town level to be supplied to households)
Dunga Galli	Surface water	1894	2.27	1.702	0.766	0.936	0.374	0.561
Dharjawa	Spring water	1974	3.5	2.625	1.181	1.443	0.577	0.866
Masoot	Spring water	1985	0.57	0.427	0.192	0.235	0.094	0.141
Shawalla	Spring water	1993	0.3	0.225	0.101	0.123	0.049	0.074
Khanitak	Spring water	2009	0.38	0.285	0.128	0.156	0.062	0.094
Total	–	–	7.02	5.265	2.369	2.895	1.158	1.737

G, gross; N, net; RCU, residential and commercial uses.

Table 5. Water availability in Havellian.

			Million litres per day (MLD)			
Access point	Water source	Date of commission	Daily supply (G)	After deduction of 10% conveyance losses (N)	Commercial (at 40%)	Residential (at 80%) (water availability at town level to be supplied to households)
Tube well 1	Groundwater	1978	0.567	0.510	0.102	0.408
Tube well 2	Groundwater	1993	0.3785	0.340	0.068	0.272
Tube well 3	Groundwater	2003	0.567	0.510	0.102	0.408
Tube well 4	Groundwater	2010	0.227	0.204	0.040	0.1632
Total			1.74	1.565	0.313	1.252

Table 6. Water availability, consumption and sufficiency ratio.

Indicators	Murree	Havellian
A. Water availability at town level (litres per capita per day)	101.70	74.43
B. Water storage at household level (litres per capita per day)	77.85	52.33
C. Difference of town level and household level water availability (litres per capita per day)	23.85	22.1
D. Water consumption (litres per capita per day)	53.0	83.0
E. Water sufficiency ratio ($S_1 = A/D$)	1.92	0.89
F. Water sufficiency ratio ($S_2 = B/D$))	1.47	0.63
G. *Water sufficiency ratio ($S_3 = A/150$)	0.68	0.50

*Ratio estimated taking into account the national standards of water consumption (150 litres per capita per day).

(150 litres per capita per day), estimated WSR shows that both towns are facing water insufficiency ($S_3 = 0.68$ and 0.50, respectively, in Murree and Havellian) (Table 6).

Factors affecting water availability

Climate change, inadequate governance and demographic changes are key challenges affecting the water availability at both household and town levels. A majority of households in the study towns perceived changes in climate in the last 20 years (Table A1 in Supplementary Materials). Almost one-quarter of households in each town reported that changes in rainfall and temperature patterns are resulting in changes in water availability (Table 7). Among governance and management factors, more than 90% of households in both towns reported that inadequate institutional support to protect water resources is resulting in either a decline or fluctuation in water availability. Almost 30% of households in Havellian and one-fifth in Murree

Table 7. Household perception of reasons of change in water availability.

Reasons for change	Murree (% households)	Havellian (% households)
Decrease in rain and increase in temperature	24.1	22.7
Overexploitation of groundwater resources	18.3	28.5
Mismanaged water resources	18.8	12.6
Poor and inefficient water infrastructure	9.9	6.4
Inadequate institutional support to protect water resources	95	92.5
Others (increase in population and urbanization)	2.9	3.5

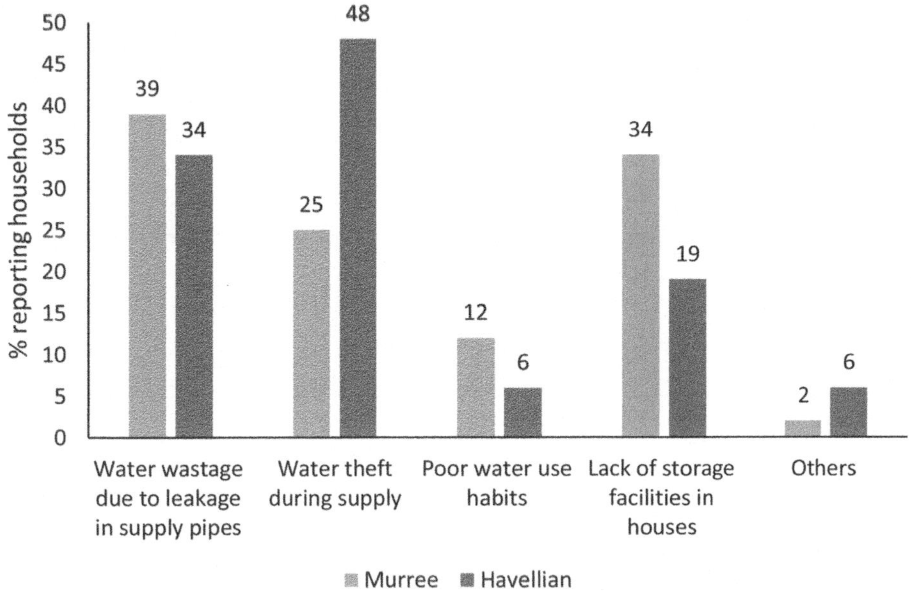

Fig. 5. Household perception of losses to available water.

perceived that overexploitation of groundwater resources results in changes in water availability. Around 10% in Murree and 6% in Havellian reported poor and inefficient water infrastructure as the reason of change. In both towns, around 3% of households reported that demographic changes such as increasing population and urbanization are impacting the household water availability (Table 7).

Households also reported some key factors resulting in losses to available water in the towns (Figure 5). These factors may be among the possible reasons for the gap between quantities of available water at town and household levels (Table 6). In Murree and Havellian, around 40% and 35% of households, respectively, reported that leakage in pipes, while supplying water from sources to households, is resulting in water losses. In Murree, one-quarter of households reported water theft as a reason of water losses during supply. However, in Havellian, nearly 50% of households reported such a reason (Figure 5).

In Murree, around 35% households reported that they do not have adequate water storage facilities at home. In Havellian, one-fifth of households reported inadequate storage facility as the reason of water losses. In Murree and Havellian, respectively 12% and 6% households reported that poor water use habits such as direct excessive use of water for domestic purposes (without storing), kitchen gardens, washing vehicles, and bathing pet animals result in water losses (Figure 5).

Overall, household perception for reasons of change in water availability (overexploitation and mismanagement of available water, and poor infrastructure) (Table 7), and perception of losses to available water (Figure 5) reveal that available water is being mismanaged during supply and use at household level in both towns.

Factors affecting water consumption

Among the factors affecting water consumption, the most important factors are rapid urbanization, high population growth and changing socioeconomic status of households. In the past three decades,

Table 8. Population in study towns.

	Years	Murree	Havellian
Population	1981	10,744	10,378
	1998	13,975	12,016
	2017	17,079	16,821
	2030	19,938*	19,780*
	2050	26,142*	26,262*
Population growth rate** (%)	1981–1998	1.56	0.87
	1998–2017	1.06	1.79
Population density (persons/km^2)	1981	364	11,531
	1998	473	13,351
	2017	579	18,690

Source: Pakistan Bureau of Statistics 1981, 1998 and, 2017.
*Projected population.
**Average annual compound growth rates.

population and population densities have shown a significant rise in both Murree and Havellian. The growth rate of population in Murree has grown at the rate of 1.56% between 1981 and 1998 and 1.06% between 1998 and 2017. Similarly, the growth rate of population in Havellian was 0.87% between 1981 and 1998 and 1.79% between 1998 and 2017. The projections show that both towns will have a population of more than 25,000 by 2050 (Table 8). A rise in population has a strong impact on water consumption through adding more pressure on available water at household level and nearby springs as in the case of Murree. Havellian is also gaining increasing importance as a result of upcoming major transportation infrastructure projects connecting the central parts to the north of Pakistan. Hence, urban growth and expansion of the town is imminent in the near future.

In Murree and Havellian, there has been a drastic increase in urban area in the past two decades. The land use has changed significantly in Havellian, with around a 22% addition in the built-up area in the last 19 years. The scenario is similar for Murree, with about a 22% addition to the built-up area in the last 19 years (Table 9). Urban sprawl in both towns indicates the fast rate of urbanization in the two towns (Figure 6(a) and 6(b)). Urban sprawl in any area affects water consumption in such a way that it increases economic activity, which first increases non-residential water consumption such as commercial or industrial water consumption. Urbanization promotes population growth via in-migration and results in changes in lifestyle and socioeconomic status of local people. It may result in an increase in domestic water consumption.

Table 9. Land use change in Murree and Havellian.

Town	Total area (km^2)	Built-up area 1998 (%)	Built-up area 2017 (%)	Change (%)
Murree	29.52	10	31	21
Havellian	0.90	63	85	22

Source: Author's own calculation using GIS techniques and satellite images of Landsat and Sentinel.

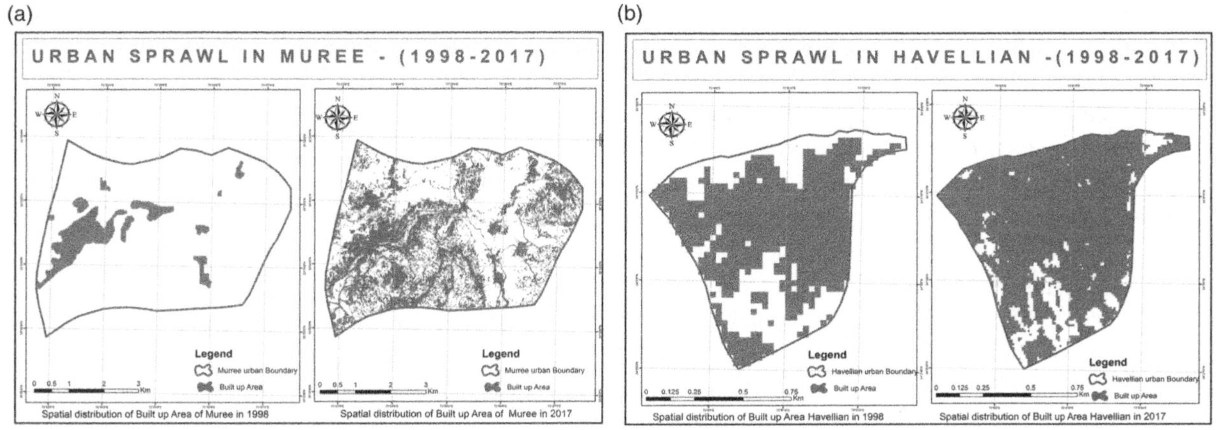

Fig. 6. Land use change and urban sprawl in Murree and Havellian 1998–2017.

Discussion

Due to increasing population, limited surface water storage and increasing impacts of climate change, pressure on groundwater has increased tremendously in Pakistan, including northern Himalayas, over the past several decades (Bhutta & Alam, 2006). Results of this study have also revealed that the populations in two Himalayan towns – Murree and Havellian – are highly dependent on groundwater resources. In Murree, around 80% of households are dependent on spring water (supplied by municipality as well as direct use). In Havellian, 86% of households are dependent on municipal water supply which is mainly fed by groundwater pumping (Figure 4). In both towns, nearly 5 to 12% of households also directly pump groundwater to fulfil their consumption requirements. In Murree, more than 15% of households perceived that spring water is not sufficient to fulfil their consumption requirements. Likewise, in Havellian, around 13% of households reported that municipal water supply is not sufficient. In both towns, 20–30% of households reported that water supplies are occasionally sufficient (Table 3). These reported/perception data are validated through quantitative estimates. Estimates of water availability, consumption and sufficiency ratio showed consistent results for Havellian, where water requirement for consumption is higher than water supplies. However, in the case of Murree, quantitative estimates showed the water supplies are sufficient to meet consumption requirements. Sufficiency ratio is based on average availability and consumption. Thus, it may not reflect the number of households who face water insufficiency. More importantly, sufficiency ratio based on national standards of water consumption shows that both towns have insufficient water supplies to meet the consumption requirement of their respective populations (Table 6).

In both towns, it is found that at household level, a significant proportion of supplied water is either being used directly without storing it or it is wasted due to mismanagement within households (Table 5). Around 20% and 30% of households, respectively, in Murree and Havellian reported that overexploitation of groundwater is impacting its overall availability. Likewise, around 20% and 13% of households reported mismanagement as the reason of change in water availability (Table 7). In the face of climate change, mismanagement of groundwater poses a serious challenge to overall sustainability of water resources. Climate change has already impacted the water availability in both towns, as reported by more than one-fifth of households (Table 7). If mismanagement and climate change continue to impact water resources, both towns may face serious water crises in the future. In particular, the gradual drying up of springs in Murree may result in a water stress situation in the town. In addition to risks to sustainability of water availability, water consumption requirements are also increasing gradually due to high population growth (Table 8) and rapid urbanization (Table 9 and Figure 6). Although most of the parts of the western Himalayas that fall in the Indus Basin are sparsely populated (Hassan, 2010), the case is not the same for some urban centres in these mountain areas, as Himalayan towns in the Indus Basin have witnessed an overwhelming growth in population over the past few decades. This growth in population is largely related to migration from rural areas to urban centres. Analysis of subsequent censuses in Pakistan establishes that the trend of migration remains from deprived and underdeveloped areas to large urban centres in the country (Gazdar, 2003; Hassan, 2010). The demographics of Himalayan towns of the Indus Basin have also rapidly changed due to the increase in trade between the neighboring countries and since most Himalayan centres lie along major trade routes, urbanization and population growth seems inevitable. This increase in population in the urban clusters of Murree and Havellian is similar to most Himalayan towns in the Indus Basin. The urban growth rate in northern Pakistan has remained uniformly high from 1981 to 2009 (Nazir & Schmitt Olabisi, 2015).

The future of water supplies to these Himalayan towns appears uncertain, while plans for sustainable management of aquifers remain a far cry (LEAD, 2016). High dependence on a singular water resource indicates the high vulnerability of Himalayan towns towards future hydro-meteorological shocks. In the absence of water conservation or development of alternative water resources in the Himalayan region, the urban water situation in the towns is set to become more precarious in the coming years. The region is warming at a rate higher than the global average (Shrestha *et al.*, 2015a, 2015b). Future projections indicate that while total precipitation will remain constant, the number of rainy days will reduce meaning lesser but more intense precipitation events (Lutz *et al.*, 2014). Furthermore, scientific literature suggests that urbanization increases surface runoff generated from precipitation and limits infiltration. Recent studies have now shown that increased urbanization can influence the timing and magnitude of precipitation, thus controlling the local climate of the urbanized area. This can be traced to 'urban heat island' effect, which affects the air circulation in urban areas. Moreover, increased carbon emissions in urban areas can also affect the local climate of the towns more adversely than rural areas (O'Driscoll *et al.*, 2010). These climatic impacts will have direct implications on the water resource situation in the urban Himalayas.

Where regulations are scanty in terms of water accessibility to inhabitants of these towns, control over urban sprawl on groundwater recharge zones is also poor. This can have serious impacts on towns like Havellian, where groundwater availability is directly dependent on groundwater recharge. City planning and development agencies across Himalayan towns lack substantial strategies to control paving of groundwater recharge zones as is the case of increasing encroachment of urban population in the spring recharge zones of Murree area, as reported by key informants and focus groups alike. Increased urbanization leads to more paved surfaces which inhibit infiltration and thus the groundwater recharge (Leopold, 1968; O'Driscoll *et al.*, 2010). The extent of urbanization in Himalayan towns, however, has soared within the past two decades, with significant losses to land cover at the hands of urbanization, as is evident from Figure 6 and Table 9.

The findings of this study are consistent with those from other urban centres in the HKH region. Urban population in all parts of the HKH region is also increasing rapidly. Singh *et al.* (2019) revealed that annual average growth rate (2001–2011) of population is 3–5% in the Himalayan urban areas of Bangladesh, India, Nepal and Pakistan. Migration from rural to urban centres for better economic opportunities is an obvious reason for the increased rate of urbanization (Rasul *et al.*, 2019). It has further widened the gap (deficit) between water demand and supply in the urban centres. Singh *et al.* (2019) also compiled data on water demand and actual supply in 13 Himalayan towns of Bangladesh, India, Nepal and Pakistan. The study shows that all towns are facing a water deficit ranging from 30% to 88% (calculated based on data presented in Singh *et al.*, 2019). Towns from the Indian Himalayas, e.g., Darjeeling, Kohima and Tansen face more than 80% water deficit, which is alarming, and likely to add more pressure on groundwater resources in these towns. Urbanization trends are almost similar to those in other countries of the Global South. By 2050, the urban population is projected to be around 66% of the total global population, with nearly 90% of the increase being concentrated in Asia and Africa (UNDESA, 2014). This increase in urban population is likely to have serious impacts on groundwater resources in urban areas.

In particular, in this study, it can be established that urbanization, climate change and lack of good governance are impacting the groundwater resources in Himalayan towns. All these factors are affecting current and future sustainability of this precious water resource. Thus, there is a dire need for effective and problem-centric groundwater policy for Himalayan towns and/or a need to have a clear directive for sustainable urban groundwater utilization and development in the National Policy on Water in Pakistan.

Conclusions

In the Hindu-Kush Himalayan (HKH) region, rapid urbanization, population growth, limited surface water storage and increasing impacts of climate change are affecting water availability, consumption and overall sufficiency, particularly in the urban centres. Findings from two Himalayan towns – Murree and Havellian – from Pakistan's Indus Basin area revealed that groundwater is the main source of domestic water, and almost all households are heavily dependent on it. In Murree and Havellian, respectively, springs and municipal water (pumped water) are the sources of groundwater. In both towns, around 35% of households reported that groundwater resources are not sufficient to fulfil their domestic water requirements. Quantitative estimates of water availability, consumption and sufficiency ratio showed consistent results for Havellian, where groundwater is not sufficient to meet the requirements for consumption. However, in the case of Murree, water supplies are sufficient. However, sufficiency ratio based on national standards of water consumption showed that both towns have insufficient water supplies to meet consumption requirements of their respective populations.

In both towns, there is a substantial gap between available water (municipal supply) and actual stored water at household level, indicating the mismanagement issues of water resources at household level. It is also evident from household perception data, which underlined that overexploitation and mismanagement were the key challenges to water sustainability. In addition, one-fifth of surveyed households reported that climate change has also affected the water availability in the towns. Conversely, factors such as rapid urbanization and population growth are likely to result in an increased requirement of water in the future. This will negatively impact the water sufficiency in the study towns.

Based on the findings, the study has suggested the following policy actions:

- Measures are suggested to reduce heavy dependency on and overexploitation of groundwater resources, and to diversify the water sources through tapping the potential of surface water resources.

- Awareness may be created among households as well as water managing authorities to promote judicious use of spring water. The share of spring water to municipal supply needs to be reduced, and surface water to be improved. This will help in protecting the spring sources from depletion.
- There is need to launch an awareness campaign to promote efficient use of water at household level, and to reduce mismanagement and wastage of water. An effective and water value-based pricing mechanism may also be introduced to regulate the water use at household level.
- There is a need to improve town level governance and administration to develop and implement strong regulations regarding urbanization (i.e., town planning and building control), groundwater abstraction and municipal water supply. These regulations may include governance mechanisms for spring water sources and water use rights. The regulations introduced for urbanization may include identification, demarcation and protection of recharge zones for major spring sources. In addition, strong regulatory action is required for setting the criteria and extent of groundwater abstraction at household or community level.

It is hoped that the findings of this study will be useful to understand the situation of water availability, consumption and sufficiency in the Himalayan urban centres of Pakistan, and will provide input to urban planning and implementation of the National Water Policy. The findings and recommendations of this study may also be useful for urban planning in other urban centres of the HKH region, and similar mountain areas from the Global South.

Acknowledgements

The study was carried out by the Himalayan Adaptation, Water and Resilience (HI-AWARE) consortium under the Collaborative Adaptation Research Initiative in Africa and Asia (CARIAA), with financial support from the United Kingdom's Department for International Development (DFID) and the International Development Research Centre (IDRC), Ottowa, Canada. This study was also partially supported by core funds of ICIMOD contributed by the governments of Afghanistan, Australia, Austria, Bangladesh, Bhutan, China, India, Myanmar, Nepal, Norway, Pakistan, Sweden and Switzerland. We would like to express our deepest appreciation for the assistance of local contact persons during the field data collection, and of all local informants and households who took precious time to share their insights and knowledge. We are ultimately solely responsible for the contents of the paper. The views expressed in this work are those of the authors and do not necessarily represent those of DFID, IDRC or its Board of Governors. In addition, they are not necessarily attributable to authors' organizations.

Supplementary material

The Supplementary Material for this paper is available online at https://dx.doi.org/10.2166/wp.2019.012.

References

Andermann, C., Longuevergne, L., Bonnet, S., Crave, A., Davy, P. & Gloaguen, R. (2012). Impact of transient groundwater storage on the discharge of Himalayan rivers. *Nature Geoscience 5*, 127–132.

Bhutta, M. N. & Alam, M. M. (2006). Prospectives and limits of groundwater use in Pakistan. *Groundwater Research and Management: Integrating Science into Management Decisions* pp. 105–114.

Bolch, T., Kulkarni, A., Kääb, A., Huggel, C., Paul, F., Cogley, J. G., Frey, H., Kargel, J. S., Fujita, K., Scheel, M., Bajracharya, S. & Stoffel, M. (2012). The state and fate of Himalayan glaciers. *Science 336*(6079), 310–314.

Climate-Data (2018). *Climate of Havellian*. https://en.climate-data.org/location/25720/ (accessed 6 July 2018).

Cochran, W. G. (1977). *Sampling Techniques*, 3rd edn. John Wiley & Sons, New York, USA.

Gazdar, H. (2003). *A Review of Migration Issues in Pakistan*. Refugee and Migratory Movements Research Unit, Bangladesh, and the Department for International Development, UK.

Gerlitz, J.-Y., Apablaza, M., Hoermann, B., Hunzai, K. & Bennett, L. (2015). A multidimensional poverty measure for the Hindu Kush–Himalayas, applied to selected districts in Nepal. *Mountain Research and Development 35*(3), 278–288.

Government of Pakistan, M.o.H.W. (1993). National Reference Manual of Planning and Infrastructure Standards. Division, E.a.U.A. (ed.), Pakistan Environmental Planning and Environmental Consultants (PEPAC), Gulberg, Lahore, Pakistan, p. 416.

Hassan, A. (2010). Migration, small towns and social transformations in Pakistan. *Environment and Urbanization 22*(1), 33–50.

Hunzai, K., Gerlitz, J. Y. & Hoermann, B. (2011). *Understanding Mountain Poverty in the Hindu Kush-Himalayas: Regional Report for Afghanistan, Bangladesh, Bhutan, China, India, Myanmar, Nepal, and Pakistan*. International Centre for Integrated Mountain Development, Kathmandu, Nepal.

Hussain, A., Sarangi, G. K., Pandit, A., Ishaq, S., Mamnun, N., Ahmad, B. & Jamil, M. K. (2019). Hydropower development in the Hindu Kush Himalayan region: issues, policies and opportunities. *Renewable and Sustainable Energy Reviews 107*, 446–461.

Immerzeel, W. W., Pellicciotti, F. & Bierkens, M. F. P. (2013). Rising river flows throughout the twenty-first century in two Himalayan glacierized watersheds. *Nature Geoscience 6*, 742–745.

Kottek, M., Grieser, J., Beck, C., Rudolf, B. & Rubel, F. (2006). World Map of the Köppen-Geiger climate classification updated. *Meteorologische Zeitschrift 15*(3), 259–263.

Laghari, A. N., Vanham, D. & Rauch, W. (2012). The Indus basin in the framework of current and future water resources management. *Hydrology and Earth System Sciences 16*(4), 1063–1083.

LEAD (2016). *Groundwater Management in Pakistan: An Analysis of Problems and Opportunities*. Learning and Knowledge Management Team, LEAD Pakistan, Islamabad, Pakistan, p. 44.

Leopold, L. B. (1968). *Hydrology for Urban Land Planning – A Guidebook on the Hydrologic Effects of Urban Land Use*. USGS, Washington, DC, USA.

Lutz, A. F., Immerzeel, W. W., Shrestha, A. B. & Bierkens, M. F. P. (2014). Consistent increase in High Asia's runoff due to increasing glacier melt and precipitation. *Nature, Climate Change 39*(4), 587.

Lutz, A., Immerzeel, W. W., Kraaijenbrink, P., Shrestha, A. & Bierkens, M. F. P. (2016). Climate change impacts on the Upper Indus hydrology: sources, shifts and extremes. *PLoS One 11*(11). https://doi.org/10.1371/journal.pone.0165630.

Miller, J. D., Immerzeel, W. W. & Rees, G. (2012). Climate change impacts on glacier hydrology and river discharge in the Hindu Kush–Himalayas. *Mountain Research and Development 32*(4), 461–467.

Nazir, N. & Schmitt Olabisi, L. (2015). *Forest Area and Land Use Change in Pakistan: A System Dynamics Approach. System Dynamics Society*. Curran Associates, Inc., Cambridge, MA, USA.

NOAA (1995). Murree Climate Normals 1961–1990, *National Oceanic and Atmospheric Administration*. ftp://ftp.atdd.noaa.gov/pub/GCOS/WMO-Normals/RA-II/PK/41573.TXT (accessed 15 August 2018).

O'Driscoll, M., Clinton, S., Jefferson, A., Manda, A. & McMillan, S. (2010). Urbanization effects on watershed hydrology and in-stream processes in the Southern United States. *Water 2*(3), 605–648.

Pakistan Bureau of Statistics (1981). District Census Reports for Murree and Havellian. Population Census 1981. Pakistan Bureau of Statistics, Islamabad, Pakistan.

Pakistan Bureau of Statistics (1998). District Census Reports for Murree and Havellian. Population Census 1998. Pakistan Bureau of Statistics, Islamabad, Pakistan.

PBS, Pakistan Bureau of Statistics (2017). *District Census Report*. Wing, C. (ed), PBS, Islamabad, Pakistan, p. 245.

Palazzi, E., von Hardenberg, J. & Provenzale, A. (2013). Precipitation in the Hindu-Kush Karakoram Himalaya: observations and future scenarios. *Journal of Geophysical Research: Atmospheres 118*(1), 85–100.

Qureshi, A. S. (2011). Water management in the Indus Basin in Pakistan: challenges and opportunities. *Mountain Research and Development 31*(3), 252–260.

Rasul, G. (2012). *Climate Data Modelling and Analysis of the Indus Ecoregion*. World Wide Fund for Nature (WWF) Pakistan, LEAD Pakistan, pp. 24–31.

Rasul, G., Saboor, A., Tiwari, P. C., Hussain, A., Ghosh, N. & Chettri, G. B. (2019). Food and nutrition security in the Hindu Kush Himalaya: unique challenges and niche opportunities. In: *The Hindu Kush Himalaya Assessment*. Springer, Cham, Switzerland, pp. 301–338.

Sharma, E., Molden, D., Rahman, A., Khatiwada, Y. R., Zhang, L., Singh, S. P., Yao, T. & Wester, P. (2018). Introduction. In: *The Hindu Kush Himalaya Assessment – Mountains, Climate Change, Sustainability and People*. Wester, P., Mishra, A., Mukherji, A. & Shrestha, A. B. (eds). SpringerNature, Dordrecht, the Netherlands.

Shrestha, A. B., Agrawat, N. K., Alfthan, B., Bajracharya, S. R., Maréchal, J. & van Oort, B. (eds) (2015a). *The Himalayan Climate and Water Atlas: Impact of Climate Change on Water Resources in Five of Asia's Major River Basins*, ICIMOD (Kathmandu, Nepal), GRID-Arendal (Arendal, Norway) and CICERO (Oslo, Norway).

Shrestha, M., Koike, T., Hirabayashi, Y., Xue, Y., Wang, L., Rasul, G. & Ahmad, B. (2015b). Integrated simulation of snow and glacier melt in water and energy balance-based, distributed hydrological modeling framework at Hunza River Basin of Pakistan Karakoram region. *Journal of Geophysical Research: Atmospheres 120*, 4889–4919. doi:10.1002/2014JD022666.

Singh, S., Tanvir Hassan, S. M., Hassan, M. & Bharti, N. (2019). Urbanisation and water insecurity in the Hindu Kush Himalaya: insights from Bangladesh, India, Nepal and Pakistan. *Water Policy*. https://doi.org/10.2166/wp.2019.215.

Smadja, J., Aubriot, O., Puschiasis, O., Duplan, T., Grimaldi, J., Hugonnet, M. & Buchheit, P. (2015). Climate change and water resources in the Himalayas. Field study in four geographic units of the Koshi basin, Nepal. *Journal of Alpine Research/Revue de géographie alpine 103*(2), 1–23.

Tiwari, P. C., Tiwari, A. & Joshi, B. (2018). Urban growth in Himalaya: understanding the process and options for sustainable development. *Urban and Regional Studies on Contemporary India 4*(2), 15–27.

Tshe-ring, K., Sharma, E., Chettri, N. & Shrestha, A. (2010). *Climate Change Vulnerability of Mountain Ecosystems in the Eastern Himalayas – Synthesis Report*. ICIMOD, Kathmandu, Nepal.

UNDESA (2014). *World's Population Increasingly Urban with More Than Half Living in Urban Areas*. United Nations Department of Economic and Social Affairs. http://www.un.org/en/development/desa/news/population/world-urbanization-prospects-2014.html (accessed 31 March 2019).

Wester, P., Mishra, A., Mukherji, A. & Shrestha, A. B. (eds). (2019). *The Hindu Kush Himalaya Assessment: Mountains, Climate Change, Sustainability and People*. Springer, Cham, Switzerland. https://doi.org/10.1007/978-3-319-92288-1.

WHO. (2003). *Domestic Water Quantity, Service, Level and Health*. World Health Organization, Geneva, Switzerland.

Wiltshire, A. J. (2014). Climate change implications for the glaciers of the Hindu Kush. *Karakoram and Himalayan region. The Cryosphere 8*, 941–958. doi:10.5194/tc-8-941-2014.

Received 19 January 2019; accepted in revised form 8 April 2019. Available online 24 May 2019

Dynamics of urban water supply management of two Himalayan towns in India

Neha Bharti[a,*], Neha Khandekar[a], Prateek Sengupta[b], Suruchi Bhadwal[a] and Ishaan Kochhar[c]

[a]*The Energy and Resources Institute (TERI), Delhi, India*

**Corresponding author. E-mail: neha.bharti@teri.res.in*

[b]*Centre for Ecology Development and Research (CEDAR), Dehradun, India*

[c]*International Centre for Integrated Mountain Development (ICIMOD), Patan, Nepal*

Abstract

Many towns in the Indian Himalayan Region (IHR) are experiencing permanent water crises due to increasing population pressure, urbanization, and poor management of existing water sources. This paper focuses on two towns – Mussoorie and Devprayag in the western IHR – to understand various aspects of the growing water scarcity and urban water management. In the current scenario of a changing climate, natural springs, their main water resource, are drying up. Mussoorie experiences an acute shortage of water in summer, precisely when the town hosts numerous tourists. In Devprayag, religious tourism and in-migration from rural areas contribute to rising demand. The reduced discharge in nearby streams has widened the demand–supply gap. An integrated management of water sources is crucial to solving water problems in Mussoorie and Devprayag. In both towns, little effort has been made towards recharging existing water sources. Detailed planning of the water supply system while being mindful of the floating population, a proper sewage and storm water management system, and rainwater harvesting schemes, are absent. There is an urgent need to adopt a comprehensive approach to solving urban water issues, covering aspects of demand, supply and water resources management in these hill towns for adaptive water management.

Keywords: Coping strategies; Indian Himalayan Region; Springs; Urban water management; Water demand; Water supply

Introduction

The Hindu Kush Himalaya (HKH) region is wide, encompassing more than 4.3 million km^2, including high mountain areas in Afghanistan and Pakistan in the west, India, China and Nepal in the centre,

doi: 10.2166/wp.2019.203

and Bangladesh, Bhutan and Myanmar in the east (Schmid *et al.*, 2015). The Himalayan range in India, being home to ten large Asian river systems, provides water to more than a 1.9 billion people. These rivers provide fresh water for drinking, domestic uses, industry, irrigation and power generation (Mukherji *et al.*, 2015; ICIMOD, 2018).

The Ganges is a transboundary river, with its headwaters originating in these mountains. The Upper Ganges Basin, in which our sites of study are situated, is located in the northern Himalaya, and encompasses an area of over 87,000 km^2 (Figure 1), with a wide variation in elevation and climate. Its elevation ranges from about 7,500 metres above sea level (masl) to around 100 masl (Bharati & Jayakody, 2010; Bharati *et al.*, 2011). Average annual rainfall ranges from 550 to 2,500 millimetres (mm), with most of the rainfall falling during the southwest monsoon (Bhadwal *et al.*, 2017).

With a glacier coverage of 60,000 km^2 (Bajracharya & Shrestha, 2011), the HKH region is referred to as the 'Water Tower of Asia' as it provides around 86,000,000 cubic metres (m^3) of water annually, feeding many of its major rivers (Rao *et al.*, n.d.). However, rapid warming has resulted in the shrinking of snow, ice, glaciers and permafrost in the region (You *et al.*, 2017). Scientific evidence indicates that glaciers in the HKH region have been receding, similar to glaciers in other parts of the world (National Research Council, 2012; Bajracharya *et al.*, 2015). As well, the decline in snowfall and erratic rainfall

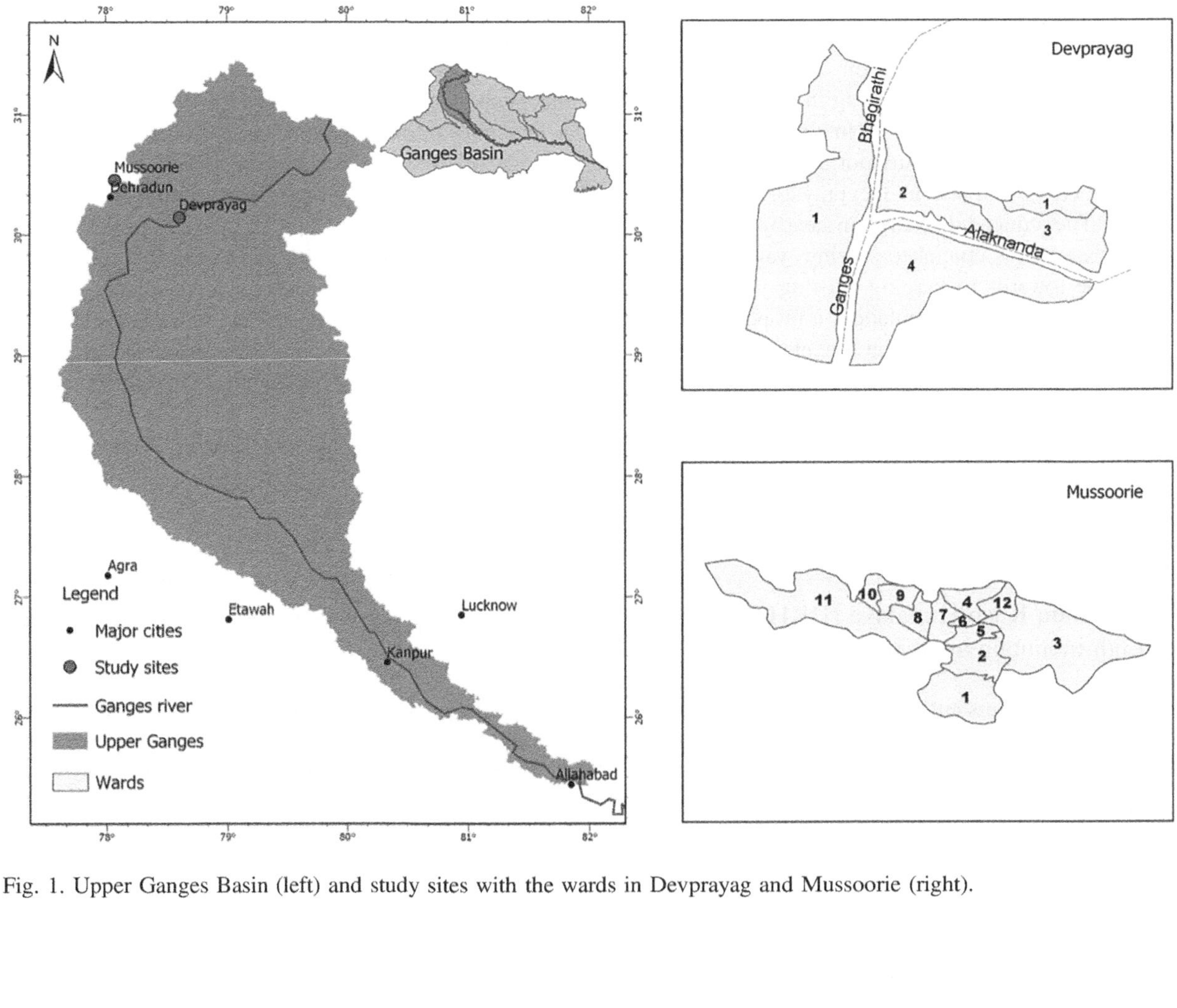

Fig. 1. Upper Ganges Basin (left) and study sites with the wards in Devprayag and Mussoorie (right).

patterns has had an impact. The Ganges Basin draws 9%–10% of its waters from glacial melt, the rest of it being largely fed by rainfall (Khan *et al.*, 2017).

Hence, a combination of factors has adversely affected water supplies. There are also issues of springs drying up in many parts of the mountains due to the lack of adequate recharge owing to changing rainfall pattern and catchment degradation. Pioneering work in these regards was carried out by Valdiya & Bartarya (1989), identifying deforestation of hill slopes as one of the primary causes. The drying up of springs will affect the flow of rivers in the region and also most of the mountain communities who depend on springs (locally known as *naula* or *dhara*) for their drinking and domestic water needs.

Conversely, an increasing demand for water, with a rising population, has added to the stress on water resources in these river systems. The current population of the HKH region is about 1.9 billion (Sharma *et al.*, 2019). In recent years, the number of cities and towns has also grown manifold, thus increasing the demand for water. Most of the increase in population is concentrated in the cities and towns in the mountain states, largely owing to the migration of people from rural and semi-urban areas seeking better access to basic facilities, including employment and education. Urbanization, therefore, presents huge challenges for meeting the water requirements of a continuously growing population.

Small towns and cities in this region also have a continuous, floating population, consisting of tourists travelling and staying in these towns for short periods, because of the region's cultural and spiritual significance. They are not generally considered part of the official census count. According to the Uttarakhand Annual Plan 2013–2014, the floating population observed in Uttarakhand during peak months ranges between 0.3 and 0.35 million. This further increases the seasonal demand for water (Government of Uttarakhand, 2014).

A physical scarcity of water is already being observed in many parts, especially the mid-hill regions (Bhadwal *et al.*, 2017). According to Government of India norms, the per capita water availability for urban areas ought to be 135–150 litres per capita per day (lpcd), but the poor living in slum areas near the urban centres receive only 72–74 lpcd (Bose & Srivastava, 2017; NIH, n.d.). Climate change is likely to add further pressure on the water resources in these regions.

With this overview in mind, this paper seeks to study the multiple factors influencing water supply and demand in Himalayan cities and towns in Uttarakhand, with the objective of suggesting possible measures to decrease the vulnerabilities of communities to water stress. We present the empirical findings from a study undertaken to investigate water access and availability, and lacunae in current water management approaches in Mussoorie and Devprayag, in the state of Uttarakhand, India (Figure 1). Both these towns vary in size considerably and have floating populations, for different reasons. We investigate the various factors compounding the situation, which increase the challenges of communities living in this region. We have attempted to document the policy, regulatory and institutional support governing water resources and unearth gaps therein. Alternative sources and coping mechanisms have been identified through field-based discussions, to better understand the fallback options that communities facing complex constraints have in dealing with resource scarcity.

Study area

Mussoorie, a medium-sized town, lies in the foothills of the Garhwal Himalayan range and is part of Dehradun district in Uttarakhand. Devprayag is a small-sized town in Tehri district. It is one of the most important pilgrimage centres in northern India for devout Hindus. The rivers Bhagirathi and Alaknanda merge at Devprayag to form the Ganges. Mussoorie has an average elevation of 2,006 metres (m) and Devprayag 830 m. Devprayag is 8.75 km^2 in area and Mussoorie 67.6 km^2 (Ramachandran & Ramachandran, 2001; Asian Development Bank, 2014). Mussoorie currently has a municipal board divided into 12 wards, whereas Devprayag has four wards, also governed by a municipal board (Nagar Palika Parishad) (Figure 1).

Mussoorie has experienced a tremendous population expansion and unregulated growth (Hewitt & Mehta, 2012). Its population grew from 18,241 in 1981 to 30,118 in 2011. In contrast, Devprayag's population has remained almost the same, rising very slightly from 2,769 in 2001 to 2,868 in 2011 (Ministry of Home Affairs, 2001, 2011).

Mussoorie's demographic spread is diverse. It has some densely packed wards, with buildings up to four floors; it also has high-rise, eight-storey apartments, in wards 8, 9 and 10. Slum areas and residential quarters in many wards are often found with common water and toilet facilities. Devprayag is sited at a height of 30 m above the water level. It is divided into three areas, separated from each other by the holy rivers. These areas are connected with each other only by the means of two suspension bridges, over which only pedestrians can pass. The houses are generally two-storeyed, and the old houses are built of large stones (AHEC, 2011).

It being one of the fastest growing hill stations in northern India, with a high tourist influx in summer and attracting around half a million tourists annually, were the pivotal reasons for having selected Mussoorie as one of the two sites of research for this study. Conversely, Devprayag is one of the most important pilgrimage centres in India, with a tourist influx mainly concentrated during the 'Char Dham Yatra' season. Additionally, the increase in in-migration from rural areas nearby is causing a very gradual rise in Devprayag's population, making it an interesting case to study. It was also unfortunately one of the worst affected areas during the Uttarakhand flood disaster of 2013, which, among other impacts, badly affected the water infrastructure.

Methodology used

The study used a mixed method approach for an in-depth understanding of the emerging water situation and adaptive mechanisms to address water-related challenges in the context of water consumption for drinking and domestic purposes. It combines quantitative data from household surveys, qualitative evidence from focus group discussions within the wards, and interviews with key informants in municipal offices, the relevant government departments, hotels/restaurants, clinics/hospitals and NGOs, among others.

The sample size for the household survey was 350 households, divided across the two towns. The sample size in each town was determined using the unitary method, in proportion to the number of households in both towns. The total households in Mussoorie and Devprayag are 6,445 and 739, respectively. The sample size for Mussoorie was calculated at 314 households in 11 wards. For Devprayag, it was 36 households in four wards.

A structured household-level questionnaire was used to survey the sample households in the two towns. The survey was conducted over January–March 2017. Households within wards were selected to represent different, relevant contexts – caste background, with or without water supply connection, elevation amplitude, whether the house was rented or owned, and differential access to primary and alternative sources of water. The quantitative data were collected using Akvo Flow tool on a smartphone and a simple analysis was conducted on the survey data exported from the tool.

Household socioeconomic profile

Most respondents are between 30 and 50 years old, with educational levels up to higher secondary. Most families in Devprayag have a stated average monthly household income below INR 30,000. In Mussoorie, it is between INR 15,000 and 50,000 for 72% of the respondent families. The average family size ranges between three (Devprayag) and five (Mussoorie) members.

Most of the respondents live in houses they own – 92% in Devprayag, 69% in Mussoorie, most in *pukka* (cemented) homes. A large proportion of respondents in both towns live in 1–2 room-set houses. Importantly, 89% in Devprayag and 72% in Mussoorie have access to private toilets. A significant proportion of the sample population in both towns has migrated there, either from nearby villages, or from neighbouring countries and regions such as Nepal and Tibet, and have been settled there for 5–10 years. Around 20–40% of the respondents in both towns are permanent residents (Figures 2 and 3).

Water sources and supply

Primary water sources:[1] *Uses, frequency, duration, location and expenditure*

All households in both Mussoorie and Devprayag rely heavily on municipal water supply for various domestic purposes like drinking, cooking, washing and bathing. Around 8% of the households in Devprayag which do not have private municipal water connections collect water from nearby public stand posts for drinking or other domestic purposes. In Mussoorie, 90% of the respondent households have access to a municipal supply. Of these, only 62% have municipal taps inside their homes; the rest 38% (which do not have the infrastructure to facilitate the supply of water) are dependent on water from public taps/stand posts for drinking and other domestic purposes[2]. The remaining 10% of sampled households in Mussoorie which do not have private water connections or access to public taps/stand posts access the closest spring source available and so that becomes the primary source for them (Table 1).

In both the towns, municipal water is supplied every day to the private water connections. Even for the public taps, almost every household in both towns said they received the water daily. Private connections in Devprayag received municipal water supply for a longer duration than Mussoorie. Households in Devprayag used to receive only 1 hour of supply per day until December 2016, but have been getting water an additional hour per day since then. Every surveyed household in Devprayag mentioned receiving more than 2 hours of water supply daily, whereas in Mussoorie, a major proportion of households (71%) received only 1 hour a day of private municipal supply. Many public stand posts in Mussoorie regularly supply around 2 hours of water daily but the number of users for each tap is absurdly high (Table 2).

Private water connections in both the towns are unmetered. Users pay a tariff decided by Jal Sansthan – the official government body dealing with drinking water supply and management in Uttarakhand –

---

[1] A primary water source is one that the households primarily have access to, and rely on, for fulfilling their domestic needs. It includes private water connections and public taps/stand posts, both supplied by the municipal authorities.

[2] These households do not have private water connections due to either of two reasons: (1) the houses are *kuchcha* (non-cemented) and do not have the necessary infrastructure to accommodate a water supply line; (2) they are located in those parts of town where there is an absence of piped water supply.

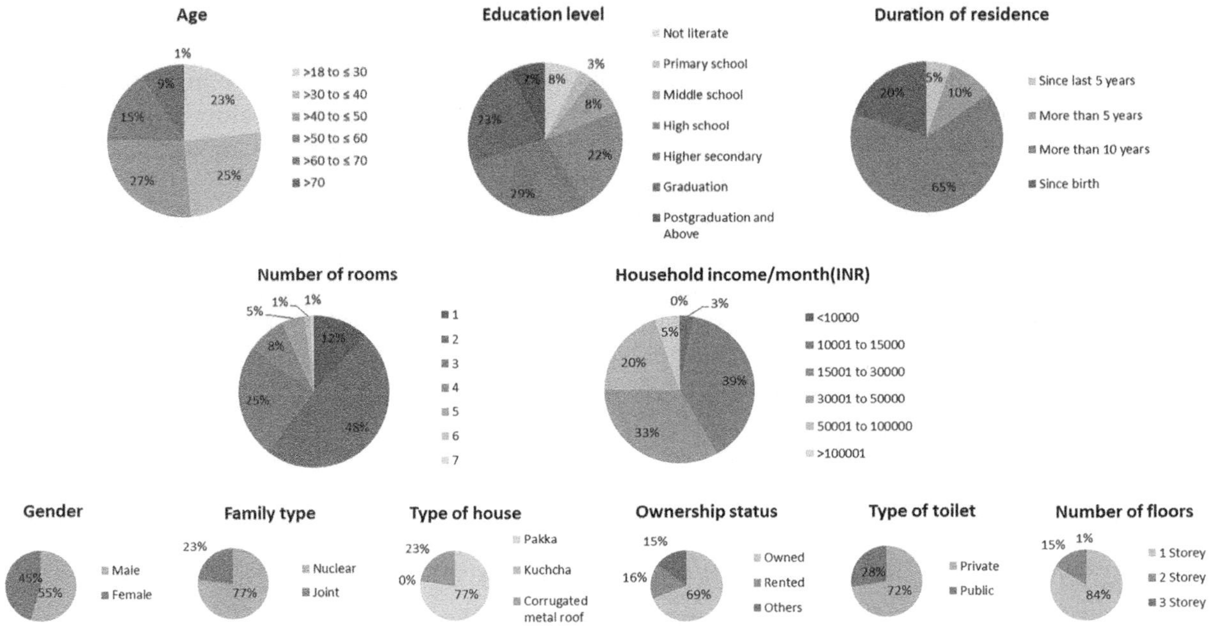

Fig. 2. Household-level information – Mussoorie.

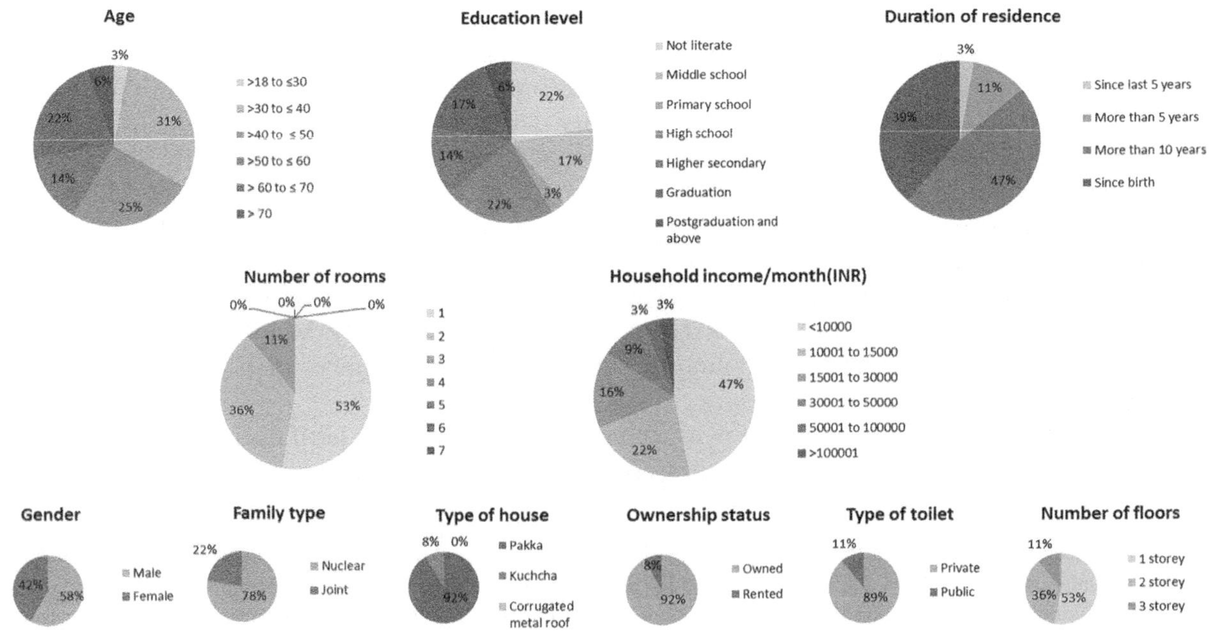

Fig. 3. Household-level information – Devprayag.

Table 1. Primary sources of water.

Primary source for domestic uses	Number of households (%)					
	Mussoorie			Devprayag		
	Public taps/stand posts	Private taps	Total	Public taps/stand posts	Private taps	Total
Municipal supply	38	62	90	8	92	100
Springs	10			NA		

Table 2. Water access: Frequency, duration and tariffs (% households).

Water access	Frequency		Duration/day		Monthly tariffs (INR)	
	Mussoorie	Devprayag	Mussoorie	Devprayag	Mussoorie	Devprayag
Private taps	Daily (100)	Daily (100)	1 hr (71) 2 hrs (27) >2 hrs (2)	> 2 hrs (100)	Up to 300 (89) 301–600 (4) >600 (7)	301–600 (100)
Public taps	Daily (97)	Daily (100)	Up to 1 hr (28) Up to 2 hrs (67) >2 hrs (6)	> 2 hrs (100)	Free	Free

based on a flat rate that takes into account two factors – the size of the property and its circle rate, and the height (or distance) to which the water is pumped. Most holders with private connections in both towns pay between INR 300 and 600 a month. A few households in Mussoorie, mainly those in high-rise apartments, pay a tariff of more than INR 600 per month (Table 2).

Piped water supply infrastructure

In both the towns, springs are the source of the piped water supply system. In Mussoorie, the municipality taps 20 spring sources to generate 9 million litres per day (MLD) of water, transported by gravity and pumping systems. There is a 98 km-wide network of distribution pipelines, with 4,065 domestic and 1,206 commercial tap connections, covering approximately 90% of Mussoorie's area. Conversely, Devprayag, being a small town, has only two spring fed streams (called *gadheras* locally), Randigad (13 km away) and Diwanigad (7 km away), supplying water to the town by the gravity system (Tables 3 and 4).

The total volume of water currently supplied in Mussoorie is 9.1 MLD against the total local demand of 6.9 MLD (Table 3). However, in reality, the town is always bursting with tourists and thus the water requirement exceeds the water availability. For example, in the summer of 2014, Mussoorie's total demand, at 14.4 MLD, was almost twice that of its supply (7.67 MLD) at the time (Ramola, 2014).

Despite geographical constraints in Mussoorie, Jal Sansthan has attempted, through the pipeline system, to reach most of the settlements in the town requiring water. There is still however a huge gap in the number of households having private connections (62%) and households not having access to the water supply system and depending either on public taps or directly accessing springs (38%). In Devprayag, the total water supply before 2014 was 0.39 MLD, far less than the demand of 0.7 MLD. What is more, the Uttarakhand floods of 2013 severely damaged the town's water supply

Table 3. Municipal supply of water to Mussoorie.

S. no.			Name of source	Volume of water supplied (MLD)	Type of source
A.	**Pumping schemes**				
1.	Murray Pumping Scheme	1	Khanalty	0.648	Spring
		2	Undercliff	0.155	Spring
		3	Bansi	0.288	Spring
		4	Kandighat Upper	0.288	Brooklet
		5	Kandighat Lower	0.691	Brooklet
		6	Rikhauli Gad	0.360	Brooklet
		7	Kandighat Stream	0.360	Brooklet
2.	Mackinnon Pumping Scheme	8	Newby	0.115	Spring
		9	John Mackinnon	0.216	Spring
		10	Chalmer Khud	0.115	Spring
3.	Bhilaru Pumping Scheme	11	Bhilaru	1.296	Spring
4.	Jincy Pumping Scheme	12	Jincy	2.419	Spring
5.	Kolti Pumping Scheme	13	Koltikhala	0.864	Brooklet
6.	Dhobighat Pumping Scheme	14	Dhobighat	0.763	Spring
B.	**Gravity schemes**				
1.		15	Company Khud	0.086	Spring
2.		16	Brookland	0.129	Spring
3.		17	Nalapani	0.036	Spring
4.		18	Pargakhala	0.158	Spring
5.		19	Douglas Dale	0.129	Spring
6.		20	Sentipani	0.072	Spring
	Total			9.188	

Source: Jal Sansthan (2017).

Table 4. Municipal supply of water to Devprayag.

S. no.	Name of source	Volume of water supplied (MLD)	Type of source
A.	Gravity schemes		
1	Old Randigad	0.21	Stream
2	Old Diwanigad	0.18	Stream
3	New Randigad	0.50	Stream
4	New Diwanigad	0.36	Stream
	Total	1.25	

Source: Jal Sansthan (2017).

system, adversely affecting the already stressed supply. In addition, the scheme was quite old and facing production problems due to high turbidity in the monsoons. To address these issues, in 2014, Jal Sansthan, with support from the Asian Development Bank (ADB), started work to reconstruct and rehabilitate the town's damaged water supply system. It also had the objective of augmenting the water supply components by adding new infrastructure in order to meet the prospective demand of the town for another ten years. The new distribution line started in December 2016 from the above two sources, supplementing the existing supply system with an additional 0.86 MLD of water. Thus, according to Jal Sansthan, the current volume of water supplied is around 1.25 MLD, almost double that of local demand, to meet the needs arising out of in-migration from surrounding rural areas.

Alternative water sources

If the primary source of water is not available, the alternative sources to rely on in the basin, mentioned by the respondents, are Ganges River water, springs, public stand posts (also supplied by municipal water) and water sourced from public and private water tankers.

In Mussoorie, the public stand post is the most preferred alternative option; 43% of the surveyed households use it for all purposes when regular private water supply is disrupted. Private water tankers and direct access to springs are next, used by 17% and 11% of the households, respectively. Many households using public stand posts, springs and water tankers also curtail their consumption due to limited water availability (Figure 4).

Water shortages are highest during May–July due to high tourist demand at that time of year combined with it being the lean discharge period of water sources. Spring water is used by 29% of the households surveyed in Mussoorie. This includes people's direct access to springs and municipal water sourced from springs, such as stand posts and municipal water tanks. However, as many as 91% of the households use it only during a water crisis (when the primary source of water is not available). Around 73% of Mussoorie's spring users access springs only in summer and the pre-monsoon period, and the remainder throughout the year. Almost all users use water supplied by tankers only during a crisis, most of it during the summer and pre-monsoon periods (Table 5). Private water tankers are usually called by a group of households (20–30 or more) in the same locality. The water is collected in large containers and carried back home. The price of the water ranges from INR 0.40–1.5 per litre, depending on the distance travelled by the tanker from the source to the locality, and the quality of the water.

In Devprayag, 44% of the households collect water from the Ganges River for all purposes when municipal water is not available. This usually happens during the monsoons, when the supply line gets cut off due to landslides or the destabilization of slopes. Those who have other options do not drink it due to quality concerns, but use river water only for washing and bathing. Around 17% of households close to spring sources use it for drinking and for other uses switch to river water (Figure 5). Of those who access springs, most use them only during periods of crisis during the monsoon, when there is a regular interruption in the supply due to the breakage of pipelines (Table 5).

There are no private water tankers that supply water to Devprayag, only municipal tankers which supply water during water shortages or when the pipelines are cut off. The municipality does not charge anything

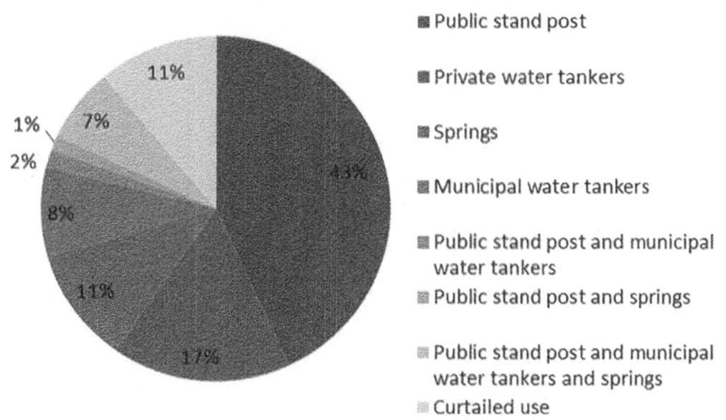

Fig. 4. Alternative water sources for Mussoorie.

Table 5. Alternative sources of water: Frequency, season used and costs.

Alternative source	Users (%)		Frequency of use (%)		Season used (%)		Cost/litre (INR)	
	Mussoorie	Devprayag	Mussoorie	Devprayag	Mussoorie	Devprayag	Mussoorie	Devprayag
Spring	29	19	Very often (9)	Very often (14)	Throughout the year (27)	Throughout the year (14)	NA	NA
			Only during crisis (91)	Only during crisis (86)	Summer and pre-monsoon (73)	Monsoon (86)		
Water tanker	31	22	Only during crisis (100)	Only during crisis (100)	Summer and pre-monsoon (100)	Summer and pre-monsoon (100)	0.4–1.5 INR	NA

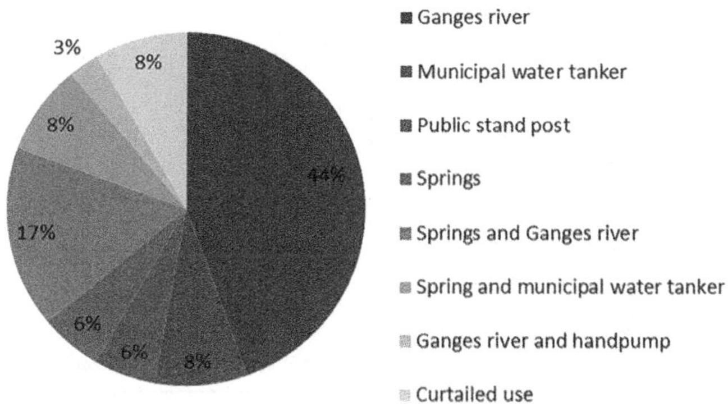

Fig. 5. Alternative water sources for Devprayag.

for the water supplied in crisis through tankers. Yet, only 26% of the households use water tankers as an alternative in Devprayag, as the town is situated on a hill slope and the tanker cannot travel easily to many houses because of the absence of a drivable road. It is difficult for people in households located down the hill to go up and collect water from the tanker. It is also easier for them to access the river.

Perceptions regarding water quality and seasonal variance

Overall, the water quality of all sources is perceived to be acceptable in both towns, except for the water provided by water tankers and river water. About 33% of users surveyed in Devprayag reported that tanker water was not clear and was of unacceptable taste. In Mussoorie, although its odour and taste were deemed acceptable, 10% of the respondents using water from tankers found it not to be clear. In Devprayag, some of the respondents found the river water neither to be clear nor taste right. A few municipal water users reported odour issues. They attributed this to the water being sourced from one of the springs having a village upstream. The spring upstream is used by livestock, for drinking and washing, which might be the cause of the perceived 'foul smell' by the respondents in Devprayag. The water quality of spring sources, in general, is perceived to be most satisfactory by users in both towns (Figures 6 and 7).

However, the quality of water from all the sources is said, in both towns, to vary seasonally. It dips most during the monsoons. For instance, most users in these towns say that the quality of municipal water – which is the major source for drinking – is merely average in Mussoorie and poor in Devprayag during the monsoon. Respondents in Devprayag even find the quality of the Ganges water poor and not potable during the monsoon, but the quality of spring water is good for half of the respondents even then. However, in Mussoorie, a majority of the respondents questioned the quality of tanker water and even springs during the monsoon.

The reasons cited for the fluctuating quality and supply of water included the supply pipes getting damaged during the monsoon by debris from landslides or slope destabilization. Users situated some distance from the pumping stations have mentioned that during the monsoon, rainwater gets mixed with their supply due to problems in the pipe network. Also, tanks and reservoirs at the source receive a great deal of debris during the monsoon, which enters the main lines and contaminates the supply. Spring sources are also contaminated, especially during the monsoon season, due to human settlements in the catchment area. Since tankers are filled directly from these sources, the water provided by them is of average or poor quality. Contamination is found to be highest in water from tankers in Mussoorie.

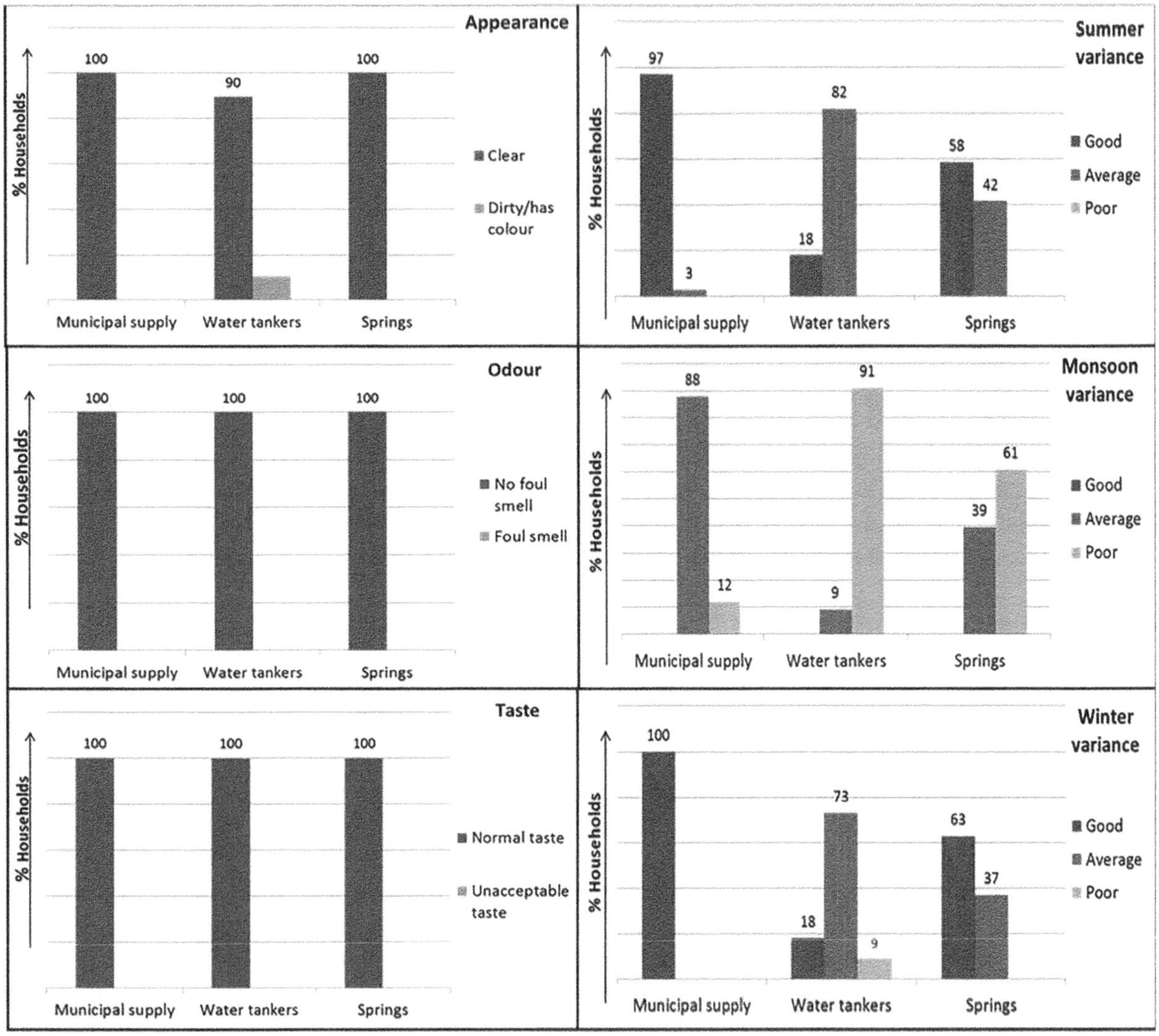

Fig. 6. Water quality and seasonal variability – Mussoorie.

Water consumption

Comparative analysis

Data collected during our survey suggest that most people in different wards of Mussoorie receive between 46 and 80 lpcd (Figure 8) and those in Devprayag even less, between 27 and 35 lpcd (Figure 9). When compared against the prescribed standards of 135 lpcd (CPHEEO manual) or the optimal access of 100–200 lpcd (WHO), these figures clearly show that the residents of Mussoorie and Devprayag only have intermediate levels of access to water.

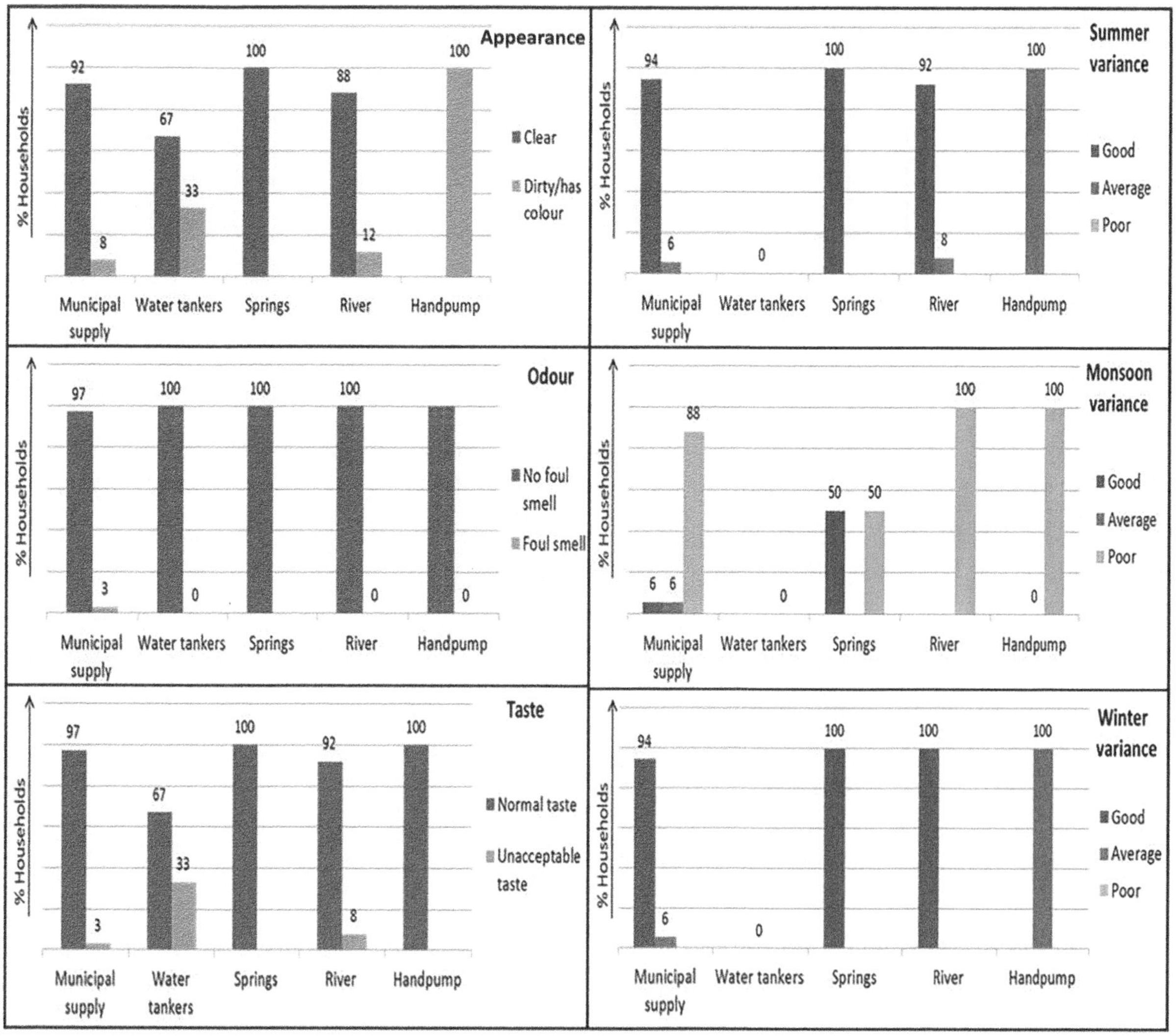

Fig. 7. Water quality and seasonal variability – Devprayag.

We now compare water consumption patterns across wards in each town. For Mussoorie, interviews with government officials revealed that even though wards 3 and 11 are considered to be 'peri-urban' and have the least number of roads, water consumption there does not differ much from the densely populated urban wards 5 and 8. This is because wards 3 and 11 receive a large number of tourists.

Interestingly, in Devprayag, water consumption is higher in cases where additional sources of water are available. For instance, per capita consumption is the highest in wards 2 and 4, despite them having the lowest population, as water is readily available not only from a primary source but also secondary sources – the River Ganges and spring water.

Conversely, respondents in ward number 2 in Devprayag consume less water than other wards as other sources are not as readily accessible to them. Many households in this ward do not have any private water connection and rely on the public stand post and river water.

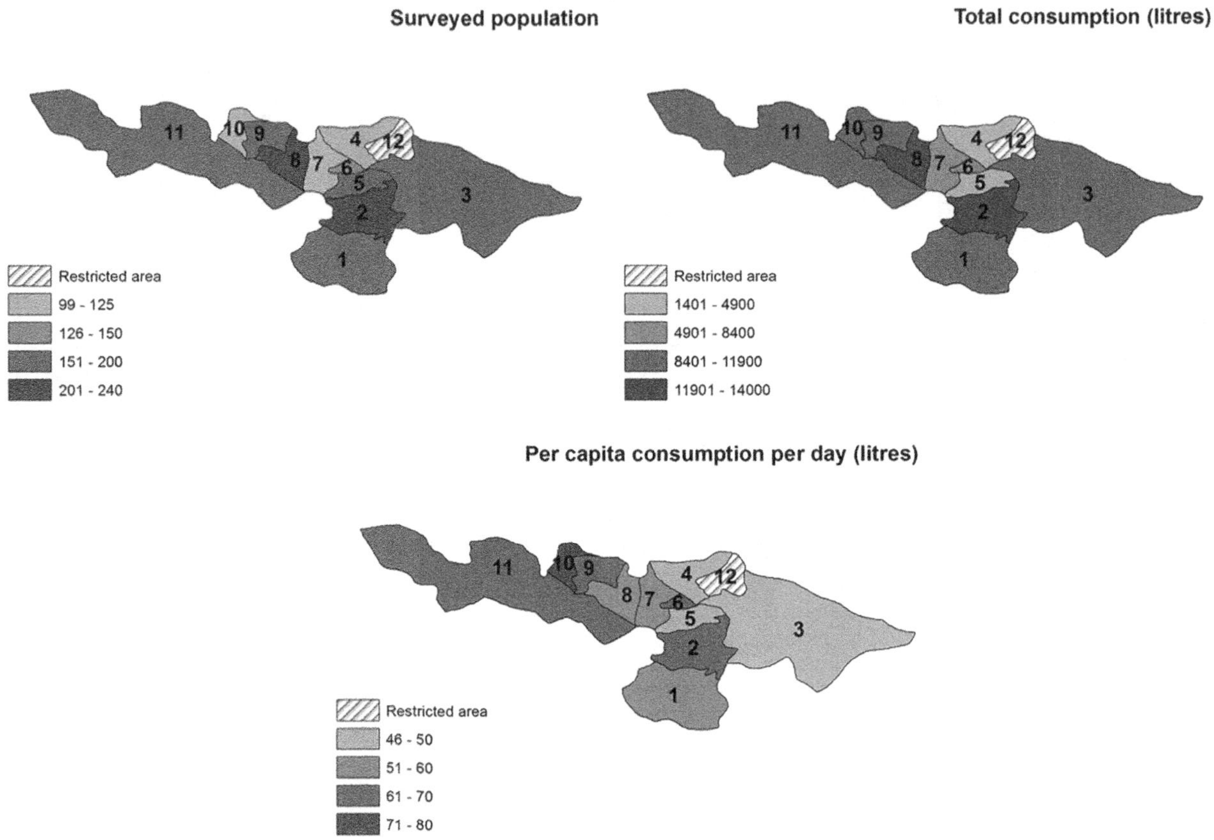

Fig. 8. Ward-wise water consumption in Mussoorie.

Thus, it is clear that having alternative sources of water in the form of springs and river water besides a municipal supply ensures greater availability and is reflected directly in the volume of water consumed by the residents interviewed.

Water collection practices

Received wisdom regarding social norms is that women are primarily responsible for water collection in households located in mountain towns. Our analysis found some variation between the two towns. Around 10% of families surveyed in Mussoorie and 69% in Devprayag stated that only adult females take the lead in water collection (Figure 10). The lower figure for Mussoorie may be due to differences in sociocultural setting; Devprayag appears to be more conservative and less urban.

In the remaining households surveyed in Mussoorie that collect water, responsibility is shared equally. In a crisis situation, all members of the family are engaged in water collection, resulting in loss of time required for other purposes such as education for children, jobs for earning males and females, and reproductive roles for females.

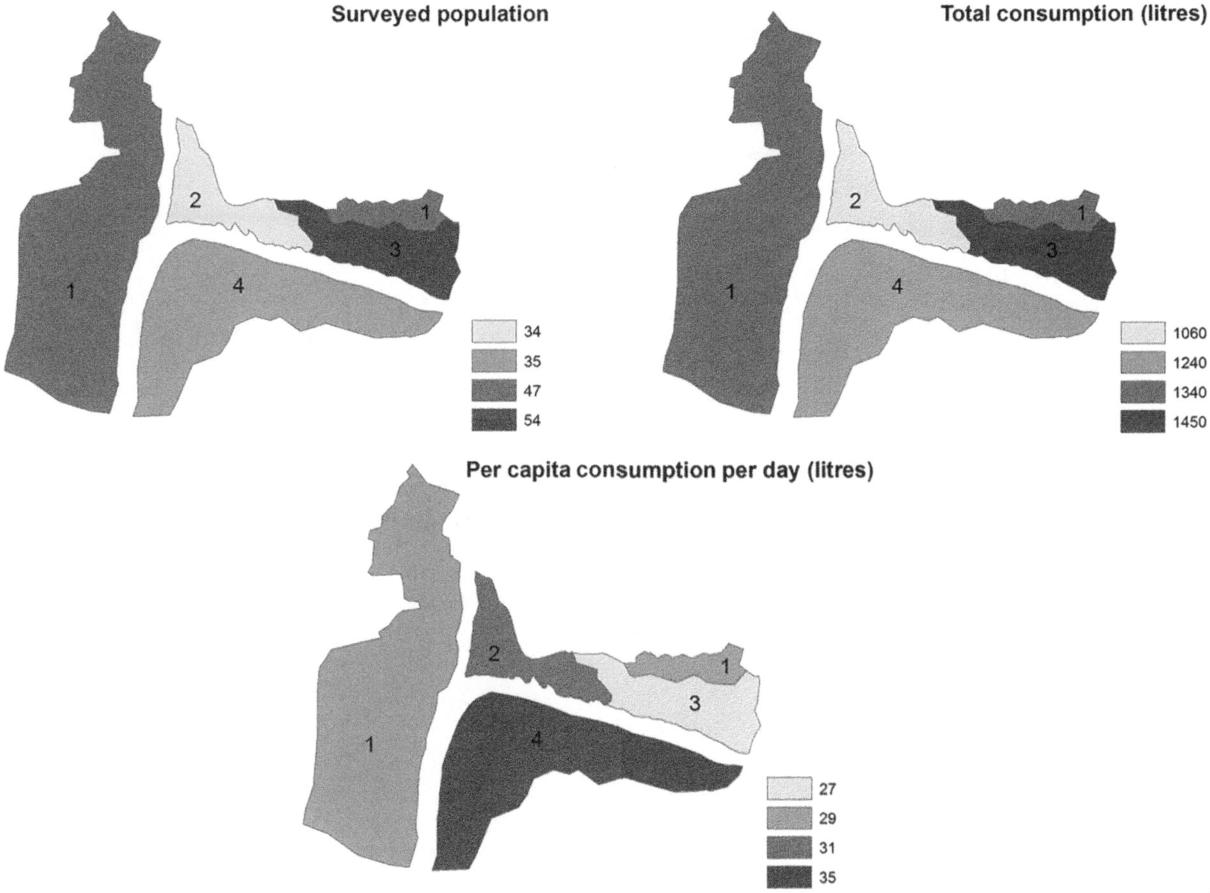

Fig. 9. Ward-wise water consumption in Devprayag.

Climate change scenarios in both towns

Perceptions of respondents

In both towns, elderly residents were asked about their perceptions regarding changes in climate over the last 30 years. A majority of the residents in both towns perceived rising winter and summer temperatures, and delayed arrival of the monsoon. They also perceived an increase in the intensity of 'short duration heavy showers' and a resultant reduction in the number of rainy days (Figure 11).

Modelling projections for climatic variables

We used Worldclim datasets (http://www.worldclim.org/) to understand the current trends and future projections of the following climatic variables for Mussoorie and Devprayag: (a) annual mean

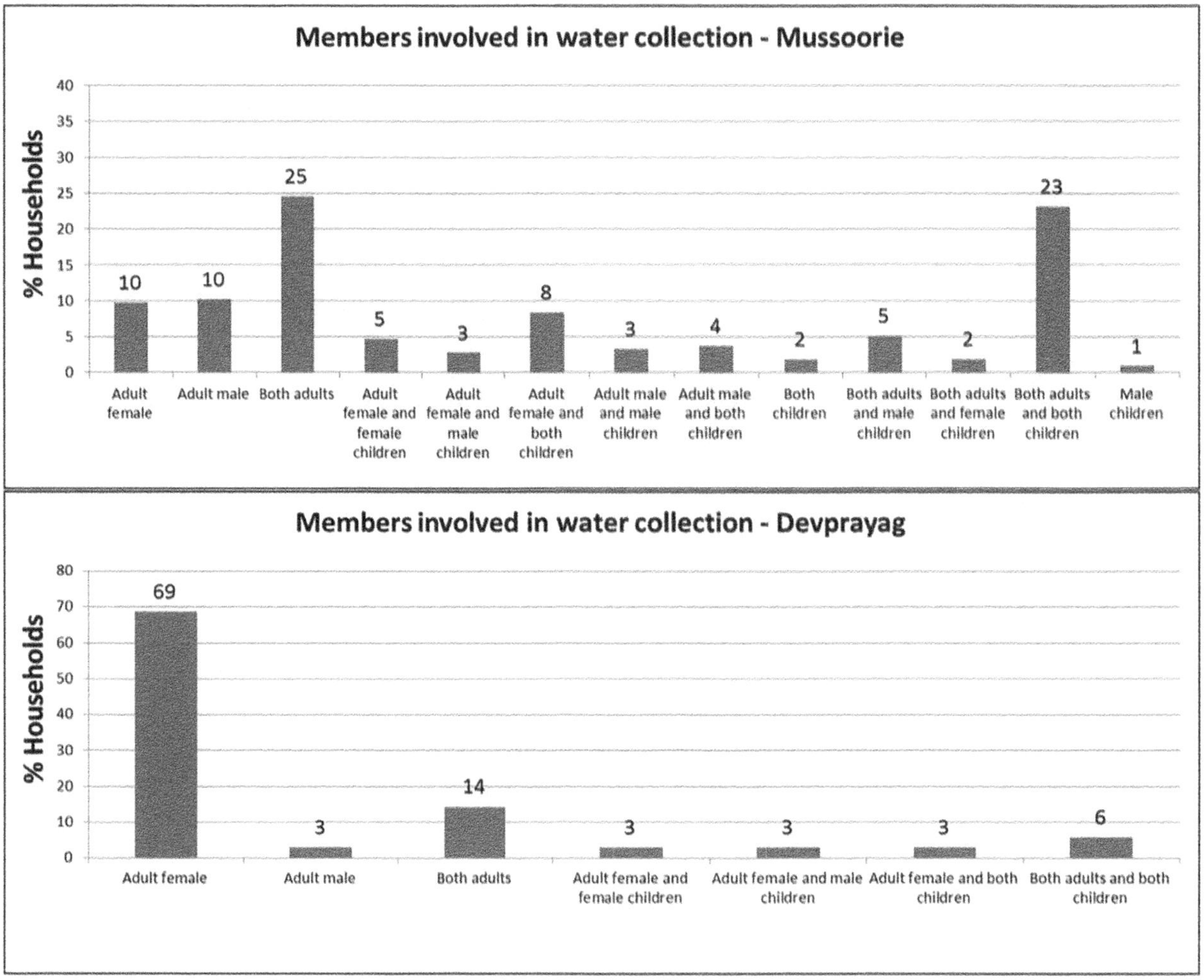

Fig. 10. Members involved in water collection in Mussoorie and Devprayag.

temperature; (b) maximum temperature of the warmest month; (c) annual precipitation; (d) precipitation during the wettest quarter; and (e) precipitation in the driest quarter. The current climatic conditions represent the period 1960–2000. The future conditions were downscaled global climate model data from CMIP5.

Tables 6 and 7 show a range of temperature rises in the mid-term future. As temperatures rise, people and animals need more water to maintain their health and thrive. Increased water demand can be expected due to the rise in temperature.

Tables 6 and 7 also show different variables of precipitation to understand the effects of climate change. Overall, annual precipitation is expected to increase from 2,003–2,728 mm to 2,063–2,859 mm in Mussoorie and 1,366–1,787 mm to 1,419–1,865 mm in Devprayag. The variability in precipitation holds great importance for seasonal water consumption.

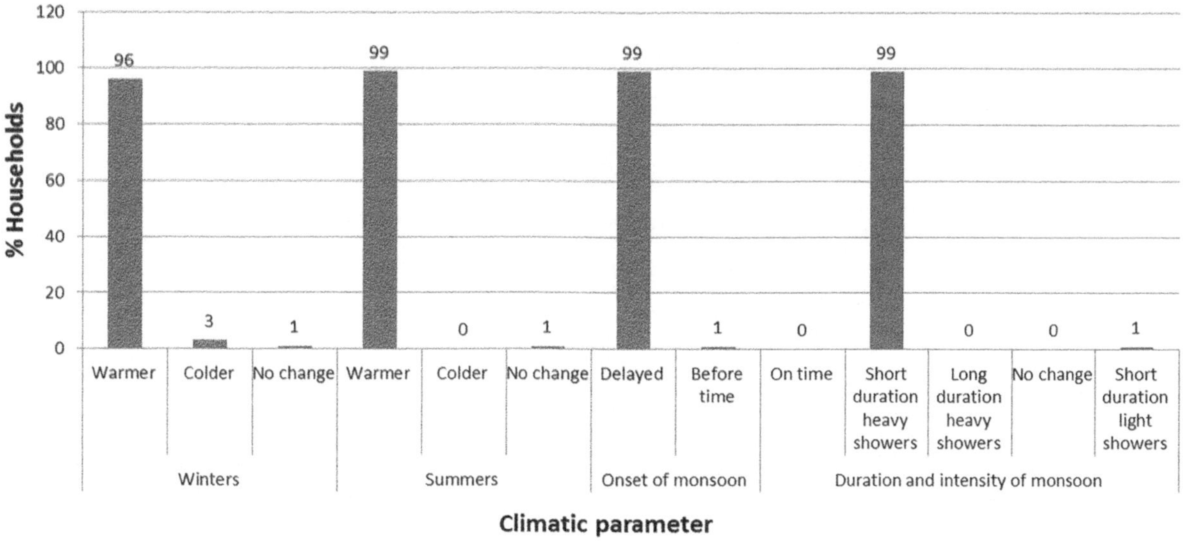

Fig. 11. Perceptions regarding the climate.

Table 6. Projected climatic variability – Mussoorie.

Variable	Current trends	RCP 4.5 (2050)
Annual mean temperature (°C)	13.1–20.7	15.3–22.8
Maximum temperature of the warmest month (°C)	23.6–34.6	25.9–36.9
Annual precipitation (mm)	2,003–2,728	2,063–2,859
Precipitation in the wettest quarter (mm)	1,412–2,184	1,490–2,316
Precipitation in the driest quarter (mm)	111–123	97–110

Table 7. Projected climatic variability – Devprayag.

Variable	Current trends	RCP 4.5 (2050)
Mean annual temperature (°C)	17.1–22.6	19.3–24.5
Max temperature of the warmest month (°C)	29.5–37.4	31.8–39.7
Annual precipitation (mm)	1,366–1,787	1,419–1,865
Precipitation in the wettest quarter (mm)	942–1,286	996–1,370
Precipitation in the driest quarter (mm)	88–107	81–90

Perceptions regarding changing water availability and solutions

Challenges in accessing water (Figure 12)

At the basin level for both towns, the biggest challenge perceived in terms of accessing water by 70% of respondents is the limited hours of supply. Participants indicated that they have observed a reduced discharge over their lifetimes in springs from which they source water, and a reduction in municipal water

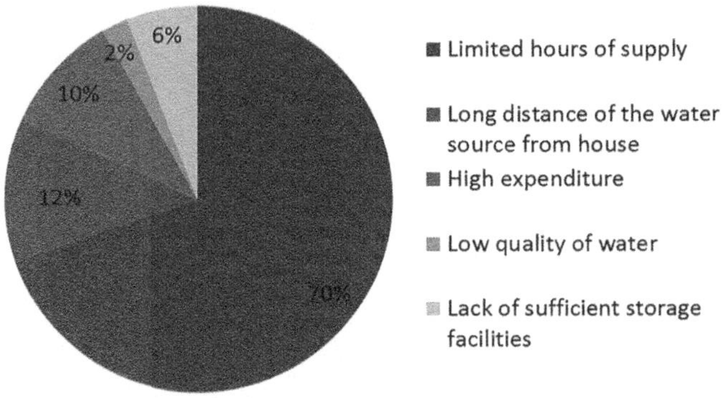

Fig. 12. Most critical challenges in accessing water (basin level).

supply, which is also sourced from springs. They reported witnessing events of 'no water supply' at a stretch of 3 days at least twice a year in Mussoorie and 15–20 days in Devprayag during the monsoon season. Thus, when the municipal supply gets disrupted during the lean season, they are unable to fall back on spring sources closer to the town as their discharge is low. This leads to competition for access to water. Residents also have to invest additional time in standing in long queues at public taps for fetching water.

Second, 12% of the respondents perceived 'distance' as a barrier in accessing water. This includes both the distance from which water is sourced, and the distance that residents have to travel in times of crisis to collect water. In Devprayag, the next major nearby source is the river Ganges but that itself carries mud and debris. There is also the risk of flooding during the monsoon and therefore fears of approaching the river then.

According to the respondents, the overall costs incurred presently on accessing water are high. This is the case not only because the water is being sourced from distant sources, but also because the system is vulnerable to natural hazards occurring in the mountainous terrain. As well, additional costs are incurred in purchasing water from private tankers. Not just quantity, but the deteriorating quality of water was also highlighted. Respondents complained of an occasional unacceptable odour and taste. Silt and even insects can be seen in the water during the monsoon.

Reasons for changing availability of water (Figure 13)

Both the towns are rapidly urbanizing – if one accounts for their floating populations as well – and hence their demand for water is increasing. Also, while Devprayag receives pilgrimage tourists, Mussoorie receives tourists on leisure trips, which further adds to the demand for water. The pressure on the water supply is compounded during the pilgrimage and tourist seasons.

The concretization of green slopes and deforestation in nearby areas is impeding the recharging of aquifers. Rainfall is becoming erratic and the number of rainy days is perceived to have reduced. Thus, collectively, the 'depletion of resources due to high demand' (44%), 'decrease in rainfall and increase in temperature' (17%) and 'mismanaged water resources' (22%) are seen as the major reasons for the water crisis.

In Mussoorie, Jal Sansthan is still running machinery installed during British rule and some parts require urgent replacement. The damage faced by such infrastructure is aggravated by extreme climatic events. The flood disaster event of 2013 severely damaged the water supply system in Devprayag,

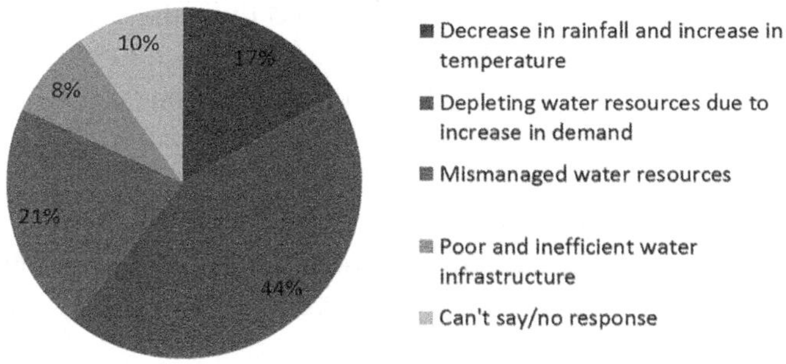

Fig. 13. Reasons for the water crisis.

disrupting the supply for 10–15 days. Deforestation, in order to accommodate more people, for making space for houses and roads, has led to a loosening of the earth, thus making the region more susceptible to landslides which can damage pipes.

Solutions for the water crisis

Since issues of water crisis in both towns are being attempted to be solved by bringing water from newer sources, almost half of the respondents identified with that as a solution (Figure 14). In Devprayag, residents suggested bringing an additional supply through a water uplift scheme from the Ganges River as they see its water as a surplus nearby source.

About 14% suggested 'rejuvenation of springs' as a solution. Another 12% felt that installing rainwater harvesting structures could help to some extent.

Coping strategies undertaken

Residents in both Mussoorie and Devprayag cope by diversifying their sources and accessing water directly from natural sources in times of crisis. With climate change bringing additional uncertainty, having multiple options to rely on seems all the more desirable, and even necessary.

People in the two towns have varied ways of coping, owing to differences in geographical features and sociopolitical conditions. In Mussoorie, residents fall back on accessing water stored in public

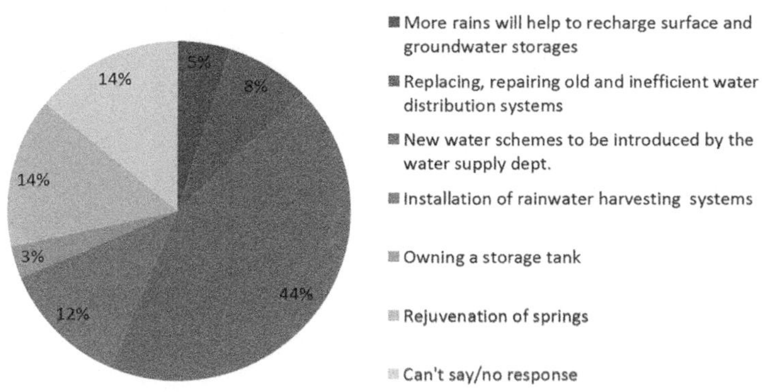

Fig. 14. Solutions offered to deal with the water crisis.

tanks or springs, whichever is closer. Many respondents said they also cope majorly by storing water in as many containers as they can. Families in Mussoorie are also forced to curtail their water consumption to a major extent. This can give rise to issues of health and hygiene, especially in settlements with a common toilet facility, as there is limited water available per person. Using private water tankers is another important measure adopted (Figure 15(a)).

In Devprayag, around half the surveyed families fetch water from the river flowing nearby. Nearly 22% rely on spring sources near their settlement. Collecting water from municipal water tankers is another way to cope (Figure 15(b)). However, as discussed earlier, houses located down the hill find it difficult to access private tankers in Devprayag.

Discussion and conclusions

Urban water management systems in hill cities and towns currently focus only on augmenting the water supply system from different sources. However, there is a need for multiple solutions or practices with the potential to meet the water needs of the present and future, without altering the urban hydrological cycle. Therefore, this calls for reforms in the urban water sector and for developing a framework that takes into account the topography, climate, ecology and the socioeconomic, institutional, administrative and political characteristics of an urban area before evolving sustainable options for it. A combination of soft, hard, structural and non-structural measures can be introduced and implemented within such a framework, including water supply and demand management (Kumar, 2014). Examples of some interventions could be: water conservation; the use of non-conventional sources such as storm water, roof run-off and waste water; storm water management; leakage reduction; water meters and pricing; building institutional capacities to manage water; and community and private sector participation in urban water management. Having analysed the problem in previous sections, we conclude that to improve the urban water system in Devprayag and Mussoorie, working on both the demand side and supply side management is needed. With a changing climate posing new challenges, multidisciplinary strategies need to be incorporated in the urban water management system.

This following section discusses measures that can be adapted within a framework to decrease the vulnerabilities of the hill communities to water stress without compromising environmental sustainability.

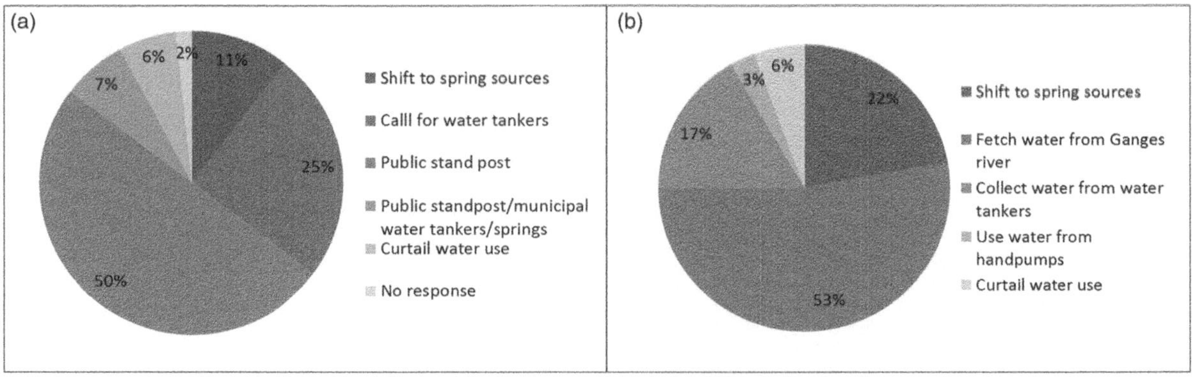

Fig. 15. Coping strategies undertaken: (a) Mussoorie and (b) Devprayag.

Spring protection programmes

Both towns are heavily dependent on spring sources for their water supply and drinking. This is the case for the entire Himalayan region. Around 90% of the drinking water supply in the mountainous part of Uttarakhand is spring-based (Niti Aayog, 2017). However, the springs are in crisis and their discharge is declining due to environmental degradation, changing land use patterns and a warming climate. A survey in Uttarakhand found that 30% of the springs in the region had almost dried up and an additional 45% were on the verge of going dry (Chakraborty, 2013). Due to the changing climate, water supply schemes falling under Devprayag's jurisdiction have recorded a decrease in discharge, at the source, of more than 50% over the last three years (Sharma, 2016). Similarly, spring sources in and around Mussoorie have displayed a noticeable reduction in discharge during the last decade. Two sources, Jinsy and Bhilaru, responsible for the largest portion of total water supply in the town, have reduced by 20%, from 450 litres per minute (lpm) in 2008 to around 365 lpm in 2017 (Jal Sansthan, 2017).

In this context, state-wide springshed management programmes and science-based and community-led springshed management interventions are extremely relevant for growing mountain cities in the IHR, and an effective way to enhance discharge in springs, even during the lean seasons. There have been many community-centric initiatives regarding springs in the states, but they are largely focused on distribution rather than regeneration.

An exception is the Dhara Vikas programme in Sikkim. It aims to catch surface run-off water and use it to recharge groundwater sources by digging staggered contour trenches and percolation pits, after identifying the specific recharge areas based on principles of geohydrology. The Rural Management and Development Department (RMDD) of the Government of Sikkim conceptualized and led the Dhara Vikas initiative to revive dying springs in the state. It is supported by different stakeholders, including NGOs, private organizations and the local communities. The communities are themselves engaged in this process of mapping springsheds, and aid the revival of springs at the village level. End-users were identified and trained as parahydrogeologists for monitoring and managing springs. This helped rejuvenate many springs in the region (Tambe *et al.*, 2012; Niti Aayog, 2017). The Niti Aayog report recognized that in order to replicate successful examples such as Dhara Vikas in other places, the demand for initiating regeneration must come from the communities themselves, and that state departments can support such initiatives with the help of NGOs. However, for the uptake of such a programme at the regional scale, the participation and support of state and local governments is required. The fact that some of the success stories, such as Dhara Vikas in Sikkim, have been led by state governments clearly demonstrates the important role that governments can play in the effective implementation of springshed management programmes. NGOs can also be involved in generating awareness within the community and demystifying the science of springshed management (Niti Aayog, 2017). Similarly, a formalized methodology in this regard has been recently produced by a consortium of practitioners which takes into account all the above points and can be referred to for future spring revival initiatives (Shrestha *et al.*, 2018).

Storm water management

Run-off is high in mountainous areas due to the incline. Rising urbanization accentuates run-off generation, due to the increase in impervious surfaces, leaving less space for infiltration. In traditional urban water systems, storm water is considered a nuisance and is not tapped and used. If storm water were collected and treated, it could be used to augment local storage or local water bodies.

There are open drains in both Mussoorie and Devprayag, and there is no provision to collect and treat storm water before its outlet into the river. Storm water management interventions can be structural and non-structural. Structural measures include actions that tap and reuse storm water; non-structural measures are practices that reduce the amount of run-off generated. Mountain communities can supplement their water supplies through rainwater harvesting. Harvesting one's roof water can reduce the dependence of households on municipal water and also make it cost-effective. It can also minimize the use of municipal treated water for non-potable use (Che-Ani *et al.*, 2009). There are no existing non-structural storm water control measures in Mussoorie and Devprayag. However, certain measures undertaken by the PWD, the Mussoorie Nagar Palika and the local public can qualify for structural storm water management measures. These include the construction and maintenance of culverts, drains, aqueducts and retaining walls, locally called *pushtas* in different locations. These structures are built and maintained primarily to avoid waterlogging and flash floods in the town.

Wastewater treatment

If treated wastewater is reused, it reduces the fresh supply needed at source, and increases the availability of future water supply. However, both Mussoorie and Devprayag are devoid of wastewater treatment plants. In Devprayag, a sewage treatment plant (STP) with a capacity of 1.4 MLD began operating in 2017, but only one ward out of four is connected to the STP. As even in that ward only 20% of houses are connected to the sewer lines, the STP is not receiving enough wastewater and is running at one-fourth capacity.

In Mussoorie, STPs exist but are not functional. As per the information from key informant interviews (KIIs), construction began in 2012 and was completed in 2–3 years. However, connections were not allotted to all the households and the households with the connections are waiting for these STPs to begin functioning. According to local opinion, these STPs cannot function as the volume of water needed to run the STPs is not available in the town. Therefore, the present practice of sewage disposal at the individual and community level continues to be the use of septic tanks.

Regularization of tariffs

Water utilities have displayed no focus on volumetric pricing or the metering of municipal supply water. The tariff for water supply needs to be regularized, and should include production and supply costs. Research has shown that domestic tariffs are heavily subsidized in Indian cities (ADB, 2007).

There is no metering system presently in either Mussoorie or Devprayag, and no analysis as to whether the costs of production and supply of water are being met. Experts have argued that metering domestic water supply will help reduce unaccounted water losses in utilities (ADB, 2014). The pricing of water on the basis of volume used, for which metering is a precondition, will encourage urban water users to reduce wastage, thereby also reducing their monthly water bill (Bassi & Kumar, 2012). Whereas the main reason for distribution losses is poorly maintained infrastructure, unbilled and illegal connections are the biggest causes of unaccounted for water losses. Tariff regularization or water metering would also improve the financial condition of the water utilities as it would help increase their revenues.

Capacity-building

The capacities of the line departments or institutes involved in water management should be strengthened so that their ability to manage water resources in a sustainable manner also increases. In keeping with this, the capacities of various government departments linked to springshed usage or management in both towns should be built, as springs are the primary source of water in both. The capacities of consumers should also be raised, and there should be initiatives stressing the involvement of communities, especially women, leading to the increasing role of communities in inclusive urban planning. Inter-state cross-learning and support of experts from across the states should also receive focus for enhancing capacities.

Improving the governance structure

Besides the inefficient management of water supply systems, multiple institutional arrangements and their isolated style of functioning add to the problem (Bassi & Kumar, 2012). During KIIs in Devprayag, it was found that the tehsil office, which keeps a record of population, had little communication with the Jal Sansthan in planning and designing new initiatives. Thus, an augmented supply line with support from the ADB, planned for ten years, was set up not taking into account upcoming educational institutions in the area, which will house 2,000 students.

Convergence with existing policies

Several existing development programmes like watershed ought to include the protection of springs from the externalities of infrastructure development, including roads. Such an approach would tie well with the Government of India's strategic focus on sustaining the Himalayan ecosystem in the context of climate change. A working group formed by the Niti Aayog identified 'Inventory and Revival of Springs of Himalaya for Water Security' as one of their themes for working on sustainable development in the Himalayan region. It recommended mainstreaming springshed programmes through a convergence with existing development programmes such as the National Mission for Sustaining the Himalayan Ecosystem (NMSHE), the Integrated Watershed Management Programme (IWMP) and the Mahatma Gandhi National Rural Employment Guarantee Act (MGNREGA), and activities under climate change cells in various states, among others (Niti Aayog, 2017).

Replicating the institutional framework

Efforts at creating and strengthening the institutional framework must be replicated across states. A policy framework is needed whereby governments play an enabling role in partnership with civil society and the private sector to facilitate management of springs at a localized level.

Database

Besides developing an online database, it would be useful to tag vulnerable villages and springs; upload discharge data and recharge measures if taken to rejuvenate those springs. Data on population (including the floating population), spring discharge, water quality, metering and municipal infrastructure need to be made available to keep a check on trends in demand, water withdrawals and availability (McKenzie & Ray, 2009; Niti Aayog, 2017).

In the context of the IHR, the important driving forces will be: (i) enhancing institutional/organizational/local capacities; (ii) the availability of finance to carry out the required training and recruitment; and (iii) the desire for making the water supply infrastructure (both physical infrastructure and human resources) more efficient and resilient; and (iv) developing public–private–civil society partnerships. The challenges to achieving the above could range from the lack of political interest and the lack of a policy framework to the lack of skills and knowledge to carry out the required training. However, efforts should focus on strengthening the driving forces, and articulating clear strategies, structures and arrangements for carrying out the required interventions, so that it may ultimately lead to a better and more efficient performance of the urban water supplying utility.

Also, the ignorance of ecological upkeep in the larger context of rivers, watersheds and aquifers has led to large gaps in practice and policy in developing a strategic and integrated response to water management in Himalayan towns. Thus, there is the need for a paradigmatic shift in water governance, from focusing on developing water sources to the management of water as a resource. More integrated and adaptive management approaches and innovations are key to enhancing urban water management systems of the Indian Himalayan Region. This also helps achieve global commitments of the country in terms of enhancing water security (Goal 6.1 SDG (UN Sustainable Development Goals), INDC (Intended Nationally Determined Contribution), 2012).

Acknowledgements

This work was carried out by the Himalayan Adaptation, Water and Resilience (HI-AWARE) consortium under the Collaborative Adaptation Research Initiative in Africa and Asia (CARIAA) with financial support from the United Kingdom's Department for International Development (DFID) and the International Development Research Centre (IDRC), Ottawa, Canada.

References

ADB (2014). *IND: Uttarakhand Emergency Assistance Project*. Uttarakhand Jal Sansthan, Jal Bhawan, Dehradun. https://www.adb.org/sites/default/files/project-document/148887/47229-001-iee-07.pdf.

AHEC (2011). *Assessment of Cumulative Impact of Hydropower Projects in Alaknanda and Bhagirathi Basins*. The National River Conservation Directorate, Ministry of Environment and Forests, Government of India, Delhi. http://www.moef.nic.in/downloads/public-information/CH-1.pdf.

Bajracharya, S. R. & Shrestha, B. (2011). *The Status of Glaciers in the Hindu Kush–Himalayan Region*. ICIMOD, Kathmandu, Nepal.

Bajracharya, S. R., Maharjan, S. B., Shrestha, F., Guo, W., Liu, S., Immerzeel, W. & Shrestha, B. (2015). The glaciers of the Hindu Kush Himalayas: current status and observed changes from the 1980s to 2010. *International Journal of Water Resources Development 31*(2), 161–173.

Bassi, N. & Kumar, M. D. (2012). Addressing the civic challenges: perspective on institutional change for sustainable urban water management in India. *Environment and Urbanization Asia 3*(1), 165–183.

Bhadwal, S., Ghosh, S., Gorti, G., Govindan, M., Mohan, D., Singh, P., Singh, S. & Yogya, Y. (2017). *The Upper Ganga Basin: Will Drying Springs and Rising Floods Affect Agriculture? HI-AWARE Working Paper 8*. HI-AWARE, Kathmandu.

Bharati, L. & Jayakody, P. (2010). *Hydrology of the Upper Ganga River*. International Water Management Institute, Colombo.

Bharati, L., Lacombe, G., Gurung, P., Jayakody, P., Hoanh, C. T. & Smakhtin, V. (2011). *The Impacts of Water Infrastructure and Climate Change on the Hydrology of the Upper Ganges River Basin*. IWMI Research Report 142. International Water Management Institute, Colombo. doi:10.5337/2011.210.

Bose, P. & Srivastava, P. (2017). *Water Supply for Urban Poor in India*. Water Aid, New Delhi.

Chakraborty, S. (2013). Restoring springs combats water scarcity in India's Himalayas. *Thomson Reuters Foundation*, 4 November. http://news.trust.org/item/20131103221719-dcup3/.

Che-Ani, A., Shaari, N., Sairi, A., Zain, M. & Tahir, M. (2009). Rainwater harvesting as an alternative water supply in the future. *European Journal of Scientific Research 34*(1), 132–140.

Government of Uttarakhand (2014). *Uttarakhand Annual Plan 2013–14. Presentation Made to the Planning Commission*. http://planningcommission.gov.in/plans/stateplan/Presentations13_14/uttarakhand_2013_14.pdf.

Hewitt, K. & Mehta, M. (2012). Rethinking risk and disasters in mountain areas. *Revue De Géographie Alpine/Journal of Alpine Research 100*(1). doi:10.4000/rga.1653.

ICIMOD (2018). *Strategy and Results Framework 2017*. ICIMOD, Kathmandu.

Jal Sansthan (2017). *Annual Report 2017*. Jal Sansthan, Government of Uttarakhand, Mussoorie.

Khan, A. A., Pant, N. C., Sarkar, A., Tandon, S. K., Thamban, M. & Mahalinganathan, K. (2017). The Himalayan cryosphere: a critical assessment and evaluation of glacial melt fraction in the Bhagirathi basin. *Geoscience Frontiers 8*(1), 107–115.

Kumar, M. D. (2014). *Thirsty Cities: How Indian Cities Can Meet Their Water Needs*. Oxford University Press, New Delhi.

McKenzie, D. & Ray, I. (2009). Urban water supply in India: status, reform options and possible lessons. *Water Policy 11*(4), 442–460.

Ministry of Home Affairs (2001). *2001 Census Data*. Government of India, New Delhi. http://www.censusofindia.in.

Ministry of Home Affairs (2011). *2011 Census Data*. Government of India, New Delhi. http://www.censusofindia.in.

Mukherji, A., Molden, D., Nepal, S., Rasul, G. & Wagnon, P. (2015). Himalayan waters at the crossroads: issues and challenges. *International Journal of Water Resources Development 31*(2), 151–160.

National Research Council (2012). *Himalayan Glaciers: Climate Change, Water Resources, and Water Security*. The National Academies Press, Washington, DC. https://doi.org/10.17226/13449.

NIH (n.d.). *River Basin Information System of India*. Water Resources Systems Division, National Institute of Hydrology, Jalvigyan Bhawan, Roorkee. http://nihroorkee.gov.in/rbis/india_information/drinking.htm.

Niti Aayog (2017). *Inventory and Revival of Springs in Himalayas for Water Security*. Department of Science and Technology, Government of India, New Delhi. http://www.dst.gov.in/sites/default/files/NITI-Aayog-report-Springs-29Dec2017-FINAL.pdf.

Ramachandran, H. & Ramachandran, N. (2001). *Carrying Capacity of Mussoorie*. Concept Publishing Company, New Delhi.

Ramola, A. (2014). Drinking water crisis deepens in Mussoorie. *Tribune*, 4 June. https://www.tribuneindia.com/2014/20140604/dun.htm#6.

Rao, P., Areendran, G. & Sareen, R. (n.d.). *Potential Impacts of Climate Change in the Uttarakhand Himalayas*. http://lib.icimod.org/record/12785/files/1092.pdf.

Schmid, M. O., Baral, P., Gruber, S., Shahi, S., Shrestha, T., Stumm, D. & Wester, P. (2015). Assessment of permafrost distribution maps in the Hindu Kush Himalayan region using rock glaciers mapped in Google Earth. *The Cryosphere 9*, 2089–2099. doi:10.5194/tc-9-2089-2015.

Sharma, B. (2016). Sustainable drinking water resources in difficult topography of hilly state Uttarakhand. India. *American Journal of Water Resources 4*(1), 16–21. doi:10.12691/ajwr-4-1-2.

Sharma, E., Molden, D., Rahman, A., Khatiwada, Y. R., Zhang, L., Singh, S. P., Yao, T. & Wester, P. (2019). Introduction. In: *The Hindu Kush Himalaya Assessment – Mountains, Climate Change, Sustainability and People* (P. Wester, A. Mishra, A. Mukherji & A. B. Shrestha, eds), SpringerNature, Dordrecht, pp. 1–19.

Shrestha, R. B., Desai, J., Mukherji, A., Dhakal, M., Kulkarni, H., Mahamuni, K., Bhuchar, S. & Bajracharya, S. (2018). *Protocol for Reviving Springs in the Hindu Kush Himalayas: A Practitioner's Manual. ICIMOD Manual 2018/4*. ICIMOD, Kathmandu.

Tambe, S., Kharel, G., Arrawatia, M. L., Kulkarni, H., Mahamuni, K. & Ganeriwala, A. K. (2012). Reviving dying springs: climate change adaptation experiments from the Sikkim Himalaya. *Mountain Research and Development 32*(1), 62–72.

Valdiya, K. S. & Bartarya, S. K. (1989). Diminishing discharges of mountain springs in a part of Kumaun Himalaya. *Current Science 58*, 417–426.

You, Q. L., Ren, G. Y., Zhang, Y. Q., Ren, Y. Y., Sun, X. B., Zhan, Y. J., Shrestha, A. B. & Krishnan, R. (2017). An overview of studies of observed climate change in the Hindu Kush Himalayan (HKH) region. *Advances in Climate Change Research 8*, 141–147. https://doi.org/10.1016/j.accre.2017.04.001.

Received 7 October 2018; accepted in revised form 16 December 2018. Available online 20 February 2019

Perform or wither: role of water users' associations in municipalities of Nepal

Sreoshi Singh[a], Krity Shrestha[b], Menaka Hamal[c] and Anjal Prakash[d]

[a]*Corresponding author. 1136, Casa Serenita, Sobha City, Thanisandra Road, Bangalore 560064, India*
E-mail: sreoshi@gmail.com

[b]*Practical Action, Indira House, Panipokhari Hill, Kathmandu, Nepal*

[c]*International Centre for Integrated Mountain Development, GPO Box 3226, Khumaltar, Lalitpur, Kathmandu, Nepal*

[d]*TERI – School of Advanced Studies – Hyderabad Campus, Sy. No. 37/1/2, Gopanpally, Gachibowli, Hyderabad, Telangana 500 032, India*

Abstract

In Nepal, water institutions have played a very significant role, and in Tansen and Damauli, the presence of user groups has indicated that proper management of water can help people avert critical water shortages. However, although in both Tansen and Damauli the user groups have been operating for a long time, their performances vary. In Tansen, infrastructural constraints tend to throw up challenges, although operational hazards associated with the supply systems are no less threatening. Moreover, there is large-scale corruption in the systems' upkeep and maintenance, allowing low grade vendors to operate in place of readily available efficient institutions. In Damauli, the systems have been rather perfectly managed, except for minor glitches from time to time. Funding has been good and community bonding has paid off. This paper delves into the community-managed water systems in the two cities and how the performance varies across them and the factors that play a role.

Keywords: Damauli; municipalities; Nepal; Tansen; water institutions; Water Users' Association

Introduction

Water is essential for sustaining life and livelihoods. Over the years, continuous discussion has centred around how water should be managed, due to the inherent challenges. Economic, social and political factors have often been found to determine people's access to water (Gareau & Crow, 2006). These factors

doi: 10.2166/wp.2019.051

have primarily determined how policies are framed by the state, resulting in the poor being marginalized. At this juncture, identifying the importance of governance and more specifically 'water governance' is crucial (Huitema *et al.*, 2009). Studies across the globe have identified the role of communities in the planning and management of resources as ideal for sustainable and inclusive growth (Lopez-Gunn, 2003). This is where 'governance' as a concept becomes all-encompassing. It is broader than just the government, incorporating both state and non-state actors, both private and public (Peters & Pierre, 1998).

In Nepal, water institutions play a critical role in the irrigation sector (Roth *et al.*, 2005) and 'proved to be very popular and successful in developing awareness amongst the farmers and in making the WUAs more active and responsive to the needs of the users' (Howarth *et al.*, 2005: 1). A research project in the 1950s around the irrigation sector in Nepal laid the path for a well-developed community-managed irrigation system in the country (Pradhan, 1989). Pradhan (2015) mentioned how their research findings emphasized the need for farmer management irrigation systems (FMIS) for judicious use of water across the different irrigation users while maintaining better cooperation among the users (Bhatta *et al.*, 2006). Its key features focus on water being a community resource which must be properly allocated and distributed within the community based on water titles, implying obligations of the users so that operation and maintenance is done on that basis. While decision-making will lie with the general assembly, the implementation will remain with the executive committee, and non-adherence to terms and conditions laid down would lead to penalties. The researchers identified that these features could only be sustained through 'strong social capital development in the irrigators' community over long period of time' (Pradhan, 2015: 42). Additionally, infrastructure development without the role of social capital will be unsustainable (Ostrom, 2000).

This paper focuses on the role of water user groups in providing access to drinking water in two urban locations in Nepal. It focuses on the question of how well the urban water user groups have been performing in providing access to water for the urban poor. This paper draws from data of 350 household surveys conducted in the two selected towns of Tansen and Damauli in Nepal. The survey was conducted in 2017–18 with a view to investigating households' response to questions regarding the status of water sources, supply systems and satisfaction levels on access, quality and quantity of water and coping mechanisms wherever water is inaccessible. Based on the results of the data analysis of some of the key indicators, the paper dwells on the working of governance mechanism in the two towns. This paper is divided into the following sections. After the introduction, a section focuses on the major developments in the drinking water sector in Nepal and the evolution of drinking water user groups. Next, a brief background of the two towns is given and data from the primary survey are presented. This is followed by a section that focuses on the drinking water user groups formed in the two towns and evaluates their performance. Then, a detailed analysis of the drinking water user groups is presented to discover why they underperform and finally, conclusions are drawn.

Major developments in drinking water scenario in Nepal

In this section, we document the evolution of the water users' associations (WUAs) in Nepal through acts and regulations. The changes and legitimization of WUAs have implications on governance and management of water at a local level especially. We focus on the drinking water sector where WUAs have been functioning for more than five decades, which was further recognized by the constitution, and promulgated in 2015. The right to drinking water in Nepal has been protected by the constitution of Nepal, 2015 in Article

25, sub article 4 of Part 3 – Fundamental rights and duties (Nepal Law Commission, 2015). Also, under Nepali law, drinking water is a consumer commodity. The Essential Commodity Protection Act 1955 deems drinking water to be an essential commodity and, as such, declares that it should be strictly protected. The main legislation in relation to drinking water in Nepal is the Water Resource Act 1992, an umbrella act, governing not only drinking water, but other uses of water and overall water resource management in Nepal (Magar, 2009; Aryal, 2011). It further gives priority to the right to use water for drinking purposes over any other domestic or commercial use. There are two regulations under the Act, for drinking water purposes the Water Resource Regulation 1993 and the Drinking Water Regulation 1998.

Evolution of drinking water users' group/associations and liaison at local level

The Drinking Water Regulation 1998 specifically deals with drinking water and sanitation. Among other things, it regulates drinking water users' associations (DWUAs), known simply as water users' associations (WUA) or water users' cooperatives, the quality of drinking water and drinking water suppliers. Rule 3 of the Drinking Water Regulation 1998 provides that groups of people who wish to benefit collectively from developing and operating their own project may form a WUA. A WUA may also be registered for the operation of a project developed by GoN (Government of Nepal) (then His Majesty's Government). If GoN wishes to operate a project through a WUA as per Section 11 of the Water Resource Act 1992, persons willing to operate the project must form and register a WUA as per Rule 5 of the Water Resource Regulation 1993. Drinking water suppliers are also considered as WUAs and corporate bodies that are essentially fulfilling a state's responsibility to supply drinking water as a service and for commercial gain. They are regulated by the Drinking Water Regulation 1998 which defines a water supplier as a WUA and/or licensee holding a licence for the operation of a water supply system. Here, the emphasis was given on enabling the communities themselves to take up the lead role in the identification, selection, formulation, implementation and management of water supply programmes. It gave priority to the disadvantaged and promoted hygiene, sanitation and promised to minimize environmental impacts. As per the policy, most of the water projects were to be handed over to the user committee, municipality, water supply-related organization or community-based organization. As Mayntz (2006) puts it very rightly,

> '*this represents a shift to a new form of governing society that is more inclusive and cooperative than the traditional ordered rule of government and concept of political steering; moving from placing state governments and political authorities at the centre of action to control socio-political processes to meet socio-economic goals.* (Hill, 2012: 17–18)'

The Act of 1993 itself mentioned the role of user groups, which will perform like a corporate body, with perpetual succession and will target using water for collective benefit. While several rural projects have been successful, there have been multiple challenges due to lack of knowledge of sources, lack of participation, guidelines regarding role of state and role of users in pricing and cost recovery.

The major role of the DWUAs is to supply water while maintaining the water quality parameters. They are also responsible for maintaining and repairing the water supply system and compensating communities and households as per the country's current rules in the case of purchasing/using water sources from other areas/communities. They are not allowed to pollute the water sources and are responsible for conducting necessary source conservation activities such as regular cleanliness, among others. The regulation has bestowed the WUAs with certain powers as well; they are empowered to impose conditions of service, such as tariffs and to reduce or cease the service and levy service charge. However, these powers have to be exercised as per the agreed constitution. In the case of changes in power, the proposal may be amended in the general assembly.

The other major pieces of legislation governing drinking water are the Nepal Water Supply Corporation Act 1989. This Act establishes the Nepal Water Supply Corporation (NWSC) as a public corporation responsible for providing clean drinking water and sewerage services to the urban population. The scope of work of the Corporation has been determined by His Majesty's Government (HMG) by notification in the Gazette. This Act establishes the right of people to drinking water and sanitation and imposes a duty on the state (through the Corporation) to provide drinking water and sanitation. The Local Self Governance Act 1999, which primarily deals with the decentralization of government, also gives local bodies some responsibility in relation to the utilization, conservation and management of water resources, maintenance of sanitation facilities and waste management. The Act and its regulations make local bodies (village development committees, district development committees, and municipalities) responsible for delivering certain services, apart from providing them with the power to make policies and implement programmes related to drinking water and sanitation and to raise revenue via local taxation, fees and other means (Water Aid Nepal, 2005).

As per the Federation of Drinking Water and Sanitation Users Nepal (FEDWASUN), there are about 41,250 DWUAs in Nepal. Before 1992, these WUAs were directly registered with District Administrative offices in their respective districts. Since the promulgation of the Local Self Governance Act (LSGA) in 1999, they are registered at the municipality/local level because the local bodies have ownership rights over the water resources within their respective administrative boundaries. Hence, the LSGA gives precedence to the local bodies of the respective water resource over the ownership by the state, as depicted in the Water Resources Act.

A case of two emerging towns of Nepal: Tansen and Damauli

About 85% of the population in Nepal lives in rural areas. Recently, many people have been migrating to urban centres of the country due to economic and social factors. The urban population of the country is largely concentrated in the Terai region and the valleys of the mid hills spread across 58 municipalities. The urban population has been rising since 1971, from 4% of the total population to 13.9% in 2001, and is expected to reach 26.7% by 2021. Simultaneously, urban water demand has also been increasing at the rate of 6 to 9% (WHO, 2014). In urban areas, the data from Central Bureau of Statistics (CBS, 2011) shows that around 87% of the total population living in urban areas has access to drinking water while 85% of the population in rural areas has access to water. Similarly, if we look across the ecological regions, 84.89% of the total population in the hills has access to drinking water. The access to drinking water in the hills is higher than in mountains or Terai (CBS, 2011).

The pressure on water and sanitation services are high, although latest estimates show that 85% of the total population has access to basic water supply services and 62% has access to basic sanitation facilities (access to toilets). Nepal has already achieved the Millennium Development Goal (MDG) targets for water supply (73%) and is making progress to the MDG target for sanitation (WHO, 2014). It is estimated that 40,000 water supply schemes of multiple types and scales including rural, semi-urban and urban, gravity flow and pumping are in operation. While most of the rural and semi-urban water schemes are operated and managed by water users and sanitation committee (WUSC) – a kind of community-based organization formed by the consumers themselves, the urban schemes are operated by a water utility such as NWSC, a semi-government agency (WHO, 2014: 1).

Tansen was recognized as a municipal town in 1961 (Nepal, 2001) while Damauli became a municipality in 2011. Both the towns are considered among the most planned municipalities in Nepal. Tansen lies in Palpa District and is the administrative centre in the hills of western Nepal. Located at an altitude of 1,372 m, on the highway between Butwal and Pokhara, on the crest of the Mahabharat Range, the town overlooks the valley of the Kaligandaki River to the north. The town is famous for the Newari traders and artisans known for their metal and Dhaka (textile) industries. Tourism is also prevalent during October–December and March–April. There is also a huge influx of religious tourists during Maghe Sankranti (February) festival in the Ridi River (Portnov & Adhikari, 2006). Old records from Tansen indicate that the first water supply system through gravity flow was set up during the rule of the Ranas in Nepal. Engineered by the British, the system can supply over 150,000 litres of water per day to around 12,000–15,000 people (Joshi, 2017). The water user group in Tansen was formed in 1980 and has been officially distributing and managing drinking water supply in its urban wards. Water is supplied to all wards except Wards nos. 9, 12 and 14, as they are peri-urban wards of the municipality. Water supply in peri-urban wards is through local resources such as springs and handpumps. There are 1,760 taps registered with the users' committee, and no new taps registered after 1994. According to the users' committee, total water demand, for 15 wards, is 35,00,000 litres per day (120.29 lpcd) whereas 2,000–2,500 litres of water is supplied per household per day and the current storage capacity of reservoirs is around 13.5–14 hundred thousand litres (Poudel, 2017). Figures 1 and 2 show the spring water and storage infrastructure within the two municipalities.

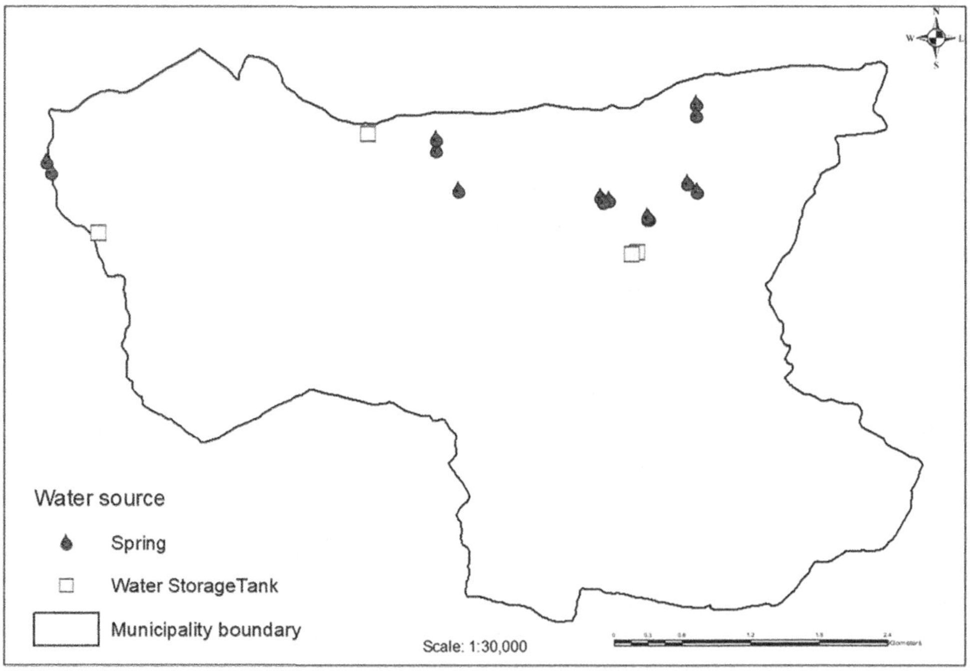

Fig. 1. Springs and tanks in Tansen municipality. (*Source*: Department of Survey, Nepal.)

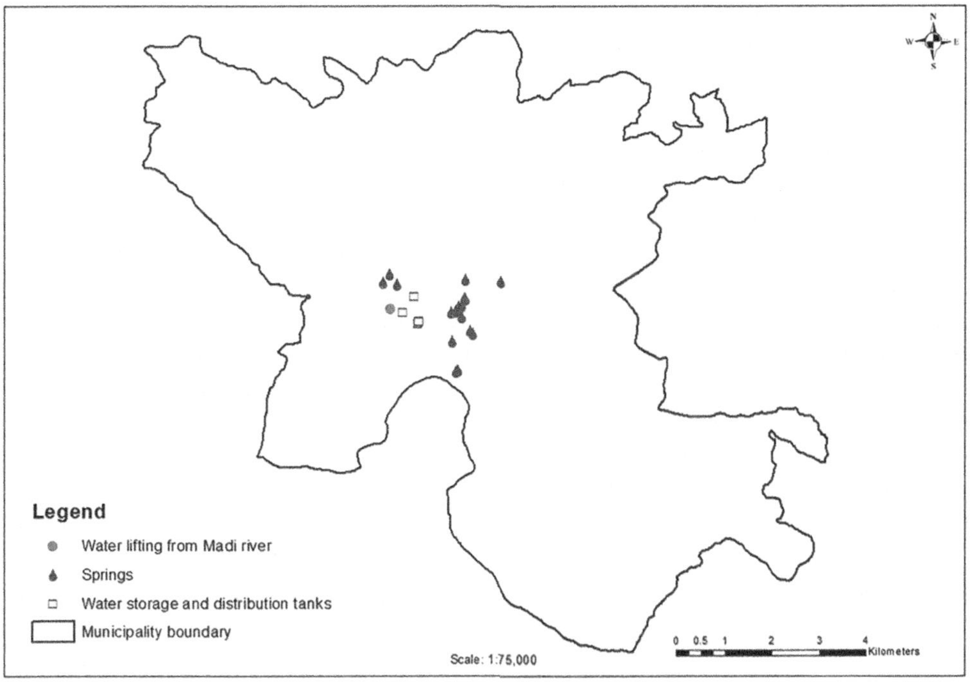

Fig. 2. Springs and tanks in Damauli, Vyas municipality. (*Source*: Department of Survey, Nepal.)

Vyas, a municipality of which Damauli town is a part, was established in 1992 after re-establishment of multi-party democracy in Nepal in 1991. The municipality is a hub of local economic activities, due to which it has witnessed huge in-migration from the surrounding villages and from other areas within the district and outside. People are settling there for business, better infrastructure, better employment opportunities and education facilities. Hence, the population of the city and demand for water is rising. There are two water supply systems providing drinking water to the residents of Damauli through 1,902 taps in the municipal areas of which 36 are public taps. Damauli Drinking Water and Sanitation Cooperative was officially registered and became operational in 2008. Latest official figures indicate that there is a demand for 2.5 million litres of water per day in Damauli, but the supply is just around 1.5 million litres (Poudel, 2017). The current storage capacity of the tank is about 1.6–1.7 million litres. To meet the deficit, the municipality is also looking for new sources, as groundwater extraction is only a temporary solution. In this process, various stakeholders have been involved, including residents of particular wards where water supply is not adequate.

A primary study was conducted in 2016–17 to understand the water budget of the two towns and the role of institutions in managing water supply. Household level data were collected from 350 households across the two towns (208 in Damauli and 142 in Tansen). In the next section, we provide the key findings of the study.

Household water use in two towns: heavy dependence on spring sources

In Tansen and Damauli, the numbers of households that originally belonged to these two towns was rather limited. From around the year 2001, there has been substantial migration, indicated by the respondents. Almost 71% and 69% of respondents have been living in the respective towns for a maximum period of 15 years with only a small proportion (12% and 23%, respectively) being born there. This is also a clear indication that water demands in these two urban centres have increased substantially owing to the increasing number of residents.

Table 1 shows that 92% of households in Tansen depend on municipal supply for drinking and only 6% on springs, while in Damauli, only 56% and 34% are dependent on these two sources, respectively. This is attributed to the fact that municipal water is actually derived from springs and, therefore, people prefer using this rather than using the spring water directly to drink. In Damauli, the low proportion of households depending on the municipality may be attributed to the fact that the municipal water systems are not able to serve the peri-urban wards as efficiently as the core areas. Households in the peri-urban areas depend largely on other minor spring sources. An interesting observation stemming from the data indicates that in Tansen and Damauli, the figure for spring water users across all uses is high (69% and 53%, respectively). Further, data for the two cities (see Table 2) show that residents in Damauli use more spring water than Tansen (34% and 6%, respectively) irrespective of the season.

Piped water supply from the municipality is available only to 60% of surveyed households in the core urban wards of Damauli with the duration of supply being around one and half hours in almost 63% of households (see Tables 3 and 4). On the contrary, duration for which water is available in 92% of households in Tansen with access to a municipal source for drinking and other purposes, is up to 1 hour, received by 74% of the households. This association has also been found to be statistically significant.

Table 1. Main source of water for drinking in Tansen and Damauli.

	Surveyed households (%)	
Sources of water for drinking	Tansen	Damauli
Municipal supply	92	55.8
Groundwater (drawn through personal borewell/tube well)	0	0.5
Groundwater (drawn through community tube well)	1.4	1.0
Stream/lake	0	0.5
Bottled water	0.7	1.5
Springs	6.2	34.0
Open well	0	6.8
Total	100	100

Source: Primary household survey, 2016–17.

Table 2. Season-wise use of spring water.

	% of surveyed households	
Use of spring water (season-wise)	Tansen	Damauli
Throughout the year	21	74
Summer and pre-monsoon	73	24
Monsoon	6	0
Winter	0	2
Total	100	100

Source: Primary household survey, 2016–17.

Table 3. Access to municipal water source in Tansen and Damauli.

	% of surveyed households	
Access to municipal water	Tansen	Damauli
Access available	92.4	60.2
No access	7.6	39.8
Total surveyed households	100	100

Source: Primary household survey, 2016–17.

Table 4. Duration of water supply from municipal source.

	% of surveyed households	
Duration of water supply from municipal source	Tansen	Damauli
Less than 30 minutes	7	6.4
31–45 minutes	7	0
45–60 minutes	74	23.4
61–90 minutes	10	7.4
Above 91 minutes	2	62.8
Total	100	100

Source: Primary household survey, 2016–17.

Focus group discussions with households reveal that the supply schedules are not satisfactory. In the case of Tansen, the most water-scarce period is during the pre-monsoon (*Chaitra-Jestha* – in local terms) when water is available for only 1 hour in 7–9 days. During the monsoon period (*Asar-Bhadra* – in local terms), there is minor improvement, but it still continues to be a cause for serious worry for residents. During the post-monsoons, water is received for 1 hour in 3–5 days and again worsens (in *Ashwin-mangsir*) with 1 hour in 5–7 days. The winter schedules are similar to those of pre-monsoons. Therefore, increasing use of bottled water and tanker water as an alternative to municipal source has been observed here (see Table 5), a practice mostly restricted to the core areas, where households from the higher income groups reside and can afford the delivery charges of commercial sources. Hence, it is a coping mechanism of the rich and elite while poor households cope with water shortages, by fetching water from nearby springs, lakes and stone spouts. Additionally, lack of large storage systems in the homes of the poorer households hinders the purchase of water from tankers.

At this point in time, where some of the key issues have been brought forth and inequities are visible, a review of the water supply system providing water to Tansen and Damauli is of utmost importance.

Table 5. Purchase of water from private tankers.

	% of surveyed households	
Water tankers use	Tansen	Damauli
Using tanker water	48	7
Not using tanker water	52	93
Total households in both municipalities	100	100

Source: Primary household survey, 2016–17.

Our purpose in this paper is also to review the larger governance issues. The piped drinking water supply system in Tansen and Damauli is actually drawn from the spring sources. In Damauli, water from the Madi River is also tapped through a pumping system. Both the systems were handed over to the community in Damauli and Tansen in 2008 and 2012. In Damauli, the handover happened rather quickly, as the town was selected for the ADB project (Small Town Drinking Water Scheme), whereby the supply system was to be augmented and infrastructure improved. Despite this, the town's demand for water continues to increase, with inequity in access. Focus group discussions in Tansen have not indicated a very positive response from people regarding the challenges faced in accessing water. We realize that the role and performance of the DWUAs have an implication for water availability in the households of these two municipalities.

Water user groups in Tansen and Damauli: an analysis

The distribution of springs and water storage in Tansen and Damauli are shown in Figures 1 and 2. Further, detailed accounts of each of the water sources for the respective user groups follows. In Tansen, water supply is drawn from five large springs and a few smaller ones. The main spring sources Bhulke, Sisne (Khane) Mul, Banjha, and partially, Holongdi and Teendhara are used by the Tansen water user group; Banjha is the only gravity system, which is currently out of use due to lack of repair and maintenance. Bartung water user group draws water from Changdi and Dhondre, and Kunsare-Kajipauwa-Prabhas drinking water users' committee draws water mainly from Holongdi and Baghchahara. In Damauli, households had been receiving water from the municipal sources, namely, the Gunadi spring from which only three wards received water until the water user committee became functional. The other source is a pumping system from the River Madi – also meant for the core urban wards. Both these sources have limited capacity. Before the municipality in Damauli was in charge, the District Division office had been managing the supply system. When it was handed over to the municipality in 1991, there were 800–900 taps in the town.

The system was unable to meet the water demands of the growing population in general and specifically the households residing in the peri-urban wards. This was mainly attributed to the pipeline system which was not well connected to all the nooks and corners of the municipality, and to limited carrying capacity of pipes and reservoirs. As a result, the water treatment capacity was also low.

The shortfall was met by households (for purposes apart from drinking) using either direct access to the river water or from other nearby springs. To manage the system more efficiently, the formation and governance by water user cooperatives, with improved water services being the foremost priority, became the need of the hour. Therefore, the Damauli Drinking Water and Sanitation Cooperative was officially registered in 2008, after which, 1,200 taps were available in the city. The ADB project for Drinking Water in Small Towns in Nepal, launched in 2000, did work as a precursor, intending to provide safe drinking water to all the wards in Damauli. The pipeline system was laid by the municipality and then taken over by the water user committees in 2008. However, the committees in the urban cores have been found to be more pro-active than the peri-urban wards. Out of 12 cooperatives, Damauli Drinking Water and Sanitation Cooperative has been mainly responsible for supplying drinking water to the core urban Wards nos. 2, 10 and 11 and the partial areas of the peri-urban Wards 1 and 8, of the city. The cooperative in Damauli is run by the 11 working committee members who are elected by the general assembly, formed by the elected committee members.

Infrastructural constraint and operational hazards

In Tansen, the municipality has 1,760 taps and is in need of more taps. The authorities are unable to take applications due to their constraints in providing new connections. This is mainly due to current storage capacity which is only 1.4 million litres and is presently operating at its peak. At the very beginning, therefore, infrastructural constraints tend to throw up challenges, although operational hazards associated with these supply systems are no less. Water from Bhulke, the most important source for the Tansen water users group is pumped up to Tansen Bazar using 15 pumps, which run 24 hours on a dedicated electricity line. In the wake of even one pump failing or an electricity snap, the entire cascade of pumps is disturbed and supply is stalled. The late arrival of the technician due to the distant location delays the repair process and distribution further. The capacity of this pumping system has also declined over the last 30 years, from 1 million litres, to approximately 0.7 million litres in 2016–17 during the monsoon, and around 0.4 million litres during the dry season, attributed mainly to inefficiency of the new pumps, which have replaced the JICA funded pumps. On the contrary, in Damauli, Rudepani Drinking Water and Sanitation Cooperative is responsible for the pipeline system supply to the community in Ward 13. Operating at a smaller scale, this cooperative is performing better because it is functioning in coordination with the municipality and division drinking water office in Damauli for funds and technical support, not only concerning repairing old tanks but also building new reservoirs near the source. The *tole* (lane in English) development committee is also active in the process and is cooperating to raise funds. As a result, any disruption in the system is handled with care and utmost urgency, although there was an instance of the dalit community complaining that when their pipes were damaged during road construction, they took over four months to be repaired.

In Damauli, responses from the field have indicated that in the core areas, water supply timing that is fixed by the user committee for an hour from 5 am to 6 am and a similar routine in the evening does not reduce the water woes. FGDs with womens' groups have indicated that these timings are not maintained and keep varying from household to household and *tole* to *tole* (*lanes/areas within the wards*) resulting in some households getting water for 4 hours, while some for 1 hour or even less. Users continue to face uncertainty regarding a tap water supply system for everyday use and therefore the dependency on springs is not only high, but also is being used throughout the year as a source. In households of the peri-urban wards, water is available almost every day from a minimum 2 hours up to 5 hours.

Declining water availability from limited sources: implication and consequences

Sisne is a major spring source, located in a peri-urban ward of the municipality, from which water has been channelled since 2006. Previously, it was a part of farmland of the upstream communities and used for irrigation. The government/committee paid money to the owner/community member and installed a water extraction system here. According to the caretaker, a 5-inch (12.7 cm) diameter pipe collects water from the source to the tank (with capacity of 1,000,000 litres). However, in recent years, he has observed the discharge reducing to half during pre-monsoons and although several storage tanks are in other locations, the situation is quite bleak. Low water discharge leads to higher prices for water. In areas within the Bartung user committee, the last two years have seen an increase in the amount of money paid by users; from NPR 20 (20 cents USD) for 6,000 litres up to NPR 80 (about 80 cents USD) for 6,000 litres of water. Moreover, no tanker services are available in these wards due to the location being away from the core city areas. Hence, people depend on the committee supply, even if it becomes more expensive, and from the nearby spring sources. Discussions in the field indicated that the supply from the DWUA is not uniform within wards. Moreover, households are split across the Bartung and Tansen water users groups and therefore water availability varies as the sources change. Households within the Bartung user group lack taps and often rely on nearby springs and water spouts. In some clusters, such as Durgabhawani/Raksaha Tole, Devidanda there is potentiality a source of groundwater, available at a depth of 1,000–1,500 m and the households in these areas completely rely on the tube wells. Households in Barbhanjyang also use tubewells. These households have installed a borewell which draws water and transports it using a pipe of 10 inches (25.4 cm), but according to households, only 4 inches (10.16 cm) of water is available. This shows that even the borewell potential is diminishing in the area. Similar scenarios have been reported from Damauli, where discharge from the Rudepani spring source, spring sources of the Kaura Pateni drinking water cooperative, has reduced substantially; along with water availability from the Kholsi drinking water lifting project since 2010, leading to an increased burden of water collection for women. In Ward 13, women often have to wait in long queues to collect water, for which they have to get up very early in the morning. Households in Ward 9, that depend on the lifting project in Damauli, are faced with water quality issues and diarrhoea is reported during summer and monsoon seasons.

The Prabhas drinking water committee, according to a senior member, is performing very well and has managed to fulfil the demand of 60,000 litres per day in the area it serves (from Baghchahara and Holongdi, which are the major sources for the system). However, the users disagree with the statement and speak of approximately 327 households owning private taps, of which 122 households in Pravash have a metered connection. Similarly, in Kunsare, 80 households share one public tap provided by the users' committee. Several other areas (mostly the peri-urban households) within the jurisdiction of this committee have a similar story to relate, and the exorbitant amount to be paid for a new connection (NPR 50,000, i.e., USD 440)) dissuades households, which are largely agrarian in nature. As a result, two to three households pool their personal funds to apply for a community private tap. In peri-urban wards of Tansen, a very interesting issue was raised by the households, who spoke about how the larger committees like Tansen water user committees are lobbying to buy the sources of water from where the smaller water user committees access water. This would lead to further inequity across the wards of Tansen, a practice that would work against the well-being of people. In Damauli, the user groups are more pro-active in the urban wards. Purandi Drinking Water and Sanitation Cooperative in Damauli is a very successful case where the communities are very satisfied with the water supply. Additionally, there are seven springs from where households fetch water for drinking, cooking, washing, bathing and livestock watering and no complaints have been reported regarding the quality during the discussions. Partial reduction in the discharge was reported by the households. Women are extremely happy as all the households have a connection and the drudgery of water collection is reduced, although power failures and breakdown of the pump drawing groundwater disrupts the supply occasionally. Only in such situations, do they fetch water from the river. The cooperative has also received ||NPR 1,100,000 (about USD 9,700) as funding from the municipality for repair and maintenance of the old infrastructure. The CBO members go to the Municipality and Division Drinking Water Office for technical support as may be required during this process.

The problem of drying up of spring sources and lowering of the groundwater table is expected to escalate over the years. During focus group discussions, various participants and key informants were asked to narrate their observations in terms of these declining trends. Increasing drought, heavy and infrequent rainfall and increasing temperatures are some of the impacts almost all the participants unanimously agreed upon. These changes are closely associated with changing climatic conditions. They also reported that winter rain is gradually disappearing and hence the dry season is prolonged, causing more water tension. These trends and impacts will only become more prominent over the years adding more woes to the water supply scenario.

Water user groups: why are they malfunctioning?

Tansen and Damauli water user groups have experienced multiple challenges in operation and maintenance. When bigger projects are the focus, a huge gap in technical capacity and governance practices has often translated into the inefficiency of the system. In Tansen, for example, Kajipauwa-Kunsare-Prabhas drinking water user committees have access to more sources and therefore experience larger issues of governance, transparency and accountability, apart from the frequent political influence. Smaller water user groups tend to be less political and more transparent, which, of course, in the case of Damauli has been in name only. These challenges lead to lack of trust between the committees and the users. Focus group discussions, for example, exposed the corruption and collusion between the private repair agencies and the WUAs to allow their businesses to flourish, whereas efficient technical institutions like Butwal Institution, which is closer to Tansen, are ignored, despite their efficiency capacities. As a result, the system crashes frequently, when the inefficient private technical agencies are called for repairing the pumps. Even during the purchase of pumps, a considerable amount of money is paid to committees by private agencies to promote low quality pumps and parts associated with them, which tend to break down frequently, and help the business of the incapable private institutions to thrive. This not only disrupts the water distribution and households' daily challenges but allows the seeds of corruption to be sown within the system, perpetrating long-term gains of a few people. Additionally, unequal access is also an issue which only those who have the power within the committee manage to deal with, while frequent protests by others (regarding the mismanagement of the system) fall on deaf ears.

The above governance issues and malpractices resonate with households' strong belief that the institution rules and policies are not inclusive and not 'community oriented'. There are issues of disparity in access and inequality in distribution. The decision-making process is unfair and operates more or less like a syndicate, because power is held by force and political support rather than through fair and inclusive processes. Therefore, users' participation in the committee is uninvited and even mandatory meetings are not organized. In households of the Bartung committee, for example, application for a new tap connection (costing NPR 50,000 (USD 440)) requires a long wait time to get one; some households have been waiting for 11 years, while those who pay double the amount (NPR 100,000 (USD 880)) get a connection faster and avoid queues. Unfortunately, the households getting new connections are getting less water as the diameter of the pipes are smaller and supply less water. These malpractices are only leading to inefficiency in performance and, therefore, the committees are incurring huge losses despite the water price hike. There are no mechanisms to ensure representation of poor, women and marginalized households or even subsidies to these communities. Lack of transparency and accountability to the people allows these practices to be sustained. Most of the committee members are affiliated to various political parties, and these political appointments have made the committee unaccountable to the users and their grievances. As per the users' group constitution, there is a mandatory election mechanism, but the communities have indicated that over the years, the same people are elected, and the communities have no influence or say in the process.

Women, who are solely responsible for collecting and managing water in most of the households, are barred from the associations, as the tap is generally registered in the name of their male counterpart. Many women in the discussions shared how they were unable to engage in income generating activities due to irregular hours of water distribution from the municipal supply and the uncertainties associated with it. In Tansen, women must spend around 6–7 hours, especially in the dry season, collecting water from stone spouts. As a result, livestock rearing as a livelihood option for women in peri-urban areas, around Tansen, is gradually declining. Women in urban as well as peri-urban areas are, therefore, impacted due to the mismanagement of water by the user committees and consequent water shortages. Unlike Tansen, Damauli has shown better representation in the associations, not only in terms of the members, but also coordination across other groups in the wards. For example, Baireni drinking water and sanitation committee have seen women at the executive level in the committee. Both male and female members of the committee go to the municipality and other relevant organizations to demand funding support and negotiate for the best deals. Additionally, political influences are unheard of and therefore practices are clean and more transparent, although issues of supply may arise from time to time. Due to the proper coordination and cooperation among members, these issues are resolved more easily than what has been observed in Tansen.

Conclusion: will WUAs survive or wither?

Much of the Global South has embraced decentralization since the 1990s, enabling efficient mobilization and allocation of resources for improved governance at the local level. This process also generated a renewed focus on community participation in water supply and sanitation projects. Studies have shown that community participation in water projects in developing countries results in ownership and promotes project sustainability. It helps settle internal differences, make better decisions, that are accepted by all, and increases technical knowledge and management experiences of the beneficiaries of these water projects (Njogu, 2014; Das, 2015).

The literature on the sustainability of community-managed water supply is, however, controversial as there seems to be a lack of agreement about the sustainability of the model. However, Marks & Davis (2012) point to the fact that the extent of involvement during the project planning phase determines, to a large extent, the feeling of ownership in the project and its success in the future. This paper has reviewed the cases of two towns in Nepal, where WUAs have been fairly successful in one, while in the other, they have fallen at the last hurdle. This is of large concern since the projected urban population in Nepal is estimated to reach 50% of the total population by 2030 and has a significant bearing on governance, programming and management for urban WASH services. The Small Towns Water Supply and Sanitation Project in Nepal, implemented initially in 29 towns in the first phase, has seen both success and failure; Damauli representing the former group, while in the second phase, it intends to target another 21 towns (ADB, 2018). There is a need to review some of the arguments that support the success of Damauli WUAs as against those in Tansen. The WUAs in Tansen are large, and have been found to be less efficient as there has been a lack of community engagement, due to political interference leading to sustained control of the resource by a privileged few, who engage in corrupt practices within the system; thus, unable to provide services to the poor and marginalized who need it most. Moreover, at every stage, efficiency of the system is compromised along with the daily struggle for water access. Our research has shown that source maintenance, quantity of water available and future planning is extremely limited in

Tansen, which is also reflected in the low per capita water consumption (41.42 lpcd). In Damauli, the ADB project, has perhaps played a role in laying the path for not only new systems to be installed, but also funds utilized wisely and in a planned manner for improved services in this municipality. The per capita water consumption of Damauli is also higher than that of Tansen (65.9 lpcd and 41.42 lpcd, respectively), although neither meet the international standards set by WHO which stands at 125 lpcd. Further, discussions with communities in Damauli regarding water supply within homes, infrastructure development, performance and overall satisfaction levels point towards the success of the WUAs, in making them socially inclusive, more transparent and accountable vis-à-vis those in Tansen.

It must be understood at this stage that over a period of time the demand for water in both the municipalities will increase further and, therefore, proper management of the systems is essential. Both Tansen and Damauli are already experiencing low water discharge from some of the major sources, but no interventions involving sustainable practices and innovations by the committees/cooperatives have been planned. This is critical as sources are drying up and efforts to rejuvenate spring sources, and springshed management is essential to meet the increasing water demands. The aspect of climate change adaptation, and resilience, seems to be lacking in the working of the WUAs. Users, especially in Tansen's urban core areas, are increasingly reliant on alternative water sources and private water tankers. In Tansen, users are paying as much as Rs 357 (about USD 3) for every 1,000 litres of water from private water tankers. The option of water tanker use is for big hotels and rich households concentrated in core urban areas of Tansen, who have adequate storage facilities. Households on rent and those living in peri-urban areas do not have access to a water supply from private tankers due to limited affordability. During discussions, the households with access to the alternative private sources revealed that they preferred these expensive alternatives against the unreliable water supply from municipal sources, which further raises concern about the diminishing trust between the WUAs and the beneficiaries thus leading to subsequent alienation.

This trend has directly led to increased disparity vis-à-vis rich and poor, urban and the peri-urban residents. Experience in Tansen shows that adaptation is only possible for the rich, through alternative sources; private water tankers and therefore WUAs have actually failed with regard to water equity.

The groups need to identify that increasing population-induced water demand, upcoming technology and changes in rainfall patterns call for a specific paradigm on spring water management (Patil *et al.*, 2018), which is the traditional source of water in these towns. Hence, development of a science-based, community-involved approach to managing the springs, if adopted as part of the WUAs' charter of activities, can make these towns more resilient to future water scarcities driven by climate change. Additionally, policies around proper land use mapping and infrastructure planning from the municipality can greatly aid in conservation of existing sources. Smaller WUAs such as Bartung in Tansen have already bought some land around their water sources to conserve the water sources. These pro-active measures by the larger groups in Tansen will lead to sustainable water supply in the core urban as well as peri-urban areas. ADB's role in the STWSSP has been increasing, through value addition built on previous ADB interventions in the sector.

> *'It will strengthen particularly (i) technical robustness, (ii) institutional capacity focusing on project municipalities and WUSCs, (iii) long-term operational sustainability, and (iv) climate-resilient approaches and smart technology application. To overcome start-up delays, the project will assist project development by supporting the detailed designs of WSS sub-projects for newly established provincial capitals and district headquarters apart from Kathmandu valley. The project is aligned with the government's sector policies, ADB's Urban and Water Operational Plans, ADB's Water*

> *for All Policy, and the Sustainable Development Goals (gender equality, clean water and sanitation, and climate change).* (ADB, 2018: 3)'

The outcomes that have been observed are, better functioning of the WSS systems, greater institutional strengthening and improved financial planning. As part of the improved process and planning, Tansen and Damauli would benefit at large, but without donor support, who will always try to push their agenda forward. At this moment, Tansen and Damauli are both functioning without donor support and there are success stories. Hence, if WUAs come together as strong institutions to find ways of better and effective governance, they can rule out the interference of donors.

The water user committees are independent bodies that are run as per their constitution and are independent of municipality or government scrutiny. However, most of the water user groups have a municipality/local government representative as the officiating member on their respective board and executive committee. As a result, there is widespread corruption and bad governance practices. The communities would like the municipality to play a positive role and mitigate these malpractices. From the larger lens of urban development and rising water demands, essential goals need to be set ahead to ensure that the water cooperatives/committees are more pro-active, accountable and inclusive in structure. Their roles must be reviewed at regular intervals and their performance monitored by the local communities, for which, municipal bodies must have written rules and regulation on how these cooperatives must operate. In this regard, experiences across the globe, particularly from Africa, highlight key aspects which help in the success of the community-managed water projects. First, capacity building, construction supervision and providing support to the community water project management committees during the first year of implementation are strongly recommended for sustained community participation. Second, a budget is essential and should be carefully prepared to ensure that there is no misappropriation of funds. It also increases transparency within the community members who are in charge of the funds and the way it has been spent by the persons responsible for each expense related to the project (Twebaze, 2010 in Njogu, 2014). Third, at the construction and implementation phase, clear communication channels need to be put in place so as to keep stakeholders informed of any modification to the project design and implementation strategies. Lastly, careful selection of members on the committee is very essential if the project is to be successful. It is essential, therefore, to involve women and youth during the implementation/project planning phase. Poricha & Dasgupta (2011) mention that through participatory processes of engaging with the local communities, NGOs and civil society organizations working in the area can help improve existing systems and also develop new site-specific suitable systems for water supply and sanitation. Moreover,

> '*practical guidelines for the managers of participation processes is available, through numerous case studies, either as elaborations of the concept of participation, or in the form of research that usually focuses on the question of what are the criteria of 'good' or 'effective' participation processes.* (Von Korff *et al.*, 2012: 33)'

The two municipalities Tansen and Damauli are growing urban centres of Nepal. With water user committees representing more and more women and marginalized groups, one would have expected considerable change in access to water in these two towns, but it has not really materialized as expected. Reducing disparity in access, inequality in services, quality of water supplied, functionality and overall sustainability of system remain key challenges of the water projects in these towns. Water governance in the form of community participation is considered most desirable, as it involves greater sense of ownership along with accountability. In this regard, the argument, which the title of the paper proposes, 'perform or wither' somewhat strongly affirms the need for change within WUAs, for their success. Across the globe, experiences have shown that the success of community water projects lies in their performance, which in turn, largely depends on the organizational composition, rules of selecting members, putting together ideas of members wisely, capacity building and learning new techniques, to design projects; planning and managing finances efficiently at every level, keeping members well informed of who is in charge of which aspect of spending, along with proper implementation and monitoring. If the chain breaks at any point, success will be skewed. Tansen and Damauli both have equal potential to develop and strengthen their WUAs. These water user groups already have a network at national level called Federation of Water and Sanitation Users Nepal (FEDWASUN) which can play the crucial role of building their capacities and improving governance.

Withering is definitely not a solution and cannot help these towns cope with future water demands. It will only result in increasing the gap between the rich/powerful vs poor/powerless with regard to water access, and allow international donors to promote their agenda. Therefore, through appropriate and corrective measures, performance/efficiency-related governance practices (within WUAs) can actually change the entire environment of these towns and breed sustainability and self-reliance. With good governance in place, performance indicators of WUAs are bound to show higher scores and will sustain and become an inspiration nationally and globally.

Acknowledgements

This work was carried out by the Himalayan Adaptation, Water and Resilience (HI-AWARE) consortium under the Collaborative Adaptation Research Initiative in Africa and Asia (CARIAA) with financial support from the United Kingdom's Department for International Development (DFID) and the International Development Research Centre (IDRC), Ottawa, Canada.

References

ADB (2018). *Proposed Loan: Urban Water Supply and Sanitation (Sector) Project, Nepal.* Report and Recommendation of the President to the Board of Directors. https://www.gtai.de/GTAI/Content/DE/Trade/Fachdaten/PRO/2018/10/Anlagen/PRO201810245009.pdf?v=1 (accessed 3 August 2019).

Aryal, R. S. (2011). The law on ownership and right to water in Nepal. Final report, Jal Shrot Vikas Sanstha/Global Water Partnership Nepal, Kathmandu.

Bhatta, K. P., Ishida, A., Taniguchi, K. & Sharma, R. (2006). Performance of agency-managed and farmer-managed irrigation systems: a comparative case study at Chitwan, Nepal. *Irrigation and Drainage Systems 20*(2–3), 177–191.

CBS (2011). *National Population and Housing Census Survey (NPHCS 2011).* Central Bureau of Statistics, National Planning Commission, Government of Nepal.

Das, P. (2015). The urban sanitation conundrum: what can community-managed programmes in India unravel? *Environment & Urbanization 27*(2), 505–524. https://doi.org/10.1177%2F095 62478155 86 305 (accessed 3 August 2019).

Gareau, B. J. & Crow, B. (2006). Governing water: contentious transnational politics and global institution building. *International Environmental Agreements: Politics, Law and Economics 6*(3), 317–320. https://doi.org/10.1007/s10784-006-9007-1.

Hill, M. (2012). A starting point: understanding governance, good governance and water governance. In: *Climate Change and Water Governance.* Hill, M. (ed.). Springer, pp. 17–28.

Howarth, S. E., Parajuli, U. N., Baral, J. R., Nott, G. A., Adhikari, B. R., Gautam, D. R. & Menuka, K. C. (2005). *Promoting Good Governance of Water Users' Associations in Nepal.* Department of Irrigation of His Majesty's Government of Nepal.

Huitema, D., Mostert, E., Egas, W., Moellenkamp, S., Pahl-Wostl, C. & Yalcin, R. (2009). Adaptive water governance: assessing the institutional prescriptions of adaptive (co-) management from a governance perspective and defining a research agenda. *Ecology and Society 14*(1), 26.

Joshi, A. R. (2017). *The Most Expensive Water in Nepal.* https://www.thethirdpole.net/en/2017/07/06/the-most-expensive-water-in-nepal/

Lopez-Gunn, E. (2003). The role of collective action in water governance: a comparative study of groundwater user associations in La Mancha aquifers in Spain. *Water International 28*(3), 367–378.

Magar, S. T. (2009). Water resources in Nepal: institutional analysis based on legal provision. *Occasional Papers in Sociology and Anthropology 9*, 199–247. http://himalaya.socanth.cam.ac.uk/collections/journals/opsa/pdf/OPSA_09_09.pdf

Marks, S. & Davis, J. (2012). Does user participation lead to sense of ownership of rural water systems? Evidence from Kenya. *World Development 40*, 1569–1576.

Mayntz, R. (2006). Governance theory als fortentwickelte steuerungstheorie? In: *Governance-Forschung: Vergewisserung über Stand und Entwicklungslinien.* Schuppert, G. F. (ed.). Nomos Verlag, Baden-Baden, Germany, pp. 11–20.

Nepal, P. (2001). Functional base analysis of Tansen municipal town. *Contributions to Nepalese Studies 28*(2), 285–293. http://www.thlib.org/static/reprints/contributions/CNAS_28_02_08.pdf

Nepal Law Commission (2015). The Constitution of Nepal, Nepal Gazette. Pg 23, Article 35, sub article 4.

Njogu, K. D. (2014). *Influence of Community Participation on Performance of Constituency Development Funded Rural Borehole Water Projects: A Case of Kerwa Sub-Location, Kiambu County, Kenya. MA thesis*, University of Nairobi. https://pdfs.semanticscholar.org/898f/a39ac21a1625189147801d4233f8f1f559ac.pdf (accessed 28 July 2019).

Ostrom, E. (2000). Collective action and the evolution of social norms. *The Journal of Economic Perspectives 14*(3), 137–158.

Patil, S., Barola, Y. & Kulkarni, H. (2018). *Springshed Restoration in Kalsi Block of Ddehradun District Under the Programme 'Strengthening State Strategies on Climate Actions Through Practicing Climate Resilient Practices'.* Technical Report of Advanced Center For Water Resources Development And Management (ACWADAM) with inputs from Central Himalayan Rural Action Group.

Peters, B. G. & Pierre, J. (1998). Governance without government? rethinking public administration. *Journal of Public Administration Research and Theory 8*(2), 223–243.

Poricha, B. & Dasgupta, B. (2011). Equity and access: community based water management in urban poor communities: an Indian case study. *WIT Transactions on Ecology and the Environment 153*, 276–285.

Portnov, B. A. & Adhikari, M. (2006). Spatial patterns of urban growth – does location matter? A Case Study of Nepal. In: *46th Congress of the European Regional Science Association: 'Enlargement, Southern Europe and the Mediterranean'*, August 30–September 3, Volos, Greece. https://www.econstor.eu/bitstream/10419/118160/1/ERSA2006_014.pdf

Poudel, A. (2017). *Water Availability Under Climate Stress in A Hilly Settlement of Nepal: A Case Study From Tansen, Palpa. MSc. Dissertation*, Central Department of Environmental Science, Institute of Science and Technology, Tribhuvan University Kirtipur, Kathmandu, Nepal. http://www.hi-aware.org/2017%20-%20%20MSc%20Ashmita%20thesis.pdf

Pradhan, P. (1989). *Patterns of Irrigation Organization in Nepal: A Comparative Study of 21 Farmer-Managed Irrigation Systems (No. 1).* International Water Management Institute, Colombo, Sri Lanka.

Pradhan, P. (2015). Role of water research institutions in Nepal and its impact on policy changes. *Journal of Water, Energy and Environment 17*, 42–43.

Roth, D., Boelens, R. & Zwarteveen, M. (2005). *Liquid Relations: Contested Water Rights and Legal Complexity.* Rutgers University Press, New Brunswick, NJ, USA.

Twebaze, J. (2010). *Community Mobilization in Rural Water Supply and Sanitation Programs: How Effective is it? A Case of Wakiso District-Uganda.* Makarere University, Kampala, Uganda. http://hdl.handle.net/10570/2159 (accessed 28 July 2019).

Von Korff, Y., Daniell, K., Moellenkamp, S., Bots, P. & Bijlsma, R. M. (2012). Implementing participatory water management: recent advances in theory, practice, and evaluation. *Ecology and Society 17*(1), 30. http://dx.doi.org/10.5751/ES-04733-170130 (accessed 8 August 2019).

Water Aid Nepal (2005). *Water Laws in Nepal-Laws Relating to Drinking Water, Sanitation.* http://lib.icimod.org/record/6408 (accessed 16 February 2019).

World Health Organisation (2014). *UN Water Global Analysis and Assessment of Sanitation and Drinking Water.* WHO, Geneva, Switzerland. http://www.who.int/water_sanitation_health/glaas/2014/nepal-10-nov.pdf

Received 19 March 2019; accepted in revised form 25 September 2019. Available online 26 November 2019

Water management systems of two towns in the Eastern Himalaya: case studies of Singtam in Sikkim and Kalimpong in West Bengal states of India

Ghanashyam Sharma[a,*], Chhayavani Namchu[b], Kalsang Nyima[a], Mahindra Luitel[a], Sreoshi Singh[b] and Chanda Gurung Goodrich[b]

[a]*The Mountain Institute India, Gangtok, India*

**Corresponding author. E-mail: banstolag@gmail.com*

[b]*International Centre for Integrated Mountain Development, Kathmandu, Nepal*

Abstract

This study examines the water supply systems, their sociopolitical dynamics, and the future of water management in two Indian towns in the Eastern Himalaya, Kalimpong in West Bengal and Singtam in Sikkim. The research was centred around issues of demand and supply, water scarcity and stress, equity, water governance, and the sustainable conservation and management of water resources in a climate change context. Methodologically based on surveys, focus group discussions, and key informants' interviews, this study finds that spring sources are drying alarmingly in Singtam, even as demand is increasing dramatically due to a floating population that is more than the number of local inhabitants. The town suffers from the lack of an adequate reservoir facility and the frequent damage of water supply pipes during the monsoon. Kalimpong faces acute water shortages all year round. The political tug of war between the state government agencies and the local government around the management and supply of water, declining water discharge in spring sources, the lack of water infrastructure for repair, maintenance, and supply, and the glaring inequity between the higher, middle, and lower income groups are the immediate issues around water in Kalimpong. The spring sources that supply drinking water to Kalimpong and Singtam need immediate conservation measures to arrest the declining state of discharge. Micro-planning at the local level, reviving drying springs, and the adoption of appropriate soil and water conservation practices on a watershed basis are all important ways forward. The development of water security plans and their strict enforcement through multi-institutional collaboration can contribute to improved water governance and socioecological restoration for sustainable water resources management.

Keywords: Dhara Vikas; Drying springs; Governance; Kalimpong; Micro-planning; Singtam; Water management; Water supply system

doi: 10.2166/wp.2019.229

1. Introduction

The countries of the Hindu Kush Himalaya (HKH) region are increasingly facing challenges in meeting their demand for water, food, and energy due to rapid socioeconomic development, population growth, and urbanization (Mukherji *et al.*, 2015). The region faces severe seasonal water shortages due to high variability in rainfall, a problem likely to be exacerbated by climate change (Vaidya, 2015). The variability in rainfall patterns, seismicity, early snowmelt due to climate change, the geological composition of aquifers in the HKH region, and unpredictable, prolonged dry periods have resulted in declining trends in water discharge in springs, rivers, and rivulets (MoWR, 2012; Gupta & Kulkarni, 2018). As a consequence, water shortages have become increasingly prevalent in cities and towns of the region, such as Kathmandu and Pokhara in Nepal; Darjeeling, Kohima, Mussoorie, and Shimla in India; Lhasa in China; Thimphu in Bhutan; and Kabul in Afghanistan (Synder, 2014). In the Sikkim Himalayan region more specifically, towns and cities are growing with unprecedented developmental and population pressure, a situation aggravated by climate change impacts. However, a scientific understanding of the socioecological situation is still lacking.

In order to fill this gap, two Indian towns, namely Kalimpong in West Bengal and Singtam in Sikkim, were chosen for a preliminary study focusing on governance and water supply systems, management, and benefit-sharing. This paper is an exploratory study of urban water supply in the two towns. It aims to expose gaps, identify potential areas of future study for the better management of water supply systems and benefit-sharing, and underlines the need for the conservation of water sources.

1.1. Historical context

This study on water and livelihoods in two Himalayan towns, in Sikkim and West Bengal, respectively, traces its narrative to the history of the region before it was defined and mapped as Sikkim and West Bengal. An understanding of this timeline is significant for a larger perspective on the changes that the region has seen – natural, anthropogenic, and sociopolitical. Basically, the challenges of water management in Singtam and Kalimpong are not a sudden outcome, but a product of many gradual changes that led to its current state of water shortage, occurring against the backdrop of climate change and increasing, unplanned urbanization. For want of space, we can touch on this history very briefly.

Historically, in terms of cultural assimilation, the Lepchas are the original inhabitants of the Darjeeling hills (which includes Kalimpong) and Sikkim. They were followed by the Limboos and later the Bhutias. The Nepalis were already settled in the western bank of the River Teesta after they conquered the land under the leadership of Prithvinarayan Shah during 1774 and later extended agriculture in the region (Pradhan, 1991). Later, during the 1960s and 1970s, the Marwari business community started settling in Sikkim and Kalimpong, along the old Silk Route, in order to trade with Tibet (Jha, 1985).

The annexation of Kalimpong by the British in the late 19th century resulted in the initiation of trade with Tibet through Jelep-la and Nathu-la (Hackett, 2008). Similarly, the merger of Sikkim into the Indian union in 1975 (Datta-Ray, 1984) resulted in several cultural and political shifts. Both of these geopolitical changes resulted in the emergence of different governance and political practices. As the 22nd state of India, Sikkim witnessed economic and sociocultural developments, political awareness and consciousness, administrative reforms, and active participation of the people through their representatives in the state legislative assembly. Kalimpong, ruled in succession by the Sikkimese and Bhutanese kingdoms, and the British East India Company during the 18th and 19th centuries, became a part of

West Bengal state after Indian independence in 1947 as a subdivision of Darjeeling district, and subsequently a full-fledged district in February 2017. It is currently ruled by Gorkha Territorial Administration (GTA) under the West Bengal government. This would have an impact on both the hill towns: Kalimpong and Singtam, now located in the states of Sikkim and West Bengal, respectively, in independent India.

During the 1990s and in 2000s, as many as 27 hydropower projects in the Teesta River and long-distance tunnels running under the mountains were planned that brought up sociocultural and socioecological issues in the sacred landscapes including the perturbation of pristine environments leading to degradation of local water sources (EPDGOS, 2015). With the increasing infrastructure development for hydropower projects, the Teesta River basin has been seriously impacted by loss of political rights, loss of indigenous religion and cultural identity, landlessness, loss of livelihoods, biodiversity loss, and environment degradation due to hydropower projects (Kerry, 2008; Huber & Joshi, 2015; Lepcha, 2018; Sharma & Pandey, 2018).

1.2. Water governance, policies, and challenges

Gupta and Kulkarni (2018), in a report for the National Institution for Transforming India (NITI) Aayog, stated that nearly half the perennial springs in the Himalayan region have dried up or become seasonal in nature, due to natural dynamism, erratic rainfall, seismic activity, and ecological degradation associated with land use change for infrastructural development impacting mountain aquifer systems. When considering the case of Sikkim Himalaya, traditional and natural wetlands and water bodies such as village ponds, *dhara/pandhera* (springs), *kuwa* (shallow wells), *khola/kholsa* (small rivers or rivulets arising from a spring), and *devithan* (water sources, abode of local deities) have all been badly degraded over the last few decades (Sharma *et al.*, 2012; Tambe *et al.*, 2011, 2012). Consequently, thousands of Himalayan villages and several towns are currently facing acute water shortages for drinking and other domestic or agricultural purposes. These structures ought to be restored, maintained, and used properly and these water bodies should not be allowed to be encroached upon for any other land use (MoWR, 2012).

The National Watershed Development Project for Rainfed Areas (NWDPRA), the Integrated Watershed Development Programme (IWDP), and soil conservation schemes in the catchments of river basins and flood-prone rivers are programmes extended by different ministries of the Government of India. India's first National Water Policy (NWP) was adopted in 1987 and revised in 2002 and 2012 (MoWR, 2002; Kumar, 2017). However, it has not categorically addressed the declining state of water sources meant to cater to the needs of mountain cities and towns and the increasing demand for water. In 2014, the Government of India adopted the National Mission for Sustaining the Himalayan Ecosystem (NMSHE), as one of the key elements of India's National Action Plan on Climate Change (NAPCC) (DST, n.d.). It is expected that marginalized and vulnerable communities in the Indian Himalayan region, which straddles 12 states, will be the beneficiaries of the mission, as it seeks to safeguard people from the impacts of climate change through evidence-based policy formulation and enhanced research and capacity.

At the individual state level, the contrast in policy is stark. Sikkim, with a stable government and not much political discord, has successfully undertaken springshed management with the first systematic and landmark initiative, *Dhara Vikas*, led by the Rural Management and Development Department (RMDD) and its collaborators. As a robust climate adaptation strategy for drought-prone areas in

Sikkim, *Dhara Vikas* (meaning, springshed development), an innovative programme was adopted in 2008 to revive and maintain drying springs, which is helping to alleviate the problem of rural water scarcity by reducing surface run-off of rainwater and allowing more water to percolate down to recharge underground aquifers, which, in turn, ensures increased discharge from springs (Tambe *et al.*, 2011, 2012; Sharma *et al.*, 2012). This scheme has been replicated and undertaken in the form of pilot projects of varying scales across Indian mountain states such as Himachal Pradesh, Nagaland, and Uttarakhand (Seidler *et al.*, 2016; Gupta & Kulkarni 2018). However, such an initiative is yet to come about in the adjoining Kalimpong district, despite its proximity to Sikkim. The reason for this lack is the political atmosphere: Kalimpong and Darjeeling Hills have been witnessing political agitations since pre-independence, based on ethnic identity, and over time this has morphed into the demand for Gorkhaland for the minority Nepali Gorkha community, demanding separation from West Bengal (Ganguly, 2005). The movement has seen outbreaks of violence, particularly since the early 1980s, and as a consequence there has been lack of coordinated initiatives towards water management by the West Bengal government and the GTA. In contrast to the coordination and collaboration between agencies on water governance in Sikkim, one study conducted in the adjacent Chibo–Pashyor area of Kalimpong has concluded that the key reasons for poor water resource governance and management were lack of coordination, convergence and engagement of local administration (PHED and the municipality) in the construction, distribution, management and monitoring of water resources, poor local-level awareness about various government schemes and programmes on water supply and systems, outdated and defunct infrastructure for overall distribution and capturing of excess run-off, and lack of properly designed schemes and projects involving community participation (Sharma *et al.*, 2019).

1.3. Climate change and reduced rainfall

According to Sikkim's *Socioeconomic Census 2005–06*, nearly 65,000 (80%) of the 80,000 rural households in the state depend on springs as a source of drinking water (Government of Sikkim, n.d.). However, these springs now face a new hazard. Although degradation of catchments has been identified as the main cause for the drying up of springs in the 20th century, climate change is now emerging as the new threat (Tambe *et al.*, 2012). In the mid-hills and low-hills, the precipitation varies from 1,000 mm annually in the subtropical agro-climatic zone to around 3,500 mm in the temperate zone. The Sikkim Himalayan region is blessed with high rainfall (annual average, 3,000 mm) but an overwhelmingly high proportion of it falls during the summer monsoon season. Sikkim has witnessed unprecedented, long spells of dryness in recent years, with 1986, 1988, 1989, 2006, and 2009 being classified as drought years. These years received 13–21% of lower rainfall than the 30-year average (1981–2010). This rainfall deficit has had adverse effects on drinking water supply, irrigation in the rabi season, the production of vegetables, oranges, and other fruit crops, and Sikkim's economy in general (Rahman *et al.*, 2012).

In the Kalimpong and Darjeeling regions, the maximum temperature showed a decreasing trend, whereas the minimum temperature showed an increasing trend. The mean temperature showed an increasing trend over the years. The annual rainfall slightly increased over the years. The extreme weather events (mainly rainfall fluctuation) increased more in recent years (after 1980) (Benerjee *et al.*, 2015). Droughts, landslides, flash floods, long dry spells, drying springs and water sources, extended summer temperatures, and increased incidences of pests and diseases in crops are some of the climate change indicators (WBSAPCC, 2010; Sharma *et al.*, 2019).

2. Study sites

Two Himalayan towns, Singtam in Sikkim and Kalimpong in West Bengal, were selected for a comparative study and to present the case of each town when it comes to shaping water resources management, demand and supply, water scarcity and stress, equity, water governance, and sociopolitical dynamics and contextual factors (such as climate, demography, historic evolutions, etc.) in a climate change context.

2.1. Singtam, East Sikkim

Singtam nagar panchayat (municipality) is in East Sikkim district and is located (27°9′0″N, 88°22′48″E) at 423 masl spread across a 0.18 sq. km area along the banks of the Teesta River and Rani Khola (Figure 1). The annual average rainfall varies between 1,500 and 1,800 mm, with the highest precipitation being in June to September. The municipality is divided into six wards (Figure 2), and has 1,144 households with a total population of 5,868 (2011 census). Of these, 3,097 are males and 2,771 females (MHA, 2011). However, its population has increased dramatically over the last 7 years. As per the Singtam District Hospital's Community Need Assessment Survey conducted during March 2018, the total headcount in Singtam municipality and Golitar was 25,665, with a floating population of more than 6,000 persons, which is rising each year. Golitar, on the left bank of the Rani Khola and Teesta Rivers, includes a population of around 5,000 that includes temporary rental occupants and East Point School campus residents. On the right flank of the Teesta River lies Manpari Busty

Fig. 1. Satellite image of Singtam. *Source*: Google Earth.

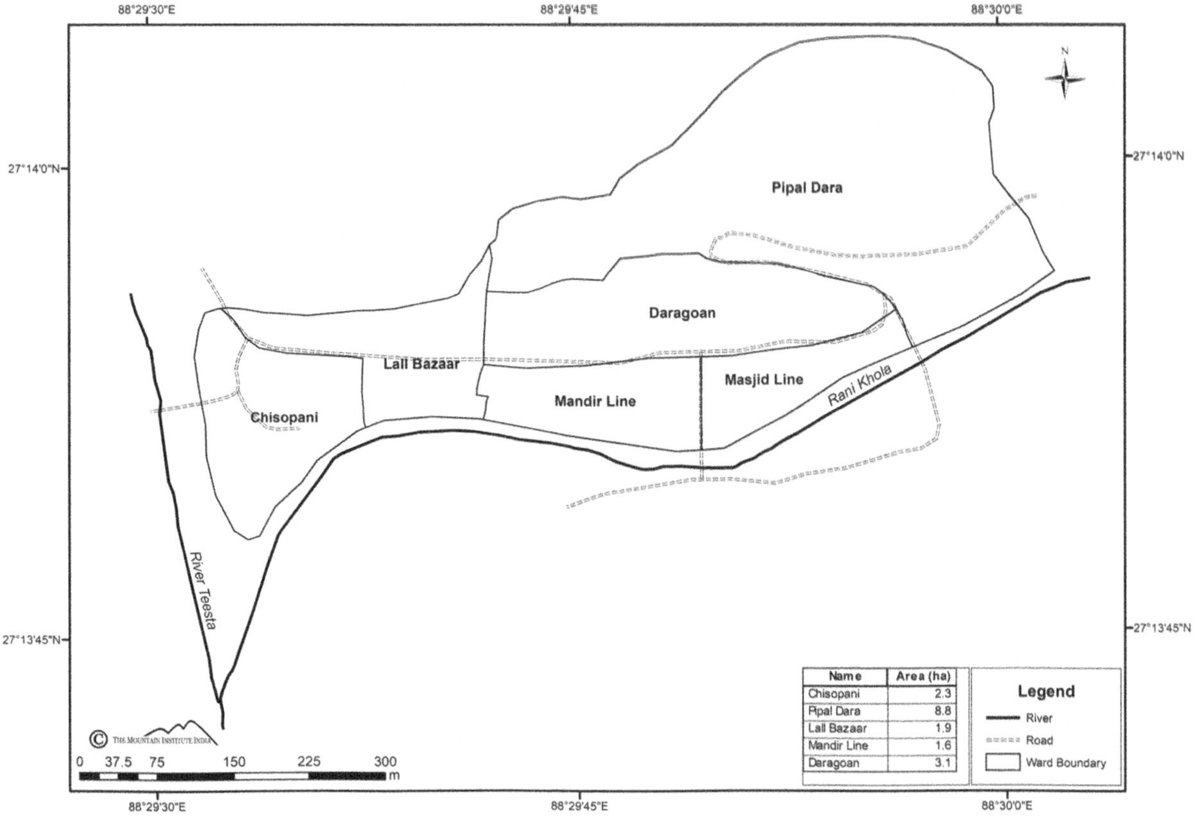

Fig. 2. Singtam town municipality and wards. *Source*: IndiKosh (2018b).

(Adarsh Gaon), inhabited by around 75 resident households and little more than 150 rental families. Both Golitar and Adarsh Gaon, an extension of Singtam, are dependent on Singtam municipality for day-to-day requirements and services.

As many as 75% of the population of Singtam are engaged in business, or employed in government or in private companies. Approximately 55% of the males and merely 14% of its female population are currently working. Around 53% of the total male population are full-time workers and 2% are part-time workers. In contrast, merely 13% of the female population are full-time workers and 2% are part-time workers (IndiKosh, 2018a).

2.2. *Kalimpong, West Bengal*

About 40 kilometres (km) away from Singtam is the town of Kalimpong (1,400 masl) in West Bengal. Kalimpong municipality (27°4′25.17″N, 88°28′37.92″E) is located between two hills – the Deolo and Durpin (Figure 3). The municipality covers an area of 9.17 sq km, has 23 wards (Figure 4) with a population of 49,403 (2011 census) (ODM, n.d.). It has grown by nearly 15% over the last decade. Around 51% of the male and merely 16% of the female population work. About 47% of the total male population are full-time workers and 4% are part-time workers. About 14% of the

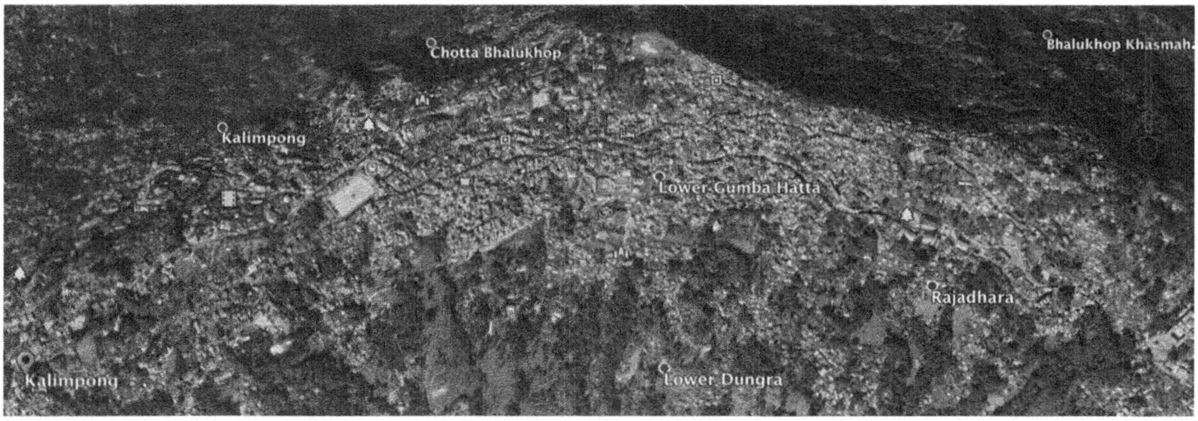

Fig. 3. Satellite image of Kalimpong. *Source*: Google Earth.

total female population are main workers and 3% are marginal workers (IndiKosh, 2018b)[1]. The annual average rainfall of the city is 2,200 mm, with the highest precipitation being in May to October.

3. Methodology

The data collected for this study were derived through qualitative and quantitative methodologies using participatory rural appraisal techniques such as focus group discussions (FGDs), key informant interviews (KIIs), and field observations. We used semi-structured and open-ended questions and conducted 18 KIIs and 7 FGDs (5 mixed and 2 female groups) in Singtam, and 23 KIIs and 12 FGDs (7 mixed and 5 female groups) in Kalimpong. As many as 276 households from 23 wards of Kalimpong and 35 households in 6 wards of Singtam were interviewed during January–December 2017 using a structured questionnaire. Data were recorded online using the AKVO Flow Android mobile application system, a data collection and monitoring tool that works in field situations, to gather reliable and geographically referenced data. The wards in Kalimpong are scattered and thus a higher number of representatives were interviewed from different locations of the town and adjacent villages within the municipality so as to reduce data bias. While in Singtam, the wards are smaller and closely located to each other with similar situations. Most of the general issues were covered in KIIs and FGDs and thus, proportionately, a lower number of the household surveys were conducted in Singtam.

The respondents' households comprised culturally similar ethnic and resident business communities (Nepali[2] 37%, Bhutia (see note 2) 19%, Lepchas (see note 2) 13%, Behari[3] 14, Marwadi (see note 3) 17%), age groups, had attained different education levels (Figure 5), and different income categories (Table 1). Around 15% of the respondents were between 60 and 80 years of age, 35% between 40 and 60 years of age, and 50% were between 20 and 40 years of age. Thirty-two per cent

[1] Typically, those who get work for more than 6 months in a year are 'main' workers; those who work for less than six months are 'marginal' workers.

[2] Nepali, Bhutia, and Lepcha are indigenous communities.

[3] Behari and Marwari are business communities.

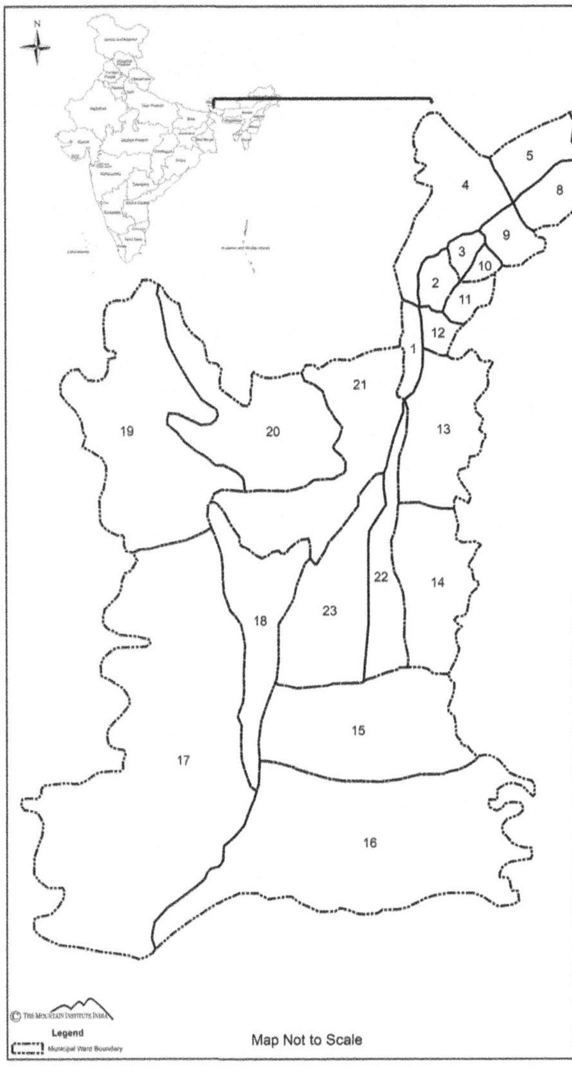

Population Data of Kalimpong Municipality (www.indikosh.com/city/322796/kalimpong)	
Ward No	Population
1	1432
2	1443
3	1260
4	2070
5	3165
6	1204
7	3772
8	2310
9	1547
10	1218
11	3142
12	964
13	3106
14	1106
15	2486
16	4737
17	845
18	1465
19	3667
20	2736
21	2551
22	1321
23	1856

Fig. 4. Kalimpong town municipality and wards.

of the respondents in Singtam were women and 56% in Kalimpong. Around 80% of the respondents belonged to households with houses they owned, while the remaining 20% were in rented apartments in both the towns. Similar and contrasting elements/aspects of water source management, transportation, distribution and supply to households, and governance for Singtam and Kalimpong are given in Table 2.

We used the inductive coding style with symbols applied to sections of text in order to easily categorize those that are related to research questions, themes, concepts, and which were easy to retrieve and organize. During and after coding, we looked for connections between codes and began with descriptive themes; we also looked for sub-categories, relationships, cause-effect, etc. and finally developed a matrix of codes and themes, and analysed the information. To ensure reliability, qualitative data were cross-checked during the FGDs, and some of them were discussed

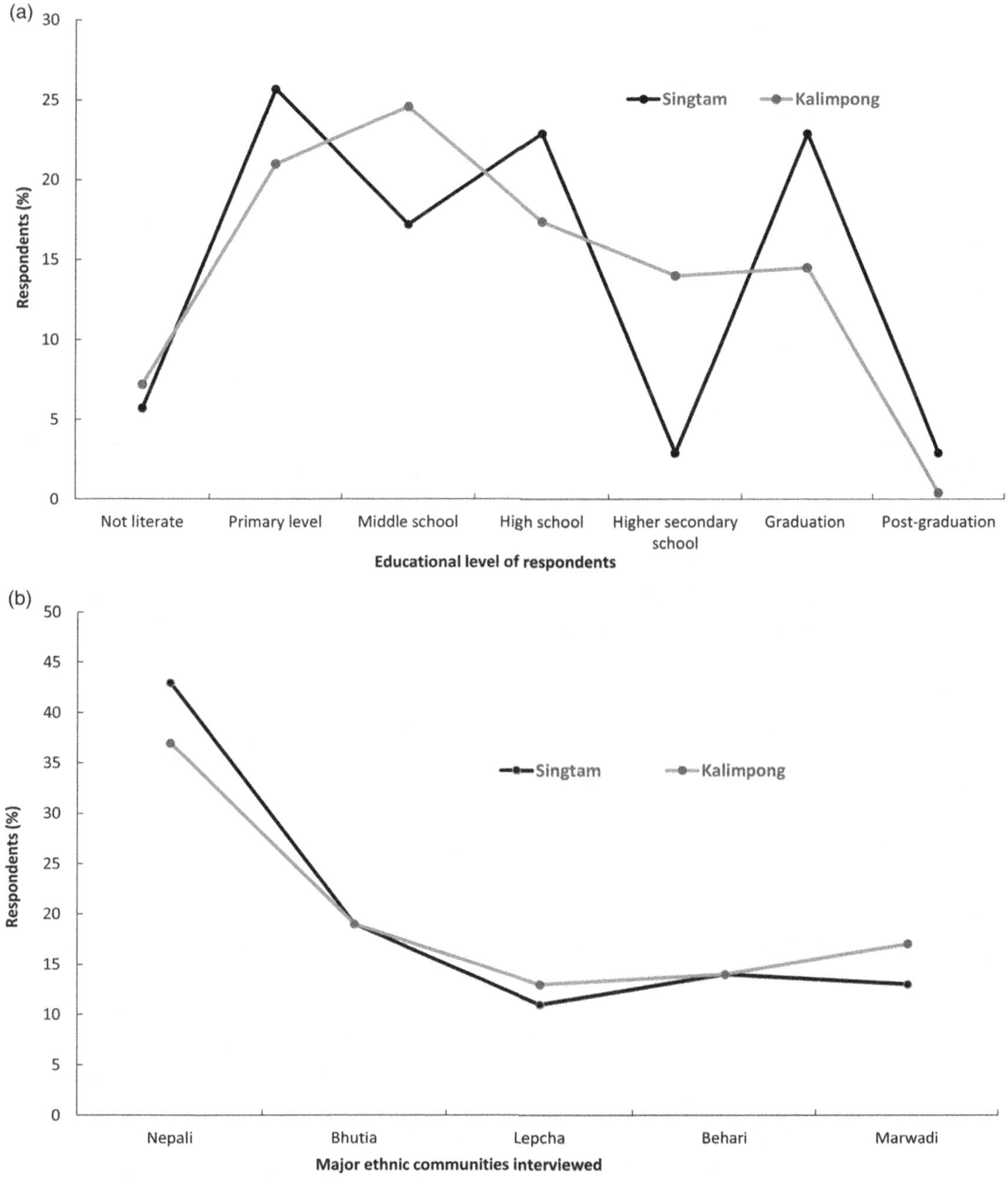

Fig. 5. (a) Educational levels of respondents in Singtam and Kalimpong. (b) Per cent respondents belonging to different ethnic and other communities in Singtam and Kalimpong.

informally with the KIIs. The participants of KIIs were members of nagar panchayats, businessmen, hotel and restaurant owners, PHED and municipal officials, pipe fitters, water vendors, household owners, and those on rentals.

Table 1. Monthly income levels of respondent households in Singtam and Kalimpong.

	Income categories (INR)						
Municipalities	> 10,000	10,001–15,000	15,001–30,000	30,001–50,000	50,001–100,000	> 100,000	Don't know
Singtam (%)	20.01	35.22	27.40	11.23	3.6	2.54	–
Kalimpong (%)	18.11	38.11	25.11	10.63	4.33	1.90	1.81

4. Water supply, distribution system and management

4.1. Drinking water sources, distribution, health, and other challenges

4.1.1. Singtam. The water distribution system in Singtam falls under the Public Health Engineering Department (PHED), Government of Sikkim. According to PHED, a total of 0.8 million gallons per day (mgd) is supplied to the town, including to the Singtam District Hospital, the police, the fire station, government offices, and public and private schools. The per capita water consumption in Singtam is only 39 litres per day (lpd), which is below the national average of 70 lpd. The total demand currently is 1.5 mgd.

The PHED transports water from Sang-chu Khola at the Martam–Nazitam gram panchayat unit (GPU), which is about 7 km away from Singtam. PHED authorities have constructed three water storage tanks, each with a 0.12 million gallon capacity, above the main town at Chisopani around 1 km away, and provide a direct connection to roughly 83% of the households. However, about 90% of the respondents' households and 70% of the KII respondents reported irregularity in water supply, receiving turbid and muddy water during the monsoon, and insufficient supply in winters. Therefore, apart from the main Sang-chu Khola source, 85% of the households also derive water from springs and streams from five other locations, most of which are located 1–3 km away from Singtam (Table 3). FGDs revealed that spring sources are tapped by the town's older residents, who have brought the water supply through pipes into their homes. A majority of the newer residents (labourers and officials of hydropower projects and pharmaceutical companies) is also dependent on the PHED supply.

Our field survey also discovered that besides PHED supply, other springs, namely, Kali Khola, Andheri Khola, Simana Khola, and Seti Khola, which are located around 1–5 km away, are significant sources of water for Singtam, Golitar, and Adarsh Gaon catering to around 1,360 families. These spring sources are 'private lines' and are poly-piped for private household consumption on payment. FGDs also disclosed that due to a warming climate and prolonged dry periods in recent years, springs have been drying. Thus, people have been shifting their source of water and looking for alternative springs. Seti Khola, located in Lower Namphing, is a major source of water for around 102 households (Figure 6(a)). In Golitar, 13 households have been drawing water from bore-wells as an alternative measure (Figure 6(b)).

PHED officials reported that they are looking for an alternative water source for Singtam that can fulfil the current total demand of 1.5 mgd. Until 2010, PHED brought water from Ghattey Khola located in Bhirkuna which dried up due to recurring prolonged dry periods coupled with anthropogenic impacts, and hence the department was forced to shift to Nazitam for a new water source.

4.1.2. Kalimpong municipality. The water source for Kalimpong municipality is the Neora forest, which is 86 km away from Kalimpong. Water brought from there is stored in a reservoir at Deolo and then supplied to the town. The responsibility for delivering water from the reservoir to the households in the 23 wards lies with the PHED, under the local government (the GTA), and the Neora Water

Table 2. Similarities and contrasting features of the selected towns on drinking water management and governance.

Particulars	Kalimpong	Singtam
Similar elements/aspects		
Location of the municipalities	A hilltop town dependent on springs, spring-fed rivers and rivulets for domestic water, for hospitals, hotels/restaurants, government offices, Indian Army and others	Town located at the bank river Teesta and Ranikhola, dependent on springs, spring-fed rivers and rivulets for domestic water, for hospitals, hotels/restaurants, government offices and others
Water sources	Sources of water are around 86 km away at Neora Forest village	Sources of water are around 7 km away, at Sang-chu, in Martam–Nazitam area
Agencies involved in water supply	The source at Neora forest is under the State government – Neora office The water is then brought in through pipelines to the reservoir at Deolo. The water distribution is then under Gorkha Territorial Administration (GTA) and Kalimpong municipality and is implemented through Public Health Engineering Department (PHED)	PHED with nagar panchayat are responsible for water distribution and supply
Contrasting elements/aspects		
Number of wards and area	Larger municipal area with adjacent semi-urban villages within the municipalities	Small town area, urbanized wards
Water distributions	Water source, connection from the source to the reservoir, and maintenance etc. are carried out by state government department. PHED under GTA is responsible for water distribution and supply. So differences in coordination in the convergence between agencies	Only PHED is responsible for water distribution and supply. The state government allocated funds for repair, maintenance and source development. PHED utilizes the fund in close coordination with nagar panchayat (elected representatives of town)
Water supply during scarcity and crises	Tankers form an alternative economy and water source for those who can buy water. Walking to nearby springs is an alternative for many. Small-scale rainwater harvesting is also carried out in many households	Around 60% household have privately owned pipes connected to natural springs in addition to PHED supply. For those do not have alternative arrangements, PHED would supply through tankers, and immediately repair and maintain distribution system
Water demand	Water demand is high as the population is a little over 50,000. The Kalimpong is comparatively dry area and receives less rainfall. Springs are drying, and the discharge is declining	The demand is still less for around 25,000 people. There are other water sources to tap water as an alternative to meet the increasing demand
Governance on water management	Political instability, frequent strikes, two different government authorities (GTA and the state government) with differences in resource mobilization and maintenance, distribution and supply of water, etc. water infrastructure set up at the time of British rule for a smaller population. No significant change in water infrastructure for about 30 years now	Stable government for the last 25 years, PHED department looking after water distribution and supply in close coordination with the nagar panchayats. As the resident, as well as floating population, is increasing, the government PHED is looking for an alternative water sources to meet the future water demand

Supply and Maintenance Division (NWSMD) at the state level. The PHED works with the Kalimpong municipality to facilitate the water distribution. As per the KIIs, water was brought to Kalimpong from 1991 onwards from the Neora River. This source required a large capital investment, which was met by

Table 3. Water sources and supply systems for Singtam.

Water source	Public/private	Number of supply pipes	Number of dependent households	Status of spring source, and effects
Kali Khola, Sirwani–Chisopani GPU	Privately owned poly-pipes	80	400	Water sources at Chisopani are drying, and as a result farmers have stopped wet paddy cultivation
Andheri Khola, West Pendam GPU	Privately owned poly-pipes	30	150	Spring catchments of the Andheri Khola have been disturbed; springs drying due to prolonged dry periods
Simana Khola, Sumin Linchey GPU/West Pendam GPU	Privately owned poly-pipes	60	300	Simana Khola originates from the Sumin Forest area, and the water sources are intact due to its wider catchment area of >500 hectares
Seti Khola, Namphing GPU	Privately owned poly-pipes	102	510	The sources of Seti Khola have been disturbed due to road-building, mudslides, and is drying rapidly
Groundwater	Bore-wells	13	78	Not known
Nazitam, Martam Khola, Martam–Nazitam GPU	Steel pipes laid by the PHED	2	1,000	The water discharge at Nazitam is gradually declining due to prolonged dry periods during November–April

Source: Authors' survey.

(a) (b)

Fig. 6. (a) Poly-pipes privately connected across the Teesta by residents of Singtam, drawing water from Lower Namphing. (b) Bore well at East Point Secondary School, Golitar.

the Government of West Bengal, with the help of the Indian Army. As a result, the Neora water was distributed between the Indian Army and the civilian population of Kalimpong.

According to the NWSMD, the Kalimpong municipality has a water demand of 2.3 mgd. Of this, Neora provides 1.6 mgd, 0.8 million gallons of which is distributed to the civilian population and 0.7 million to the Indian Army, and the remaining 0.1 million gallons to villages en route to the Deolo reservoir. The remaining 0.7 mgd used to be drawn from Thuckchok Relli. However, according to the NWSMD, and as stated in KIIs by around 95% respondents, the water level has been depleted over the course of about 5 years and the supply has fallen to about 0.4 mgd.

As per records available from the PHED office in Kalimpong, NWSMD supplies no more than 0.35 mgd for a population of over 50,000 or about 31.5 litres per person per day. This has resulted in a severe water crisis, as the current demand for water is about 1.5 mgd. Moreover, 40,000 gallons per day are to be given to the sub-divisional hospital. During the dry season, the water received from Neora Khola is only 0.35 mgd. This created a hue and cry as PHED could only supply water once in 3 days, and during the dry season once in over 5 days. Some residents collect water from natural springs, while most buy water from the markets, brought on tankers, which in turn bring water from springs from far-off places (Figure 7(a) and 7(b)). Based on KIIs with concerned officials, the differences in numbers pertaining to the volume of water supplied from the source as narrated by the two departments reveal the differences in data that exist between the two departments and a lack of good coordination.

4.2. Demand–supply gap

A basic issue that arises regarding the discussion of the demand–supply gap is the nature of the narrative about the water supplied to the households, market, and institutions. This is because the quantity of water supplied is often dependent on networking and the purchasing power of the households. It is usually difficult to derive an honest estimate in such cases, as people are conscious of what is legal and what can be problematic.

The water consumption patterns of households in Singtam and Kalimpong and their respective water sources are given in Table 4. Daily water consumption per household in both towns as per the household survey is presented in Figure 8. The majority of households requires 201–500 litres of water per day. When we triangulated the information extracted from KIIs, FGDs, and household surveys, the data on water storage and consumption mismatched, as many have illegal connections, and more than one municipal water pipeline.

KIIs revealed that 82.9% of the households in Singtam are dependent on the municipal supply; however, they do not get water in their taps throughout the day during the dry season (January–May). While household respondents reported that many receive water every alternate day, for some households, it comes every 2–3 days, for 20–30 minutes each time. About 55% of the households receive PHED water once in 3 days and even less frequently during the dry season. Consequently, around 70% of these households have to depend on alternative water sources, with the water brought through poly-pipes from different spring sources to their houses. Another 18% of the households, without PHED connections, are fully dependent on spring water transported through poly-pipes. This has been the situation in Singtam in recent years, although the water distribution system is being improved.

Fig. 7. (a) Water distributed by tankers at Kalimpong. (b) Supply pipes to Kalimpong, PHED's and privately owned.

Table 4. Uses of water in Singtam and Kalimpong (%).

	Singtam nagar panchayat			Kalimpong municipality		
Water source	Drinking/cooking	Washing	Bathing	Drinking/cooking	Washing	Bathing
Municipal supply	50.4	50	50	49.32	67.32	76.55
Rivers	1.1	3.6	4.3	3.87	6.38	5.32
Curtailed water[a]	0.7	1.1	1.1	0.92	2.43	0.89
Springs/dhara	47.5	44.9	45.6	39.23	21.45	11.78
Water tankers	0.3	0.4	00	6.75	2.51	5.56

[a]Curtailed water: Reserved water fetched from distant springs or neighbours.
Source: Authors' survey.

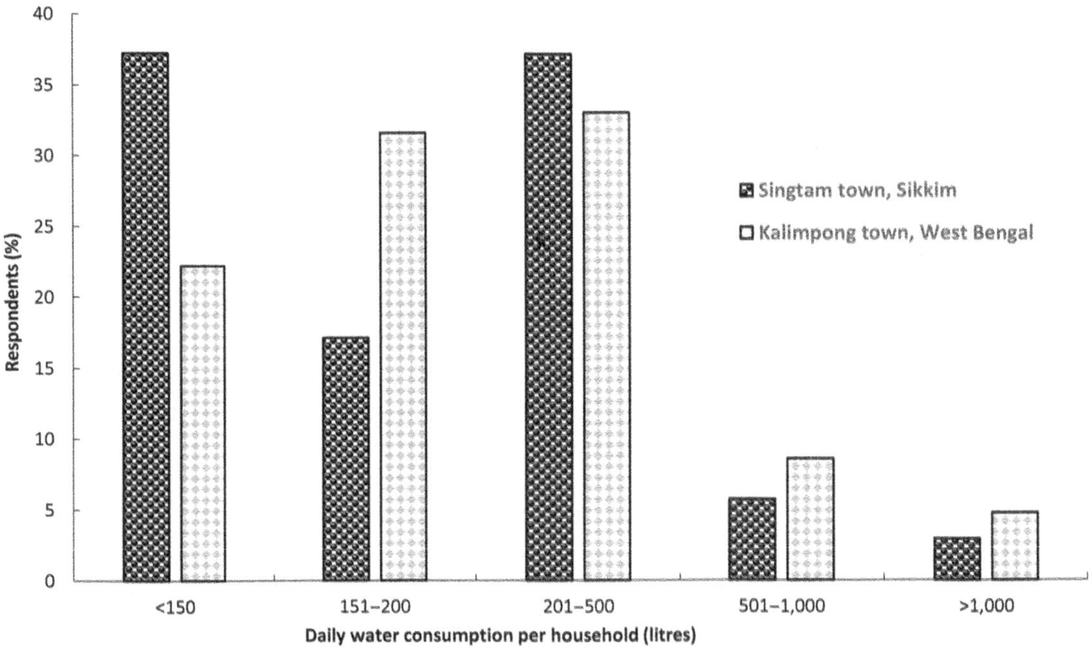

Fig. 8. Daily water consumption per household (litres). *Source*: Authors' survey.

There is no system of water collection from a common public tap in Singtam; households either have PHED connections or get water through poly-pipes, or both. As reflected in FGDs, as an alternative solution for better-off households at Golitar 13 private bore-wells have been constructed for their domestic use. The Manipal Institute of Technology at Majhitar has installed around six bore-wells, while the majority of pharmaceutical companies in Sikkim has been drawing water through bore-wells. The Singtam Fruit Preservation factory has constructed two bore-wells and bottles mineral water.

Around 91% of the household respondents reflected that some homes have two municipal pipes, or get water for longer hours in comparison with other households in addition to a poly-pipe connected to springs. Seventy-four per cent of KIIs and 95% household respondents disclosed that almost all hotels, restaurants, and elite households have more than one PHED connection, apart from water brought from springs on poly-pipes. In terms of the frequency of water supply, it varies across the seasons and between households. Interestingly, KII informants were adamant about disclosing such a situation.

There is a growing water crisis – a result of the growing population, climatic changes, erratic rainfall, and earthquakes that damage spring supply infrastructure – in Singtam. Furthermore, the Ramphu–Singtam–Dikckhu highway area is dotted with 12 pharmaceutical companies, 3 hydropower projects, and other factories and industries. A majority of their working population resides in Singtam. Therefore, the water crisis in Singtam is going to increase manifold. Water scarcity is definitely a concern for 60% of the residents, with their water supply fluctuating sharply, from getting the PHED supply daily, to receiving it once in 2–3 days or even more infrequently during the dry season.

The respondents in the KIIs, FGDs, and household surveys reflected that Kalimpong, being a larger town than Singtam and set against political assertion/instability and unrest, has become a place where the people's need for water has been sidelined in the discourse around a larger community assertion of identity politics. Partly due to this, water supply has become problematic and households are forced to take up other alternative options (Figure 9). What is ideally stated to be the system of water distribution at the top levels of governance is not what trickles down to the household level, as was evident during the course of this research with FGDs and KIIs.

In the discourse around water distribution and supply in Kalimpong, different stakeholders present varied narratives, as was expected. The responses from the PHED and NWSMD officials were analysed and they stated that adequate water is supplied by them from the source to the water reservoir. A civil society group comprising concerned citizens also reiterated this view of adequate water being supplied to the municipality. However, all the household respondents reported that with outdated infrastructure and unplanned constructions, the PHED could not distribute water proportionately to all the locations and populations.

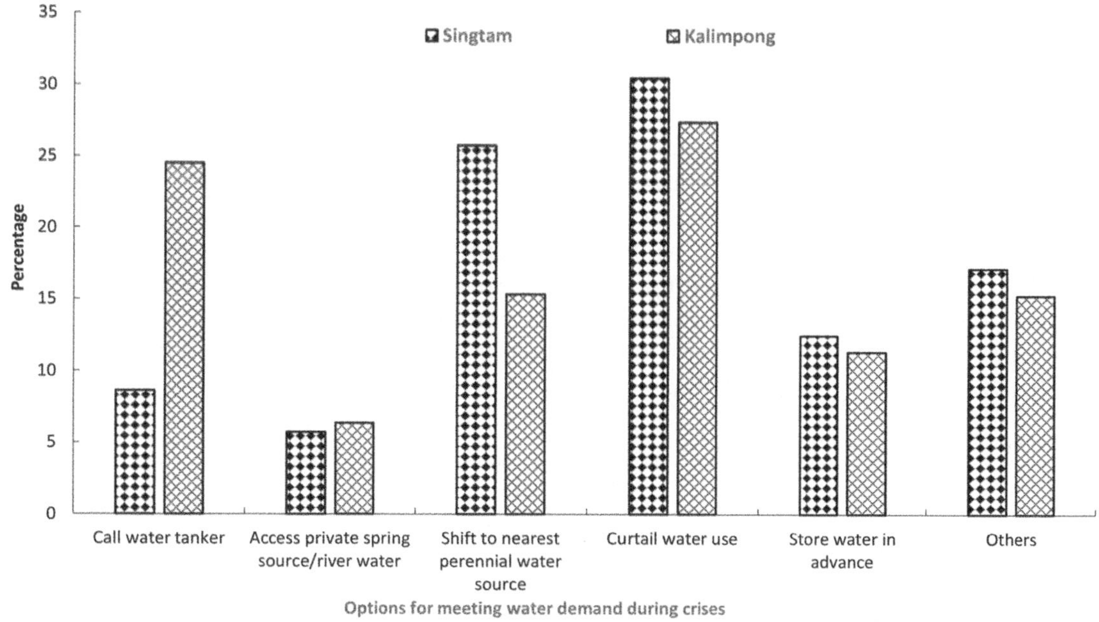

Fig. 9. Responses during water crises. *Source*: Authors' survey.

During FGDs and informal conversations during household surveys, people raised issues mainly related to poor governance and mismanaged water distribution. The main reasons household respondents of both the towns mentioned were lack of conservation, resulting in degradation due to development pressure (28%); increased demand for and overexploitation of sources (26.5%); land use change in spring catchments (24.5%); decreased rainfall and increased number of dry months (17%); and poor and inefficient water infrastructure (4%) (Table 5).

When we interacted with an elderly woman at Kalimpong about the current situation of water, she said

> '*in olden days we had several spring sources nearby with sufficient water discharge, we had lush green forests all around and had uniform rainfall year round. These days population has remarkably increased, forests have been lost, climate is altered and water in the springs have also disappeared. Bad time is approaching.*'

Kalimpong municipality comprises urban and peri-urban ward distribution. Households at peri-urban wards located away from the main roads restated the lack of municipal water connection. This is because pipelines have been cut due to poor infrastructure and the lack of planning processes. Additionally, like in Singtam, hotels, restaurants, and higher income households pay more money to get access to more water, thereby creating an imbalance and greater water shortages for people from marginalized groups and locations.

Households at Kalimpong also informed us that water supply is only around 30–40 litres per person per day, less than the per capita norm of 135 litres per day as set by the municipality. During critical moments of water scarcity, households manage to meet their water requirements through water tankers, private spring sources/river water, fetching water physically from distant springs, storing water in advance, or shifting to other perennial springs.

4.3. Health impacts due to poor water quality

According to Singtam's residents, water supplied by PHED is mostly *ledo* (muddy) and *dhamilo* (turbid) during the monsoon, when they face problems of scarcity and poor water quality. The Singtam District Hospital officials reflected that the presence of contaminants in the water has been affecting people's health adversely, including gastrointestinal illness, diarrhoea and vomiting, cholera, and dengue. The situation is yet to improve, as there is no functional water treatment plant for the storage tanks installed at Chisopani. They added that around 4,857 patients visited the hospital with diarrhoea during January–September 2018 (average 500 patients per month), confirming that water supply to the town is contaminated throughout the year. Cases of typhoid and contamination of coliform bacteria were found in the water in and around Singtam.

Table 5. Responses on rising scarcity of water.

Indicators of water scarcity	Singtam	Kalimpong
Decrease in rainfall; increase in the number of dry months	20	14
Increased demand, overexploitation of spring sources	26	27
Poor and improper water infrastructure	5	3
Land use changes in spring catchments	24	25
Lack of conservation; degradation due to development pressure	25	31

From our interaction with health officials, we know that water quality in Kalimpong too declines in the monsoon, and people face all the illnesses mentioned above, and also dysentery and the flu. Coliform bacteria were found in several sources of drinking water in Kalimpong. According to the Health Officer of the Civil Hospital at Kalimpong, the authorities had declared the water from these sources as unsafe for drinking, but acceptable for washing and other domestic purposes (Ravidas, 2012). The contaminated water in Kalimpong has increased the number of patients with bacterial infections such as typhoid, paratyphoid, cholera, bacillary dysentery, and amoebic dysentery.

4.4. Issues and challenges of water management

KII respondents argued that the key challenge Singtam faces is its floating population that is fast increasing due to the rapid industrialization along the Ramphu–Singtam–Dikchu stretch in the Teesta River basin and, therefore, water demand is going to increase manifold. Heavy traffic congestion, increasing air, water, and noise pollution, the incidence of waterborne diseases, and lack of proper management of sewerage are issues that need immediate attention. According to the KII, FGD, and household respondents, Singtam is soon going to face water shortages due to a planned road construction approved by the Sikkim government in the Lower Namphing forest, which supplies water to more than 110 households in the town, and hence it is anticipated that its supply will cease. The current water sources at Nazitam-Namphing were reported to be in a declining trend.

In the case of Kalimpong, KII respondents said that the current water supply system was designed and built around 65–80 years ago, and is incapable of catering to the needs of the increasing population of the present-day Kalimpong. Pipelines and storage tanks are not repaired periodically, and thus the feeder and supply pipes are rusted, twisted, and too weak to hold sufficient water. The water sources lack conservation and protection of their catchments. A modern metering system has not yet been established so as to maintain optimum water use efficiency and accountability and to avoid the illegal tapping that occurs at several places. There is a lack of regulation over water consumption. The town also lacks adequate water harvesting systems.

5. Governance, local bodies, and political priorities

5.1. Parallel governance vs. uniform governance

The FGD participants and KII respondents recounted that *andolan* (political unrest/movement) over the demand for Gorkhaland gained momentum again under a new political party, Gorkha Jan Mukti Morcha, and led to the formation of the Gorkhaland Territorial Administration (GTA) in 2012 in Darjeeling. The GTA has administrative, executive, and financial powers (but no legislative powers, which are vested in the Government of West Bengal). The system of parallel governance, i.e. control and authority over the same resource by both GTA and Kalimpong municipality, has resulted in a constant power struggle and mismanagement of the water supply system in Kalimpong. The constant conflict between the state government and GTA pushes issues and agendas of development and change into the background, and control and power take precedence. Transporting water from the source to the reservoir falls under the ambit of the state government. Subsequently, the distribution of water from the reservoir and to the different wards in Kalimpong falls under the PHED, which works with the local municipality. FGDs and KII revealed that this division of responsibility leads to miscommunication, with the two bodies often working in isolation, rather than in convergence. The two departments also blame each other for the shortage of water supply and scarcity. Comparably, household respondents argued that it is the local residents who suffer, as they do not know who is accountable for their grievances. Several areas do not have municipal water pipes, such as wards 15 and 17. Respondents and participants in FGDs said this was because of the *andolan* in the late 1980s, which, they say, led to the destruction of some of the water infrastructure, which was lost or sold by the agitating youth at the time.

KII respondents contemplated that in the situation of political unrest and growing unemployment in Kalimpong, the water shortage has also generated employment for many, as tank water suppliers, plumbers, and drivers. There exists a formal water market registered as the Kalimpong Water Supply Drivers' Welfare Association, which 'is operated, often in monopolistic ways, by water vendors and taxi drivers who work under the patronage of local political leaders' (Joshi, 2014). The supply of water has become less about citizens' daily needs but generates a new terrain of political tussle and control over it.

Comparatively, contrast is the system of governance in Singtam. The water distribution and supply systems have a decentralized approach involving the PHED and the Singtam nagar panchayat as per the FGDs and KIIs. Singtam is an example of uniform governance (close coordination and convergence), which operates from the government level (the PHED) to the local municipal level. Thus, the system of governance in Singtam ensured accountability, which in turn translated into more equitable water distribution, relatively regular water supply, and conflict management.

However, Singtam also has water issues; governance here is focused mainly on water supply, but it also needs to focus on policies that will cope and deal with the growing urban population, the drying of springs, and the occurrence of disasters such as landslides and earthquakes, which adversely affect water connections. Remarkably, there is a new system of private water supply, for meeting increasing demand at the households, largely unregulated poly-pipe water suppliers who are outsiders and do not own water sources. They are paid for hanging poly-pipes from the spring source to the households and monitoring the supply.

The water governance in the towns of Kalimpong and Singtam in Sikkim and West Bengal state of India shows complex socioecological dynamics with examples of cross-scalar policy challenges which particularly impact with increased environmental vulnerabilities. The policy responses failed to tackle escalating environmental and developmental challenges. It is, therefore, important to integrate approaches to address these challenges, emphasizing an emancipatory multi-scalar politics that has the potential to open up sustainable pathways in the context of dynamic social and ecological changes in the Himalayas (Satyal *et al*., 2017).

6. Socioeconomic backgrounds and correlation with water distribution

6.1. Whose issue is it anyway?

Issues of access, water shortage, its distribution, and concerns surrounding it are directly related to the socioeconomic background of households. The purchasing power of households, or lack thereof, is central to water shortages and the water economy. The officially unmentioned water economy in the region often came up in FGDs, KIIs, and even in informal discussions. This was especially evident in

Kalimpong. However, a study (Joshi, 2014) also states that access to water is 'highly differentiated and not only financial', but also influenced by a host of other intersecting disadvantages based on 'class, caste, ethnicity, and personal fortune in terms of marriage, life partners, and children'. Another major dynamic in the access to water is politics. Given the volatile political situation, local politicians have established strongholds in certain pockets and among certain groups of people, which influences the access to, and control of, resources.

Singtam is a compact urban town spread over a small area. People in Singtam mostly operate local businesses – shops, small restaurants, and canteens – or work in pharmaceutical factories and hydropower projects established along the Teesta River. A few of the migrants coming into the city are from rural areas of Sikkim. There are also an alarmingly increasing number of migrants from outside Sikkim, mostly from the plains of West Bengal, Bihar, and Uttar Pradesh.

6.2. Adaptation and coping

6.2.1. The growing water economy in Kalimpong and Singtam. During interviews and in FGDs, people said that those with high purchasing power in Kalimpong buy more than one pipeline with municipality water, and thus there is unregulated selling of water pipelines. Many residents spoke of a growing water economy, benefitting certain groups of people. It is said to work at all levels, from water tankers to local PHED sources. Many respondents blamed the inefficiency of the local government and specifically the role of plumbers working under the department. Buying a PHED line, it was said, costs roughly INR 20,000 or more. One such connection can supply water for about 20–30 minutes, daily during the monsoon, and once or twice a week during the dry seasons. However, several households had access to PHED water throughout the year, even during the dry season, as many households were capable of buying multiple PHED connections. Hence, although water shortage is a huge issue, the impacts of this shortage are felt by different households differently. During times of water shortage in Kalimpong, the most convenient option is calling for a tanker. Around 1,000 litres of water, the minimum amount of water that can be bought from tankers, costs INR 280–350. Even households that are unable to afford the tankers themselves access them by distributing the costs, and the water, among two or more households. Of the households surveyed, 36.6% admitted to calling for water tankers. However, it must be borne in mind that the households surveyed were largely from lower income groups, who are far more dependent on spring sources. Alternative sources, if any, may be available only to those with economic, social, and political resources. This leaves a high proportion of lower income households, migrants, and the slum population at a disadvantage. The fact that water can be bought by many is the primary reason why pressure has not been applied to political parties, governments, and policy-makers to implement sustainable solutions.

In Singtam, payment for water to private vendors and to the PHED started 5 years ago, although a metering system is yet to be introduced. Besides the special connection lines (with a continuous flow of water) provided to the district hospital and the fire brigade, some elite households have been given special lines, resulting in an unequal distribution of water by PHED. Those households with private poly-pipes pay around INR 1,000–1,500 per month per poly-pipe, depending on the distance and amount of water at the source. Unfortunately, and interestingly, the payment goes to illegal water vendors, mostly unskilled fitters from Islampur and other areas of West Bengal, who tap water from the forests and monitor supply to households.

This capacity of the better-off and dominant social groups to purchase extra water via PHED lines or tankers not only leaves a large section of the lower income population more vulnerable, but it also puts future generations at the risk of absolute water shortages.

6.2.2. Catering to the floating population. There is an increasing floating population (approximately 8,000–10,000 people) in Kalimpong, which is due to tourism and other businesses. Kalimpong derives much of its revenue from tourism. An important change in Kalimpong over the last 5–7 years has been the rise in homestays, new hotels, and restaurants attracting a new generation of tourists. The homestays are located in peri-urban areas as most of them are styled as local cottages and have spring sources of their own. Hotels and restaurants here cannot legally have PHED water connections as they are only meant for household residents. The representative from the hotels said that running a successful hotel in Kalimpong required about 10,000–12,000 litres of water every day, which is impossible to get out of a PHED line. Many of the hotels and restaurants located in the town deploy their own trucks to transport water for the sole purpose of running the establishment. Their owners are obliged to buy trucks, jeeps, petrol, and employ drivers and helpers. On average, about INR 0.63 million are spent a year by the hoteliers in Kalimpong municipality on getting water. The issue of the right to water becomes an issue of purchasing power.

Singtam has a floating population of over 6,000–7,000 persons per year, which is rapidly increasing, with 2,000–3,000 vehicles plying every day. Singtam is not a tourist destination, but a hub for industries, pharmaceutical companies, educational institutions, automobile workshops, factories, and hydropower projects established close by, all of which bring in a significant working population. People migrate here from all four districts of Sikkim, and even from neighbouring states, in search of employment. It also functions as a trading junction for Sikkim and is a business gateway for agricultural produce such as large cardamom, vegetables, and other cash crops. So the population in Singtam has always been 'floating' in nature and there is an increasing demand for water.

7. Lessons from Singtam

7.1. Micro-planning

Both Kalimpong and Singtam have to plan for their growing populations, ensure regular water supply systems from reliable sources, and carry out revival measures for the drying springs. But because of its small size, a uniform government, and active policies, Singtam can provide lessons to Kalimpong and other Himalayan towns on several counts. Solutions adopted locally can contribute towards meeting immediate water demands. Water shortage, although a critical issue, is not faced equitably by all households. Hence, planning solutions should be done at the micro level, in order to ensure the inclusion of the major issues and shortcomings of the existing water system. Planning and policies passed at the state level based on urban planning may not be applicable or appropriate at the local level in hill towns. An example of this is the lack of communication between the state authorities and the GTA regarding Kalimpong.

Kalimpong's 23 wards face different issues regarding access to water. Some wards lack PHED pipelines, whereas some are more dependent on springs; all this needs to be taken into account. Singtam, being a smaller municipality with six compact wards, follows a more efficient form of water governance. Not just because of its smaller size, but also the proximity of the municipality to both local institutions and the state capital, Gangtok. Local representatives have access to higher levels of government. Micro-planning is, however, required to address the increasing migration and the growth of slums, which are the most vulnerable to water scarcity and policy marginalization.

7.2. Dhara Vikas and adaptive options to revive drying springs

The RMDD, Government of Sikkim, with its partners since 2008, TMI India, Advanced Centre for Water Resources Development and Management, Pune (ACWADAM), People's Science Institute, Dehradun, and WWF India developed an atlas of approximately 700 springs, and initiated *Dhara Vikas* in the dry areas of South and West district of Sikkim for reviving drying springs. The programme has provided a number of learning experiences that can be replicated in the other parts of the Himalaya. An important aspect of *Dhara Vikas* is that local communities are involved at the village level in convergence with the Mahatma Gandhi National Rural Employment Guarantee Act (MGNREGA). Sharma *et al.* (2019) conducted a study in the Chibbo-Pashyor watershed (905–1,325 m) in Kalimpong for identifying drying springs for revival. In total, 12 out of 54 springs were identified for long-term monitoring; roughly 439 households in Chibbo-Pashyor villages depend on them (Sharma *et al.*, 2019).

8. Conclusion and ways forward

This study has focused on issues of demand and supply, water scarcity and stress, equity, water governance, and the sustainable conservation and management of water resources in a climate change context in two towns in the Eastern Himalaya, Singtam and Kalimpong. It finds that spring sources are drying in Singtam, and Kalimpong faces acute water shortages through the year. Springs, streams, and rivulets have provided water to these two towns for centuries. However, they are being gravely impacted by new stressors in the form of climate change and other anthropogenic pressures. These towns are also having to cope with a rising floating population of tourists and industrial workers.

The revival of traditional sources of water is extremely important for the sustainable conservation of water resources in the Himalayan context. In the recent past, the NITI Aayog, the NWP, and the NMSHE have all emphasized springshed development for the conservation and revival of declining springs. However, the benefits of these missions are yet to materialize on the ground in the Sikkim Himalayan region. The National Rural Drinking Water Programme Guidelines 2013 has mentioned that, by 2022, every rural person in the country would have access to 70 lpcd within their household premises or at a horizontal or vertical distance of not more than 50 m from their home. One hopes that this would benefit Kalimpong and Singtam.

The adoption of appropriate soil and water conservation practices on a watershed basis (*in situ* conservation, adopting soil conservation practices like contour bunding, terracing, construction of check dams, gully control structures, providing farm ponds, and digging trenches to collect rainwater and utilize it for supplemental irrigation) is considered to be the only way to control soil erosion and improve the environment in the catchments of the springs through successful springshed management activities. Over the past 10–15 years, safe and uninterrupted drinking water supply has become hard to find in Kalimpong due to the poor performance of water supply schemes, whereas the water supply and sanitation situation is continuously degrading even in Singtam. If governments at all levels seriously wish to counteract the plight of the urban populations and prepare for existing water-related vulnerabilities due to climate uncertainties, they will need to invest vigorously in expanding services to reduce the isolation of these towns from the main developmental agenda of the central government and the respective states. There is a need to promote coordination and cooperation among the concerned agencies, with targeted investments. The development of water security plans and their strict enforcement through multi-institutional collaboration can contribute to improved water governance and socioecological restoration for sustainable water resources management. Planning should be contextual, and a system of micro-planning is required in Kalimpong, Singtam, and other hill towns throughout the Himalaya.

Acknowledgements

The authors are indebted to the community members and key informants in Kalimpong and Singtam town for their knowledge, involvement, and information. We thank ICIMOD, Kathmandu, for funding and TMI India for facilities. We thank nagar panchayats, PHED officials, community members, and Health Department officials at Singtam and Kalimpong for information.

This work was carried out by the Himalayan Adaptation, Water and Resilience (HI-AWARE) consortium under the Collaborative Adaptation Research Initiative in Africa and Asia (CARIAA) with financial support from the United Kingdom's Department for International Development (DFID) and the International Development Research Centre (IDRC), Ottawa, Canada. This work was also partially supported by core funds of ICIMOD contributed by the governments of Afghanistan, Australia, Austria, Bangladesh, Bhutan, China, India, Myanmar, Nepal, Norway, Pakistan, Sweden, and Switzerland. The views expressed in this work are those of the authors and do not necessarily represent those of DFID, IDRC or its Board of Governors; nor are they necessarily attributable to the authors' affiliated organizations.

References

Benerjee, S., Mukherjee, A., Sattar, A. & Biswas, B. (2015). Change detection of annual temperature and rainfall in Kalimpong station under Hill Zone of West Bengal. *Indian Journal of Hill Farming 28*(2), 81–84.

Datta-Ray, S. K. (1984). *Smash and Grab: Annexation of Sikkim*. Vikas Publishing House, Delhi.

DST (n.d.). *National Mission for Sustaining the Himalayan Ecosystem*. Department of Science & Technology, Ministry of Science & Technology, Government of India. Available at: http://www.knowledgeportal-nmshe.in/

EPDOGS (2015). *Energy and Power Sector-Vision 2015*. Energy and Power Department, Government of Sikkim, India, p. 21.

Ganguly, R. (2005). Poverty, malgovernance and ethnopolitical mobilization: Gorkha nationalism and the Gorkhaland agitation in India. *Nationalism and Ethnic Politics 11*(4), 467–502. doi:10.1080/13537110500379286.

Government of Sikkim (n.d.). *Socioeconomic Census 2005–06*. Department of Economics, Statistics, Monitoring & Evaluation, Government of Sikkim. Available at: https://sikkim.data.gov.in/catalog/socio-economic-census#web_catalog_tabs_block_10

Gupta, A. & Kulkarni, H. (2018). *Report of the NITI Aayog Working Group on Inventory and Revival of Springs in Himalayas for Water Security as Part of Initiatives on Sustainable Development of Mountains of Indian Himalayan Region*. Submitted to NITI Aayog, Government of India.

Hackett, P. G. (2008). *Barbarian Lands: Theos Bernard, Tibet, and the American Religious Life*. Doctoral Dissertation, Columbia University, New York. Available at: https://library.columbia.edu/content/dam/libraryweb/locations/eastasian/Mirror_Hackett.pdf

Huber, A. & Joshi, D. (2015). "Hydropower, anti-politics, and the opening of new political spaces in the Eastern Himalayas." *World Development 76*: 13–25. doi:10.1016/j.worlddev.2015.06.006.

IndiKosh (2018a). *Singtam Nagar Panchayat*. Available at: https://indikosh.com/city/274603/singtam

IndiKosh (2018b). *Kalimpong Municipality*. Available at: https://indikosh.com/city/322796/kalimpong

Jha, P. K. (1985). *History of Sikkim (1817–1904): Analysis of British Policy and Activities*. GPS Publication, Calcutta.

Joshi, D. (2014). Feminist solidarity? Women's engagement in politics and the implications for water management in the Darjeeling Himalaya. *Mountain Research and Development 34*(3), 243–254.

Kerry, L. (2008). Lepcha narratives of their threatened sacred landscapes. *Transforming Cultures 3*(1), 227–255.

Kumar, D. N. (2017). Water management in India: the multiplicity of views and solutions. *International Journal of Water Resources Development 34*(1), 1–15. doi:10.1080/07900627.2017.1351333.

Lepcha, T. (2015). Hydropower projects on the Teesta River: conflicting against mega dams in Sikkim. In: *Water Conflicts in Northeast India*. Joy, K. J., Das, P. J., Charkraborty, G., Mahanta, C., Paranjape, S. & Vispute, S. (eds). Routledge, Tylor and Francis Group, London and New York, pp. 232–242.

MHA (2011). *2011 Census Data*. Office of the Registrar General & Census Commissioner, India, Ministry of Home Affairs, Government of India, Delhi. Available at: http://censusindia.gov.in/2011-Common/CensusData2011.html

MoWR (2002). *National Water Policy*. Ministry of Water Resources, Government of India, Delhi.

MoWR (2012). *National Water Policy 2012*. Ministry of Water Resources, Government of India, Delhi. Available at: http://mowr.gov.in/sites/default/files/NWP2012Eng6495132651_1.pdf

Mukherji, A., Molden, D., Nepal, S., Rasul, G. & Wagnon, P. (2015). Himalayan waters at the crossroads: issues and challenges. *International Journal of Water Resources Development 31*(2), 151–160. doi:10.1080/07900627.2015.1040871.

ODM (n.d.). *Kalimpong Municipality*, Office of the District Magistrate Kalimpong, West Bengal. Available at: https://kalimpongdistrict.in/kalimpong-municipality/

Pradhan, K. (1991). *The Gorkha Conquests: The Process and Consequences of the Unification of Nepal, with Particular Reference to Eastern Nepal*. Oxford University Press, Kolkata. Available at: www.himalbooks.com (Accessed 16 October 2018).

Rahman, H., Karuppaiyan, R., Senapati, P. C., Ngachan, S. V. & Kumar, A. (2012). An analysis of past three decade weather phenomenon in the mid-hills of Sikkim and strategies for mitigating possible impact of climate change in agriculture. In: *Climate Change in Sikkim: Patterns, Impacts and Initiatives*. Arrawatia, M. L. & Tambe, S. (eds). Information and Public Relations Department, Government of Sikkim, Gangtok, pp. 1–18.

Ravidas, R. (2012). Water every three days: Kalimpong supply hit, sources infected with bacteria, 23 March. Available at: https://www.telegraphindia.com/states/west-bengal/water-every-three-days-kalimpong-supply-hit-sources-infected-with-bacteria/cid/446849 (Accessed 25 October 2018).

Satyal, P., Shrestha, K., Ojha, H., Vira, B. & Adhikari, J. (2017). A new Himalayan crisis? Exploring transformative resilience pathways. *Environmental Development 23*, 47–56.

Seidler, R., Sharma, G. & Telwala, Y. (2016). Climate vulnerability, water vulnerability: challenges to adaptation in eastern Himalayan springsheds. In: *Developments in Earth Surface Processes Volume 21: Mountain Ice and Water; Investigations of the Hydrologic Cycle in Alpine Environments*. Greenwood, B. G. & Shroder Jr., J. F. (eds). pp. 280–303. doi:10.1016/B978-0-444-63787-1.00007-X.

Sharma, G. & Pandey, T. (2018). Harnessing energy potential in fragile landscapes: exploration of conflicts and emerging issues around hydropower development in Sikkim. In: *Water Conflicts in Northeast India*. Joy, K. J., Das, P. J., Chakraborty, G., Mahanta, C., Paranjape, S. & Vispute, S. (eds). Routledge, Taylor and Francis Group, London and New York, pp. 50–69.

Sharma, G., Sharma, D. P. & Dahal, D. R. (2012). *Adaptive Approaches for Reviving the Dying Springs in Sikkim*. The Mountain Institute India, Gangtok.

Sharma, G., Pradhan, N., Sharma, D. P., Barola, Y., Luitel, M., Luitel, K. K. & Nyima, K. (2019). *Conserving springs as climate change adaptation action: lessons from Chibo-Pashyor, Teesta River Basin, Kalimpong, West Bengal, India*. ICIMOD Working Paper, Kathmandu, Nepal.

Snyder, A. (2014). Shortage in the mountains of plenty: Water supply in mountain and hill cities throughout the Hindu-Kush Himalayan region. Available at: https://www.worldfoodprize.org/documents/filelibrary/images/youth_programs/2014_interns/2014_br_research_papers/SnyderAbigail_LONGReport_56ED38F157B76.pdf

Tambe, S., Arrawatia, M. L., Bhutia, N. T. & Swaroop, B. (2011). Rapid, cost effective and high resolution assessment of climate-related vulnerability of rural communities of Sikkim Himalaya, India. *Current Science 101*(2), 165–173.

Tambe, S., Kharel, G., Arrawatia, M. L., Kulkarni, H., Mahamuni, K. & Ganeriwala, A. (2012). Reviving dying springs: climate change adaptation experiments from the Sikkim Himalaya. *Mountain Research and Development 32*(1), 62–72.

Vaidya, R. A. (2015). Governance and management of local water storage in the Hindu Kush Himalayas. *International Journal of Water Resources Development 31*(2), 253–268. doi:10.1080/07900627.2015.1020998.

WBSAPCC (2010) *West Bengal State Action Plan on Climate Change*. Government of West Bengal, Government of India. Available at: http://www.moef.nic.in/downloads/public-information/West-Bengal-SAPCC.pdf

Received 5 December 2018; accepted in revised form 28 January 2019. Available online 9 May 2019

The pain of water: a household perspective of water insecurity and inequity in the Kathmandu Valley

Olivia C. Molden[a], Anoj Khanal[b] and Nita Pradhan[c]

[a]*Corresponding author. Department of Geography, University of Oregon, Eugene, OR, USA. E-mail: omolden@uoregon.edu*
[b]*Kathmandu Valley Water Supply Management Board (KVWSMB), Lalitpur Metropolitan City Ward No. 18, Sanbu, Bhaisepati, Lalitpur, Nepal*
[c]*Nepal GIS Society, Women Development Training Centre Complex, Lalitpur, Nepal*

Abstract

This paper draws on participatory research with 47 household water managers over the dry, pre-monsoon, and monsoon season, alongside expert knowledge of water management in Nepal's Kathmandu Valley. Doing so, it presents the perspectives of water managers to highlight three dimensions of water security that existing approaches often overlook. First, experiences of water security vary greatly between households over the year, even within a relatively small geographic area. Second, social connections and landownership play an important role in mediating these experiences. Third, coping with poor water supply places a burden on certain household members. This paper argues that addressing water inequities and insecurities demands research, development and policy responses to look beyond the main pipe network and engage with the variety of ways in which households secure water.

Keywords: Himalayan towns; Household water security; Kathmandu Valley; Qualitative research; Urban water

Introduction

Uma, a middle-aged woman, dreams of leaving Nepal's Kathmandu Valley to be in a village where there is flowing water. Like many residents, she summarizes the city's situation as '*paani ko dukkha*': a painful struggle for water. After her marriage she moved from the periphery of the Valley to a central urban area, Patan; often also called Lalitpur, Patan is just south of the capital, Kathmandu. Like most

doi: 10.2166/wp.2018.116

residents of the Valley, she faces growing water insecurities. Although her neighbours receive tap water, Uma does not have access to the main water grid system. About six times a day she walks to a community-run water source to fetch 15 litres of water each time; for 90 litres of water, she pays 80 Nepali rupees (NPR) (around 1 US dollar). This is a challenging task; as she explains, 'even when it rains, I have to go and get water and get all drenched. I have to carry water and walk. Now I am getting older and weaker.' Uma is worried because groundwater is increasingly scarce and she lacks affordable alternatives.

As stream flows and groundwater levels become more variable, over-tapped and contaminated (Salike & Fee, 2015; Saraswat *et al.*, 2017), urban residents like Uma worry about the future habitability of the Kathmandu Valley. Already, many households are not meeting their water needs: a recent study shows that the median total per capita water consumption in Kathmandu's households is below the United Nation's mandate of 50 litres per person per day for personal and domestic use (Raina, 2016). Moreover, a 2017 evaluation of water security in Kathmandu found that the public water company, Kathmandu Upatyaka Khanepani Limited (KUKL), provided much less than the 135 litres per person per day that the study recommends for economic growth (Thapa *et al.*, 2018). In the meantime, research continues to show that most water sources contain pathogens, heavy metals, ammonia, nitrates and other contaminants due to poor sewage and waste disposal (Udmale *et al.*, 2016). As climatic changes impact rainfall dynamics (Salike & Fee, 2015) and current landcover changes decrease the Valley's recharge potential (Shrestha *et al.*, 2012), institutions face serious water management challenges to ensure residents meet their needs now and in the near future.

Peer-review literature, policy documents and institutional reports about the Valley's water situation tend to rely on data from surveys of households and water sources. While this quantitative perspective is critical to understanding overarching patterns of water supply, use and quality, there is a lack of understanding about the variety of ways in which people address water insecurity in their everyday lives. These diverse stories of resource access in everyday life are important for understanding both the drivers of resource insecurity and potentials for transformation at multiple scales (Hulme, 2011; Wutich & Brewis, 2014; Loftus, 2015; Ranganathan & Balazs, 2015; Carey *et al.*, 2016; Furlong & Kooy, 2017).

As such, this paper presents household perspectives of water security. It does so by drawing on the experiences of 47 household-level water managers over 2017's dry, pre-monsoon and monsoon seasons. These participants all live relatively close to each other in newer and older areas of Patan. The goal in recruiting participants was to understand everyday life in an urban area across demographics. Once recruited, the lead author created 'story-maps' with participants over the year. Building story-maps requires constant participant feedback as photographs, maps and participant experiences are integrated into one story over time (see Supplementary Material). The paper relates the adaptive strategies and framings of household-level water managers within the broader context of water management and policy in the Valley.

Analysis reveals several commonalities. First, although experiences of water security vary greatly between participants over time, household water options are increasingly limited; see Figure 1 for a schematic of the types of water systems that households access. These constraints appear through either cost and/or time investments in storage, additional water sources and treatment methods to ensure water is available for different tasks each week. Although wealth, household location, social networks and landownership alleviate these burdens to an extent, the burden – and stress – of managing water falls heavily on certain household members. Over time those burdens widen inequities and constrain the capabilities of families, neighbourhoods and communities to build and sustain water security.

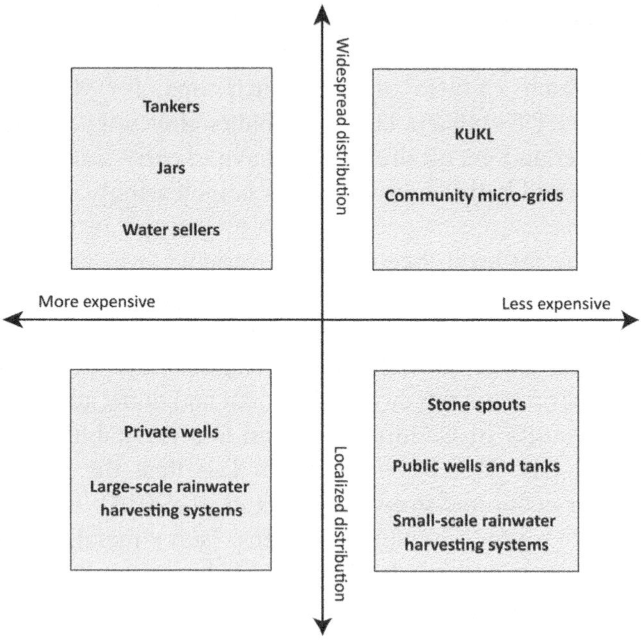

Fig. 1. Schematic of different water sources and systems available to households. © Olivia C Molden.

The following section outlines the research methodologies and approaches that inform this paper. To contextualize the study, the paper provides an overview of historic and contemporary water management issues in the Kathmandu Valley. From this background, empirical findings are presented in three sections. The first illustrates the inequities of water access by detailing the ways in which households in the study secure water over the year, focusing on an example of piped water access. The second highlights the role of social connections and landownership in household water security. The third presents the burdens placed on households and particularly on certain people within households when coping with water access. Between these findings, the growing vulnerabilities and inequities between households, especially as water access becomes more variable and expensive, are emphasized. The paper concludes with policy responses to these insights.

Methods: qualitative research insights

Prior to conducting this research, the lead author reviewed research publications on domestic water issues in the Kathmandu Valley. That review found that literature often describes patterns of water security based on the aggregated findings of large-scale surveys of households and assessments of water sources. However, there is a lack of information on the spatial and temporal variability of water security. Here, knowledge is produced within municipal and sometimes ward boundaries in-between either the dry or wet season. More fundamentally, there is a lack of in-depth research that investigates the drivers and experiences of water security in the Kathmandu Valley. Yet, examining complex environmental issues, such as water security, demands greater communication and integration between knowledge communities through multiple research approaches (Carey *et al.*, 2013; Zwarteveen & Boelens, 2014; Klenk & Meehan, 2015; Loftus, 2015; Jepson *et al.*, 2017). In the Kathmandu Valley, there is a need for more discussion based on qualitative research insights, in addition to shared knowledge production between communities, institutions and researchers.

This paper primarily draws upon insights from the lead author's year of ethnographic research in 2017 in the Patan area of Lalitpur Metropolitan City. During this time, over 150 semi-structured and informal interviews were conducted with 47 household-level water managers living in Patan during the dry, pre-monsoon and monsoon seasons. Interviews with these water managers took place over a series of interactions in 2017 in February–March, April–May and between June–August. Through these interactions, the lead author built a spatial and visual database of how water sources and storage practices changed for each household between meetings to create individualized story-maps. These story-maps synthesize visual, textual and spatial information about how participants understand, feel and respond to changes in water quality, quantity and availability. The lead author designed these story-maps to encourage participation and reflection and, moreover, to facilitate comparison, transparency and reflexivity in the research process.

To contextualize these experiences in broader institutional changes, this paper is also informed by insights from interviews with 31 experts and water managers working for governmental, non-governmental, private, community organizations during the lead author's fieldwork in 2017. The second author's expert knowledge of the governance of water and the hydrogeology of the Valley informs the insights and policy directions for this paper.

Using convenience sampling, household managers were recruited from different neighbourhoods in Patan as a case study of water insecurity in the Valley; it is important to recognize that it is likely the dynamics of water insecurity vary greatly across urban areas in the Kathmandu Valley due to the diversity of cultural histories and features in the physical landscape. The study area selected roughly encompasses a 1.5 km radius from the centre of Patan's Durbar Square outwards to Kumaripati, Balkumari, Gwarko and Shankhamul. Some participants in the study are quite wealthy while others are poor; some own their homes ($n = 33$) while others rent ($n = 19$); and some live in temporary housing due to the 2015 earthquake ($n = 4$); most participants are women ($n = 40$). (Five participants who rented provided information only for the dry season as they moved after the first or second round of interviews; consequently, only 47 participants of the 52 remained in the study throughout).

Background: situating water insecurity

Historically, cities in the Kathmandu Valley were built around cascading water distribution networks of canals, ponds (*phuku* or *pokhari*), wells (*inar*) and stone spouts or stepwells (*hiti* or *dhunge dhara*) (Becker-Ritterspach, 1995; Tiwari, 2001; Spodek, 2002; Colopy, 2012). The dynamics of these water systems vary across the landscape but in general the system catches, stores and transfers various sources of water – from rainfed springs in the Valley's surrounding forests to aquifers in the Valley floor. In addition to their cultural heritage value, stone spouts and wells still function today as vital water sources for the urban population (NGO Forum for Urban Water & Sanitation, 2009).

The longevity of these water networks and their diverse engineering techniques constitute knowledge about the natural environment, particularly of the local geology, given the Valley's zones of high and low recharge (Tiwari, 2014). In other words, the ancient water system incorporated the physical landscape and ecosystem processes. For example, people tell stories of how the worship of frogs, snakes and fish icons at water sites reflect the integration – and celebration – of life cycles in ecosystems. Here, the construction of these water systems both shaped, and were shaped by, the practices and beliefs of surrounding Newar societies (Tiwari, 2001; Shrestha & Shrestha, 2013). (Newar peoples are the historic inhabitants of the Kathmandu Valley or Nepal Mandala civilization; Newar rule over cities such as Patan or Yala in the Kathmandu Valley ended with the Gorkha invasion in 1768).

The maintenance of the ancient system relied (and to some extent still relies) on natural and cultural systems such as the Newar festivals like *sithi nakha*, where people clean their water sources in worship of water deities (Tiwari, 2001, 2002). Many communities are organized around these ancient water sites and revitalizing wells, ponds and stone spouts for both their cultural significance and utility (Regmi, 2005; Welle, 2006; UN-HABITAT, 2008; Molden *et al.*, 2016). As these practices continue today, it is important to recognize that water and water infrastructure in the Valley are more than a resource or means of distributing a resource: they embody cultural histories and changing social relationships (Tamang, 2016; Clark *et al.*, 2017).

The ancient cascading water system was radically transformed during the Rana period of government in the 1900s. Influenced and supported by the British Empire, Rana rulers imported metal pipes for elite households and stand posts for the public. In 1911, Prime Minister Chandra Shamsher Rana built the Pharping Reservoir which continues to feed pipelines supplying the Patan area. To lay down that infrastructure, which today's grid system extends from, the Rana regime set in motion a process of modernization that fragmented the infrastructure and governance of pre-existing Newar water systems (UN-HABITAT, 2008), exerting some degree of authority through the water system: stand posts appear directly adjacent to stone spouts, wells and ponds, and in other places these sites were demolished for new buildings. After the Shah dynasty regained power from the Ranas in 1951, the government expanded the piped water system for more public access (Colopy, 2012). From the 1900s to the present day, efforts to build new pipelines, roads, schools, government buildings and other infrastructure, both government and private actors have demolished and/or damaged many canals, ponds, wells and stone spouts. These developments not only diminish water availability in historic wells and stone spouts but also lead to displacement and the loss of cultural heritage (Molden *et al.*, 2016).

The metal waterpipe system initially laid down by the Ranas quickly degraded due to poor maintenance and management. Moreover, extensions and replacements of that system have been unable to keep up with increasing demands from rapid urban growth and changing lifestyles. Demands for water have increased from 35 million litres a day in 1988 to 155 million litres a day in 2000, and then to 370 million litres a day in 2015 (Thapa *et al.*, 2018). In the meantime, the extent of the built up area increased by 412% from 1989 to 2016 (Ishtiaque *et al.*, 2017). Growth in the Valley has largely been driven by rural to urban migration due to unrest during the civil war period (1996–2006) and a lack of development in rural areas. In 1990 and 2017, migrants accounted for 40% and 36% of population growth, respectively (Ishtiaque *et al.*, 2017).

Currently, the KUKL supplies 19% of demand in the dry season and 31% in the wet season, due to changes in water availability at KUKL's water pumping sites and reservoirs (Thapa *et al.*, 2018). What this means is that few households can rely on piped water. A recent study (Raina, 2016) found that, on average, households in the study area of Kathmandu only receive 1.5 hours of water every 5 days from KUKL pipes; the same study also demonstrated that poor households systematically obtain less piped water than wealthier households.

Plans to improve water supply came into motion in the early 1990s with an internationally funded scheme to privatize the water system and expand infrastructure with the Melamchi Water Supply Project, referred to commonly as 'Melamchi'. Although the Melamchi project now involves many other projects, its key feature is a 27.5 km-long tunnel to transfer water from the Melamchi basin to the Kathmandu Valley. The project promises to bring in an added 170 million litres a day, year round, from the snow-fed Melamchi River until the completion of the second phase of the project which will add 340 million litres a day from the Yangri River. Even so, models predict a 124 million litre daily deficit with the first phase of the Melamchi project, based on growing demands (Udmale *et al.*, 2016).

The Asian Development Bank approved a loan for the Melamchi project in 2001; however, it did not materialize and institutionalize until after the civil war ended (1996–2006). Moreover, due to protests at both ends of the pipeline (Domènech *et al.*, 2013) tunnel construction in Melamchi did not begin until 2009, missing its first 2007 deadline. Decades later and around US$360 million spent, the project missed its second deadline extension. In April 2018 the contractor finished building the tunnel but kilometres of pipe laying remained incomplete in the city.

In addition to building the tunnel that will divert water from mountain streams into the city's piped water system, the project required the restructuring of infrastructure and management systems within the Valley (Domènech *et al.*, 2013). After changes to the scope of the project to expand and improve piped water distribution in the Valley, institutional reform and foreign investment shifts took place in 2007 and 2008, involving the creation of three new water management entities based on a public–private model (Domènech *et al.*, 2013). These institutions are the operator of water and wastewater services (KUKL), the asset owner of those services (Kathmandu Valley Water Supply Management Board; KVWSMB) and the economic regulator (Water Supply Tariff Fixation Commission; WSTFC).

The KVWSMB is the sole authority responsible for water supply management within the valley area. It was formed from an ordinance in 2005 that was later accommodated within an Act of Parliament, the Water Supply Management Board Act, 2006 (the Act). The Act led to the formation of the KVWSMB as the authority responsible for regulation and monitoring of water supply business within the whole valley. The KVWSMB is the apex body for water supply management in the municipal areas of the valley and has issued a 30-year lease to KUKL, the main operating company responsible for piped water supply within the valley. Since its establishment, the KVWSMB has worked in the field of water supply management, groundwater regulation, identification of new water sources, artificial recharge, licensing and monitoring of small water vendors (such as tanker vendors), on social awareness campaigns and on the rapid water quality assessment of different water sources. One major goal of the KVWSMB is to ensure equitable drinking water access to all residents of the valley.

Findings

The temporal and spatial variability of insecurity

Within the span of many people's memories, water has become a highly contested and expensive resource to access. As public water sources run dry and become contaminated people must rely more on buying water. Costs include payments to KUKL, to tanker and jar companies, to neighbours, or through the installation of new technology and infrastructure, such as pumps, tanks and filters. Together, these sources and systems form a complex meshwork where boundaries of formal and informal, legal and illegal, public and private, all become entangled (Schwartz *et al.*, 2015). To help make sense of this meshwork, Figure 1 provides a schematic of these various sources based on household methods of access. As the schematic shows, households have a choice between tapping into water brought in through pipes, tanks, and jars or sources available in their compounds and neighbourhoods. For example, since Uma harvests rainwater from her roof in buckets and goes to a nearby community run stone spout, she accesses sources that are comparatively less expensive and more localized.

Due to unreliable KUKL supplies, households and communities have invested in new water infrastructures for storage, treatment and supply. Small-scale rainwater harvesting is common in households and the government now promotes rainwater harvesting and recharge on a large scale (Pandey *et al.*, 2012). Some neighbours have worked together to create small-scale water supply and distribution systems. On a slightly larger scale, several communities have created water supply systems. Depending on availability and cost, community leaders collect water from deep bores, shallow wells and tankers, and from ancient stone spouts or wells. Then, depending on membership status, people pay to collect water from public tanks and taps, from hose connection points to micro-grids, or from private tap connections to these micro-grids. Some communities also provide free water access to the broader public at certain times in the week.

Households, especially wealthier ones (Raina, 2016), increasingly rely on water brought in from the surrounding hills in tankers and jars; research in 2005 estimated that tanker and jar companies fulfil half of demands (Dixit & Upadhya, 2005). The costs of these non-grid sources are estimated to be around US$12 a month, or around 2–3% of household income (Gurung *et al.*, 2017); however, there are many other hidden costs in time spent and the cost of pumping, storing and treating water that make this a much greater burden (Pattanayak *et al.*, 2005). Cost is an important factor for research participants. Although some would like to pay more for a constant flow of clean water from taps, they do not want to keep paying high prices for water from sources like jars and tankers.

Groundwater from private wells (for homes, schools, hospitals, hotels, industries and tankers), KUKL wells, and historic community wells supply the needs of the Kathmandu Valley (Shrestha *et al.*, 2012). Industries and private groups (including community organizations) with land and capital are tapping deep aquifers through deep boring technology. Overall, 21.56 million m^3 a year are being extracted, exceeding the recharge rate of 9.6 million m^3 (Pandey *et al.*, 2010).

Research into Kathmandu's water use tends to portray households by dimensions of source access and less as a meshwork, or entanglement, of different sources that shift spatially and temporally (Schwartz *et al.*, 2015). This study finds that access to water varies because of the subjective ways people distinguish quantities and qualities of water for certain tasks, for example using jar water only for drinking, and rain water and well water for washing clothes. However, these use and storage dynamics transform, for example, in the rainy season when some people store their well water with tap water. Moreover, the amount of water households access varies greatly: some struggle daily to find enough water for drinking, washing dishes, and using the toilet. Other households worry about calling a tanker every month for their non-consumptive needs. During the dry season one participant had plenty of water in her well but the water in her neighbour's well became yellow and then ran dry.

Similar differences apply in the case of piped water. Two neighbours, Sonika and Shanta who live next to the main KUKL water supply line in Patan experience water insecurity very differently. Sonika rents a room and has access to her neighbour's KUKL connection and to the public well. Once a week she waits in line with other renters in the area and collects 100–150 litres of water from her neighbour, paying her neighbour US$2 a month for that water. Sonika feels like she never has enough water. Across the street, Shanta has two KUKL connections because he expanded his family home into two lots and receives around 8,000–9,000 litres a week for US$7 a month. Shanta only buys water from tankers after the 2015 earthquake and during the rice planting month when farmers divert water from the Pharping source to their fields. However, Shanta is concerned about the sustainability of the piped water connection as he does not trust the government or the Melamchi scheme. Shanta's experience shows that even with abundance insecurity persists due to a lack of trust and autonomy.

For households further down the piped water line, the water that comes is not as clean or as voluminous. Here, households living upstream of the pipeline and directly along the main line receive the most benefits, which coincides with more central areas such as along the road from Pulchowk to Mangal Bazar. In March, only five participants could rely solely on water from the pipes. Most participants who have KUKL access ($n = 12$) received less than 400 litres a month. However, most lack easy access at all: 11 participants lack a working KUKL connection and 19 lack direct access to a KUKL line.

As Hari explains, 'When it is Saugal's turn to get tap water [Saugal is a community downstream of the pipe], if the force of water is high then we can also get water … But to send water depends on staff of KUKL. Sometimes they do not open the tap key but usually they open the tap keys on time. Then we would use tanker water. But once a week there will be water.' There are also many areas which lack piped water connections because they are in newly developed areas and/or too far from the main line.

As the pipeline pressure is weak, many households invest in pumps and remove KUKL meters to collect more water. As such, people who can afford stronger pumps can effectively suck up more of the water that would extend further through the network. Although such practices are illegal, they are widespread and deregulated. However, for Nanichhori, a poor homeowner who lives adjacent to the main line, these forms of 'cheating' mean that she has to wait all day around the tap: 'It [KUKL water] should come thrice a week but when there is electricity the people use machine and they get water. But we don't get [any].' She says even if she could access the tap water she doesn't have a place to store the water. Overall, even with growing inequities in access between families, as for Nanichhori and Hari, the current pipe network is a source of stress for all families, albeit in different ways.

Between these diverse experiences, feelings of security and insecurity do not necessarily change between seasons of overall water abundance and scarcity. Rather, feelings of insecurity depend upon the ease of access to water sources, which includes perceived financial costs of access and storage, time and physical burdens (e.g. carrying, hand washing, filtering), and social acceptability (e.g. changing gender relations, beliefs, community membership). These costs increase as groundwater levels drop, or as public spouts run dry and taps run black or dry. As a result, families must constantly negotiate with external forces – companies, stores, syndicates, community leaders, neighbours and government officials who control water flows – which in turn reinforces power disparities.

The importance of social networks and homeownership

Homeownership is a powerful dynamic of water security. Registering a home provides residents with the ability to connect to the water mains and, if they have the means, to invest in underground or rooftop storage. Having this storage is particularly critical as it means they can pump well water, tap water or tanker water and store it for future use. Moreover, landownership often provides added income as homeowners rent out spare rooms and space or sell their land. However, not all homeowners have such benefits. Some homeowners cannot afford to move to a better area, repair earthquake damaged homes, or renovate and expand. Conversely, landless residents – such as renters, squatters and people living in informal settlements – have a more challenging time directly accessing, treating and storing water.

Whether homeowners and renters, participants were constantly adapting and changing their strategies: trying out different technologies, investing in new storage tanks, talking to neighbours about their strategies, or watching where other people collect their water. What this shows is that it is not just the ability to attach a hose to a tap or pump water from a pipeline that provides them with water security, but, more fundamentally, social connections. For example, renters like Rupa find friends that watch each other's buckets or children as they wait for water. Homeowners like Dev ask neighbours for the phone number of a more reliable tanker, while Prem invests with friends in revitalizing a community well to create new water possibilities. Or, Shanti, for example, talks about how her life has improved since 'the brothers of the community' asked her to manage the newly installed community tank. Her daily life is different now: instead of buying water and carrying buckets home, she makes a small profit from selling water and spends her time getting to know the renters and homeowners in the area.

Migrants and homeowners who have shifted to a new area explained that they feel more insecure at first because it takes time to learn about water systems. Observing other residents and making social connections improves water security. Doing so, families learn new strategies, such as using a tarp to collect rain; creating or joining a collective water collection and storage system; and acquiring more knowledge about local water sources and their management systems, such as when and where the 'key man' comes to open pipes. However, as some migrants discover, access to certain public water sources are more exclusive, favouring 'locals' or historic homeowners in the area who have membership of community water systems.

Even for household managers who do not have to leave their home to fetch water because they can call a tanker, pump water from their well, or rely more on KUKL tap water, social connections are still vital. For example, Sangita had to talk to neighbours and relatives to find a tanker that would deliver to her home after the earthquake when KUKL stopped working for a month. Gautam's street has a strong KUKL connection because one of his neighbours used to work for the government. Gautam can also drive his scooter to his ancestral home near the main KUKL line to collect water when his tap and well run dry.

Investments in certain sources, technologies and social networks provide many participants with a greater sense of autonomy and security. When asked if they would stop using non-grid sources after Melamchi, many participants said that they will continue to support their neighbourhood well and stone spout water systems and/or maintain their personal wells, but stop purchasing water from tankers, jars and other vendors.

Several participants rely on community leaders to make water management decisions. For these households in community cooperatives, the community system is an extension of their household as they are socially, economically, politically and culturally invested in their local system. For example, Samrat says it is the community's decision to incorporate Melamchi into their neighbourhood water system. Systems like the one Samrat relies on provide inspiration for other participants who lack local connections. Prem and his neighbours are investing in a tank and filter system for his community's well: 'We have to use the source that we have. We have to think how can we use the well effectively. We have a strong well where about 30,000 to 40,000 litres of water is pumped daily. That water is enough for us if we can collect properly. But we don't use it well. People hesitate if they have to spend money. We have to treat it with chemicals, medicines, filter plant and we have to clean it regularly. But in case of tap water we don't have to do all that. We just have to pay 150 [rupees]. But imagine if Melamchi comes and we have to pay 1,000 to 1,200 [rupees] per month then how difficult it will be. If someone gave us the medicines to keep in water I think this water will be okay to drink.'

In some cases, strong cultural ties have enabled community groups or 'clubs' to have a greater degree of autonomy (Molden & Meehan, 2018). As a result, some household water managers have given up on the government attempts to fix the grid. Santa Maya, a resident in a historically low caste neighbourhood, feels betrayed by the government: 'I don't care [what the government does]. We get water from the club. The government won't give you water.' Samrat, similarly expressed, 'the government had said that the Melamchi water was coming years ago. If we were just sitting and waiting for water then people would have died of thirst. But the community people has helped getting water. I don't know much about government. I don't trust in government.'

Nevertheless, most participants complain that people in their area act too individually and selfishly. Participants wish for more local unity – to have meetings to discuss what the problems are, develop strategies, go in mass to the KUKL office to complain and make demands. Participants explained that these strong social ties grant them power but acting as an individual they are powerless, especially when making demands at the KUKL office or local government offices.

Yet, even for participants who are part of strong community systems, only a few groups have been able to successfully negotiate with the KUKL office, and still they complain that the results are short lived. Take the neighbourhood of Chyasal where locals claim the government responded to their demands by building a deep boring plant next to the Bagmati river and installed new pipelines to households in that neighbourhood. Despite this investment, residents do not think water will not flow through the system anytime soon. One resident complained that it would not come until after Melamchi.

These strategies indicate that insecurity persists due to rapid changes and the increasing fragmentation of social ties and physical water flows. All participants expressed worries that water is increasingly scarce and hard to find or more expensive to buy. Here the burden falls heavily on the people who secure water for their families and communities as most lack the ability to fully trust a larger system.

The burdens of coping

Institutions often praise the resourcefulness of residents as they wait for the completion of Melamchi. Here, institutional narratives view the use of multiple sources and water saving techniques as coping mechanisms, or temporary alternatives, until the grid supplies demand (Molden *et al.*, 2016). As Ching (2018) argues through her research on public narratives of piped water supply in Kathmandu, there is a paradox of resilience: the act of coping creates a false sense of resilience. Coping thus becomes part of a cycle of inaction where the public's perceived resilience feeds into bureaucratic incentive towards blame-aversion (Ching, 2018).

However, what does coping look like in everyday life? This study finds that the burden of coping falls heavily on households as families take management into their own hands. More specifically, that burden of managing water falls on certain household members who bear responsibility for gathering, paying, storing, treating and using water. In most cases that burden falls on women, yet, with changing social relations, participants explain that men are more involved in water tasks, to varying degrees.

Take the following two experiences of household water management, from Rupa and Dev. Rupa, a grandmother, who migrated with her children to the Valley 10 years ago, says she spends her entire day 'touching water.' She must walk 10 minutes down the road to a stone spout and then wait in line, sometimes for an hour, to collect one bucket of water. When water comes and the homeowner lets the renters take water, Rupa and her neighbours take turns collecting buckets from the KUKL line. She stores the tap water in clear jars under her bed and uses it for drinking. Then, when the homeowner has unlocked the well and it is not smelling of sewage, she takes turns pulling buckets with her neighbours. Rupa's children and grandchildren will sometimes help her when they get back from school or work. Rupa says it is always hard work securing water but it is easier when there is more water available through the KUKL tap or well.

Dev, a grandfather who is long retired, has built a new home on the outskirts of Patan. He has two large storage tanks on his roof and on the ground floor, and has built a system to channel rainwater directly into the toilet cistern. Dev orders a tanker twice a month and collects one large bottle or jar of water every three days. He lets his renters use the household well.

While in completely different situations, both Dev and Rupa cannot rely on systems beyond their household to provide them with water. Dev is constantly arguing with tanker suppliers and Rupa is constantly figuring out which sources will be available. While these experiences of insecurity vary between Dev and Rupa and their households, both managers feel stress and frustration. Both wish that water management could be less stressful, just like the old days when Dev's previous household in Patan Dokha received a constant supply of water, or in Rupa's village where clean water was always available from springs.

To build security, household water managers like Rupa and Dev have invested in storage: for Rupa this consists of dozens of 5–20 litre buckets and drums stacked in her small room, while Dev has several 500–1,000 litre tanks and an underground reservoir. Looking out at the water tank-lined skyline of Kathmandu, investment in different water storage units is clearly a key mediator of water security.

Participants long for a working KUKL connection where they can just turn on the tap and not have to worry about filling all their tanks when they have a chance. Both Dev and Rupa are optimistic that they will soon stop buying other water and only use tap water as they see pipelines being laid. However, other participants feel like 'Melamchi will never come' or when it does they are sceptical of how long it will last.

Already the unfolding of Melamchi is amplifying the insecurities of some households more than others and increasing the multiple costs of coping. This is because, for the many households with unused and broken KUKL connections, household managers and community managers made a choice to either continue paying the bills despite not getting water, to ignore the bills and accumulate a large fine, or to 'cut the line.'

However, several other participants, such as Santa Maya and Samrat, opted to cut the line because they transitioned to rely on other systems instead and did not want to keep paying for water that they do not receive. As Muna explains, 'After we built this house we cut the pipelines. So now we don't have [a KUKL line]. We have cut the pipes here, but we haven't gone to the office to inform them. So, I think if they send Melamchi water then we will have to pay a lot of fees. It's been 12–13 years that we build the house … But water hasn't come for 18–19 years.'

After seeing Melamchi finally materializing in the city, these participants have realized that they need to reapply for a connection or pay up all their fines, some of which have accumulated over many years. Sarita owes around US$500 to unlock her connection to Melamchi water. However, Sarita also needs to rebuild her house which was damaged in the 2015 earthquake, so she said she will see what Melamchi looks like before paying. Currently, she collects buckets from the community well and pays the mothers' group that manages it 50 cents a month. While homeowners, like Sarita, grapple with the costs of becoming reconnected to the piped network, renters wonder how much they will have to pay to access the landlord's tap. This is because current policy governing the piped water system systematically marginalizes residents who lack certain kinds of property in the city, as seen in the disparities of access between Sonika and Shanta or Sarita and Dev.

Discussion

Urban residents are paying for the costs of ineffective and at times absent water management systems. As one participant expressed, 'let people realize that it is the government duty to provide water. Even if people don't pay, the government needs to supply water. It is a basic human right!' This paper argues that meeting these rights demands research, development and policy responses to look beyond the main pipe network and engage with the variety of ways households secure water.

When asked, participants prefer water from KUKL and/or community water sources such as stone spouts and wells. The reasons for these preferences vary but participants often emphasize perceived cleanliness, convenience, reliability and social value, especially around community systems. However, without regulation and enforcement, water sources are vulnerable to contamination and over extraction from upstream or surrounding users. Thus, not addressing the drivers of inequities and vulnerabilities surrounding water access risks perpetuating the disparities or creates new forms of insecurities, as the stories of Sarita, Samrat and Rupa demonstrate. At the very least, addressing water insecurity requires the upholding of existing laws and policies to ensure water resources are available and accessible to the public for domestic needs through better regulation, public outreach and coordination between national, municipal, community and household management systems. Addressing inequities demands a greater understanding of the social, political and economic factors that shape spatial and temporal disparities in water access between families living in a city.

Following local, national and international laws and policies, the State must ensure that water resources are available for public use. Indeed, following Nepal's 1992 Water Resources Act, the State owns all water resources 'for the rational utilization, conservation, management and development of water resources', and prioritizes domestic use above all others. While Article 35, Section 4, of Nepal's Constitution states that 'every citizen shall have the right of access to clean drinking water', those rights can be addressed through sources beyond piped water from Melamchi and KUKL. Yet, so far, the main way in which the State and development efforts have attempted to address that right is through a piped water network. However, as current practices in Kathmandu and other developing cities demonstrate, privileging the grid over other sources risks perpetuating inequities and insecurity (Bakker, 2010; Meehan, 2013; Ranganathan, 2014; Anand, 2015; Furlong & Kooy, 2017).

Policy responses

In its effort to better manage available water resources in the Valley, the KVSWMB is working on new policies and regulations. It has issued a 30-year lease (license) to KUKL with the terms of reference of operation and maintenance of the piped water supply network in the Valley.

Recently, KVWSMB also issued about 500 licenses (to date) to private tanker water vendors. Each water vendor must submit a water quality report from the desired water source. After studying the physical, chemical and microbiological characteristics of the submitted report, KVWSMB classify the water as drinking water, household use water and other water. After the categorization of the water,

KVWSMB issues the license with colour codes, e.g. blue for drinking water meeting all national standard values of physical, chemical and biological parameters, green for household use water meeting the national standard for physical and chemical parameters, and yellow for other water meeting only one of the physical, chemical or biological parameters. All license holders must put a colour sticker on their tanker according to the quality of water being supplied and this colour-coded classification for tankers has helped users to identify the water quality of tanker water they have been using.

Similarly, the licensing of more than 400 deep tube wells (to date) has helped formulate conjunctive water management strategies for incorporating good groundwater sources as a viable alternate water source to piped water.

KVWSMB has issued regulating guidelines with the aim of regulating tanker water business and groundwater abstraction in the Kathmandu Valley. *Licensing Underground Water Extraction and Use Guideline* (*2014*) and *Tanker Water Business Guideline* (*2017*) are two major policy documents that KVWSMB has been following for better management of the water nexus in the valley. KVWSMB issues groundwater abstraction license to all major groundwater users within the valley depending on their water demand on the groundwater potential of the area, water quality and on water consumption per day. Licenses thus issued have to be renewed every year and, for renewal, submission of data regarding water quality, water abstraction/day and rainwater harvesting schemes in the premises of the license holder is mandatory.

KVWSMB has additionally been working on a comprehensive and interactive database with detailed attributes about all available stone spouts, ponds, tube wells, community schemes and spring sources within the Valley. Overlapping this information on one single map will help the public and even policy makers to formulate a plan for equitable water access to all user group within the Valley.

Conclusion

This paper examines the ways in which households address water insecurity in the Patan metropolitan area of the Kathmandu Valley, Nepal. It draws on insights from institutional expertise and participatory research with 47 household water managers over the dry, pre-monsoon and monsoon season in 2017. By presenting the experiences of urban residents like Uma, Rupa, Shanta and Dev over the year as they negotiate different water challenges, this paper questions common assumptions in water security research. As such, this paper calls for greater attention to the diverse experiences of different people living in the city over time.

First, this paper highlights spatial and temporal disparities of water access across household demographics and water sources, for example between the neighbours Shanta and Sonika. These disparities demonstrate that commonly applied spatial and temporal boundaries in discussions of water security, such as the dry season or the municipal border, inadequately reflect lived experiences and conceal inequities. Here, in-depth qualitative research of household experiences helps to reveal systemic forms of marginalization that mediate experiences of water security (Loftus, 2006; Sultana, 2011; Truelove, 2011; Meehan, 2013; Buechler & Hanson, 2015), such as the difference in the ways that renters and homeowners, wealthy and poor, central and peripheral residences struggle to access piped water from KUKL.

Second, this paper highlights the role of landownership and social connections in mediating experiences of water security. Registering a home in the right areas enables access to a KUKL tap connection, and also grants the homeowner extra space for storage, potentials for well-building and a possible means of income from renters. Conversely, migrants who cannot afford to build and register a home, such as

Rupa and Sonika, remain vulnerable and insecure as they lack these options. Nevertheless, as stories like Sarita's and Uma's demonstrate, not all landlords and homeowners are more water secure or privileged. For homeowners like Sarita, Uma, Shanti, Samrat and Sanu Maya, social connections and local membership of community water systems are vital lifelines and support systems. More generally, it is through socializing more than by direct access to infrastructure that people come to build security. As this study and others show, prevailing narratives about water management often overlook these social and cultural drivers of security in favour of metrics of access and basic needs (Molden *et al.*, 2016; Tamang, 2016; Jepson *et al.*, 2017).

Third, this research emphasizes the burdens of coping with poor water supplies on particular household members. In the Kathmandu Valley, ineffective governance means that households must manage their own water carefully. However, within a household, there are often several families and, within these families, there are particular individuals who bear the responsibility of securing water. The costs of coping are high, not just financially but also in terms of time, labour and stress, as Uma and Rupa explain. Moreover, the prevailing standards of measurement of basic needs, through metrics such as litres per day per person, poorly align with household management strategies by missing critical dynamics of water storage, social networks and belief systems that affect the security or vulnerability of a family (Wutich, 2009; Jepson *et al.*, 2017; Molden & Meehan, 2018).

With these three findings, we argue that policy and development efforts must look beyond the piped water grid to fulfil residents' rights to water and build water security. Although residents dream of a functional water grid, current efforts to fix and supplement the existing piped water network will not adequately address water insecurities or inequities in access.

Acknowledgments

This work was carried out by the Himalayan Adaptation, Water and Resilience (HI-AWARE) consortium under the Collaborative Adaptation Research Initiative in Africa and Asia (CARIAA) with financial support from the United Kingdom's Department for International Development (DFID) and the International Development Research Centre (IDRC), Ottawa, Canada. In addition to expressing gratitude to the people in Nepal (both participants and friends) who made this research possible, the lead author would like to thank Nita Pradhan for her fieldwork assistance, Dr Katie Meehan for her advice, and the Nepal GIS Society, the Nepal Fulbright Commission and the University of Oregon for their institutional support. We would all also like to acknowledge the journal editors, anonymous reviewers, members of the Glacier Laboratory (Augustine Beard, Hayley Brazier, Dr Mark Carey, Becca Marshall, Holly Moulton, Alina Motschmann and Mackenzie Myers) and graduate student writing group (Dylan Brady, Belén Noroña, Dr Zack Thill and Yi Yu) for their comments and edits. Research for this paper was supported by a Fulbright-Hays Doctoral Dissertation Research Abroad fellowship. The authors have declared no conflicts of interest.

Supplementary material

The Supplementary Material for this paper is available online at https://dx.doi.org/10.2166/wp.2018.116.

References

Anand, N. (2015). Leaky states: water audits, ignorance, and the politics of infrastructure. *Public Culture 27*(2 76), 305–330. http://doi.org/10.1215/08992363-2841880.

Bakker, K. (2010). *Privatizing Water: Governance Failure and the World's Urban Water Crisis*. Cornell University Press, Ithaca, NY.

Becker-Ritterspach, R. O. (1995). *Water Conduits in the Kathmandu Valley*. Munshiram Manoharlal, Delhi.

Buechler, S. & Hanson, A.-M. (eds) (2015). *A Political Ecology of Women, Water and Global Environmental Change*. Routledge, New York.

Carey, M., Baraer, M., Mark, B. G., French, A., Bury, J., Young, K. R. & McKenzie, J. M. (2013). Toward hydro-social modeling: merging human variables and the social sciences with climate-glacier runoff models (Santa River, Peru). *Journal of Hydrology 518*(PA), 60–70. http://doi.org/10.1016/j.jhydrol.2013.11.006.

Carey, M., Molden, O. C., Rasmussen, M. B., Jackson, M., Nolin, A. W. & Mark, B. G. (2016). Impacts of glacier recession and declining meltwater on mountain societies. *Annals of the American Association of Geographers 4452*, 1–10. http://doi.org/10.1080/24694452.2016.1243039.

Ching, L. (2018). The paradox of social resilience: explaining delays in water infrastructure provision in Kathmandu. *Water Alternatives 11*(1), 61–85.

Clark, J., Gurung, P., Chapagain, P. S., Regmi, S., Bhusal, J. K., Karpouzoglou, T., Mao, F. & Dewulf, A. (2017). Water as 'Time-Substance': the hydrosocialities of climate change in Nepal. *Annals of the American Association of Geographers 107*(6), 1351–1369. http://doi.org/10.1080/24694452.2017.1329005.

Colopy, C. (2012). *Dirty, Sacred Rivers: Confronting South Asia's Water Crisis*. Oxford University Press, New York.

Dixit, A. & Upadhya, M. (2005). *Augmenting Groundwater in Kathmandu Valley: Challenges and Possibilities*. Nepal Water Conservation Foundation, Kathmandu, Nepal.

Domènech, L., March, H. & Saurí, D. (2013). Contesting large-scale water supply projects at both ends of the pipe in Kathmandu and Melamchi Valleys, Nepal. *Geoforum 47*, 22–31. http://doi.org/10.1016/j.geoforum.2013.02.002.

Furlong, K. & Kooy, M. (2017). Worlding water supply: thinking beyond the network in Jakarta. *International Journal of Urban and Regional Research 41*(6), 888–903. http://doi.org/10.1111/1468-2427.12582.

Gurung, Y., Zhao, J., Kumar KC, B., Wu, X., Suwal, B. & Whittington, D. (2017). The costs of delay in infrastructure investments: a comparison of 2001 and 2014 household water supply coping costs in the Kathmandu Valley, Nepal. *Water Resources Research 53*(8), 7078–7102. http://doi.org/10.1002/2016WR019529.

Hulme, M. (2011). Meet the humanities. *Nature Climate Change 1*(4), 177–179. http://doi.org/10.1038/nclimate1150.

Ishtiaque, A., Shrestha, M. & Chhetri, N. (2017). Rapid urban growth in the Kathmandu valley, Nepal: monitoring land use land cover dynamics of a Himalayan city with landsat imageries. *Environments 4*(4), 72. http://doi.org/10.3390/environments4040072.

Jepson, W., Budds, J., Eichelberger, L., Harris, L., Norman, E., O'Reilly, K., Pearson, A., Shah, S., Shinn, J., Staddon, C., Stoler, J., Wutich, A. & Young, S. (2017). Advancing human capabilities for water security: a relational approach. *Water Security 1*, 46–52. http://doi.org/10.1016/j.wasec.2017.07.001.

Klenk, N. & Meehan, K. (2015). Climate change and transdisciplinary science: problematizing the integration imperative. *Environmental Science and Policy 54*, 160–167. http://doi.org/10.1016/j.envsci.2015.05.017.

Loftus, A. (2006). Reification and the dictatorship of the water meter. *Antipode 38*(5), 1023–1045.

Loftus, A. (2015). Water (in)security: securing the right to water. *Geographical Journal 181*(4), 350–356. http://doi.org/10.1111/geoj.12079.

Meehan, K. (2013). Disciplining de facto development: water theft and hydrosocial order in Tijuana. *Environment and Planning D: Society and Space 31*, 319–336. http://doi.org/10.1068/d20610.

Molden, O. & Meehan, K. (2018). Sociotechnical imaginaries of urban development: social movements around 'traditional' water infrastructure in the Kathmandu Valley. *Urban Geography 39*(5), 763–782. http://doi.org/10.1080/02723638.2017.1393921.

Molden, O., Griffin, N. & Meehan, K. (2016). The cultural dimensions of household water security: the case of Kathmandu's stone spout systems. *Water International 41*(7), 1–16. http://doi.org/10.1080/02508060.2016.1251677.

NGO Forum for Urban Water & Sanitation (2009). *Traditional Stone Spouts: Enumeration and Mapping*. NGO Forum for Urban Water & Sanitation; WaterAid.

Pandey, V. P., Chapagain, S. K. & Kazama, F. (2010). Evaluation of groundwater environment of Kathmandu valley. *Environmental Earth Sciences 60*(6), 1329–1342.

Pandey, V. P., Shrestha, S. & Kazama, F. (2012). Groundwater in the Kathmandu Valley: development dynamics, consequences and prospects for sustainable management. *European Water 37*, 3–14.

Pattanayak, S. K., Yang, J. C., Whittington, D. & Bal Kumar, K. C. (2005). Coping with unreliable public water supplies: averting expenditures by households in Kathmandu, Nepal. *Water Resources Research 41*(2), 1–11. http://doi.org/10.1029/2003WR002443.

Raina, A. (2016). *Equity in Urban Water Service Delivery and the Role of Informal Water Vendors*. National University of Singapore.

Ranganathan, M. (2014). Paying for pipes, claiming citizenship: political agency and water reforms at the urban periphery. *International Journal of Urban and Regional Research 38*(2), 590–608. http://doi.org/10.1111/1468-2427.12028.

Ranganathan, M. & Balazs, C. (2015). Water marginalization at the urban fringe: environmental justice and urban political ecology across the North–South divide. *Urban Geography 36*(3), 403–423. http://doi.org/10.1080/02723638.2015.1005414.

Regmi, A. (2005). Six complexities of water governance: rise and fall of groundwater for urban use. In: *Liquid Relations: Contested Water Rights and Legal Complexity*. Zwarteveen, M., Roth, D. & Boelens, R. (eds). Rutgers University Press, New Brunswick, NJ.

Salike, I. P. & Fee, L. (2015). *Cities and Climate Change Initiative: Kathmandu Valley, Nepal*. UN-Habitat.

Saraswat, C., Mishra, B. K. & Kumar, P. (2017). Integrated urban water management scenario modeling for sustainable water governance in Kathmandu Valley, Nepal. *Sustainability Science 12*(6), 1037–1053. http://doi.org/10.1007/s11625-017-0471-z.

Schwartz, K., Luque, M. T. & Rusca, M. (2015). (In)formality: the meshwork of water service provisioning. *WIREs Water 2*, 31–36. http://doi.org/10.1002/wat2.1056.

Shrestha, B. K. & Shrestha, S. (2013). Water management in rapidly urbanizing Kathmandu valley: balancing structural linkages among water, society, and settlement. In: *Globalization of Water Governance in South Asia*. Narain, V., Goodrich, C. G., Chourey, J. & Prakash, A. (eds). Routledge, New Delhi, India.

Shrestha, S., Pradhananga, D. & Pandey, V. (eds). (2012). *Kathmandu Valley Groundwater Outlook*. Asian Institute of Technology (AIT), The Small Earth Nepal (SEN), Center of Research for Environment Energy and Water (CREEW), International Research Center for River Basin Environment-University of Yamanashi (ICRE-UY), Kathmandu, Nepal (2012).

Spodek, J. C. (2002). Ancient Newari water-supply systems in Nepal's Kathmandu Valley. *APT Bulletin 33*(2/3), 65–69. http://doi.org/10.2307/1504758.

Sultana, F. (2011). Suffering for water, suffering from water: emotional geographies of resource access, control and conflict. *Geoforum 42*, 163–172. http://doi.org/10.1016/j.geoforum.2010.12.002.

Tamang, M. S. (2016). Water connection: everyday religion and environments in Kathmandu Valley. *Himalaya 36*(2), 82–85.

Thapa, B. R., Ishidaira, H., Pandey, V. P. & Bhandari, T. M. (2018). Evaluation of water security in Kathmandu valley before and after water transfer from another basin. *Water 10*(224), 1–12. http://doi.org/10.3390/w10020224.

Tiwari, S. R. (2001). *The Ancient Settlements of the Kathmandu Valley*. Centre for Nepal and Asian Studies, Tribhuvan University, Kirtipur, Kathmandu.

Tiwari, S. R. (2002). *The Pit Conduit Water Supply System of Kathmandu*.

Tiwari, S. R. (2014). *Construction of Traditional Water Supply System in Bhaktapur*.

Truelove, Y. (2011). (Re-)Conceptualizing water inequality in Delhi, India through a feminist political ecology framework. *Geoforum 42*(2), 143–152. http://doi.org/10.1016/j.geoforum.2011.01.004.

Udmale, P., Ishidaira, H., Thapa, B. R. & Shakya, N. M. (2016). The status of domestic water demand: supply deficit in the Kathmandu Valley, Nepal. *Water 8*(196), 1–9. http://doi.org/10.3390/w8050196.

UN-HABITAT (2008). *Water Movement in Patan with Reference to Traditional Stone Spouts in Nepal*. Water for Asian Cities Program Nepal, United Nations Human Settlements Programme.

Welle, K. (2006). *WaterAid Learning for Advocacy and Good Practice: Water and Sanitation Mapping in Nepal*. Retrieved from https://washmatters.wateraid.org/publications/wateraid-learning-for-advocacy-and-good-practice

Wutich, A. (2009). Estimating household water use: a comparison of diary, prompted recall, and free recall methods. *Field Methods 21*(1), 49–68. http://doi.org/10.1177/1525822X08325673.

Wutich, A. & Brewis, A. (2014). Food, water, and scarcity. *Current Anthropology 55*(4), 444–468. http://doi.org/10.1086/677311.

Zwarteveen, M. & Boelens, R. (2014). Defining, researching and struggling for water justice: some conceptual building blocks for research and action. *Water International 39*(2), 143–158. http://doi.org/10.1080/02508060.2014.891168.

Received 1 June 2018; accepted in revised form 1 August 2018. Available online 28 September 2018

Conundrum or paradox: deconstructing the spurious case of water scarcity in the Himalayan Region through an institutional economics narrative

Rinan Shah[a,b,*] and Shrinivas Badiger[a]

[a]Ashoka Trust for Research in Ecology and the Environment, Royal Enclave, Srirampura, Jakkur Post, Bengaluru 560064, India

[b]Manipal Academy of Higher Education, Madhav Nagar, Manipal 576104, India

**Corresponding author. E-mail: rinan.shah@atree.org*

Abstract

Water scarcity in mountain regions such as the Himalaya has been studied with a pre-existing notion of scarcity justified by decades of communities' suffering from physical water shortages combined by difficulties of access. The Eastern Himalayan Region (EHR) of India receives significantly high amounts of annual precipitation. Studies have nonetheless shown that this region faces a strange dissonance: an acute water scarcity in a supposedly 'water-rich' region. The main objective of this paper is to decipher various drivers of water scarcity by locating the contemporary history of water institutions within the development trajectory of the Darjeeling region, particularly Darjeeling Municipal Town in West Bengal, India. A key feature of the region's urban water governance that defines the water scarcity narrative is the multiplicity of water institutions and the intertwining of formal and informal institutions at various scales. These factors affect the availability of and basic access to domestic water by communities in various ways resulting in the creation of a preferred water bundle consisting of informal water markets over and above traditional sourcing from springs and the formal water supply from the town municipality.

Keywords: Domestic water scarcity; Eastern Himalayan Region; Mountain towns; Urban water governance; Urbanization

Introduction

Most water scarcity studies typically provide a very narrow view of the contextual reality, describing them as physical water shortages, perhaps distinguishing them as either physical, economic or

doi: 10.2166/wp.2018.115

institutional scarcity (Falkenmark & Lundqvist, 1998; Seckler *et al.*, 1999; Government of West Bengal, 2012; Hoekstra & Mekonnen, 2012; Molden *et al.*, 2014; Vaidya, 2015). Locally, scarcity almost always manifests through a combination of all three typologies both spatially and temporally, and is also differentiated based on social class and geographic location (Bandyopadhyay & Gyawali, 1994; Ohlsson & Turton, 1998; Anand, 2001, 2004, 2011; Wolfe & Brooks, 2003; Mehta, 2006; Mehta *et al.*, 2011; Srinivasan *et al.*, 2013; Badiger *et al.*, 2014). Concerns related to domestic water scarcity often overlook social–institutional drivers which influence access. Translation of physical water availability into accessibility for communities depends on the current state, scale and ability of the water infrastructure to overcome seasonal changes and reduce the burden of water collection. This process requires a detailed understanding of the functioning and organization of multiple water institutions (Agrawal & Yadama, 1997; Anand, 2001, 2004; Badiger *et al.*, 2014).

Institutions are rules, norms, and practices which govern decision making in a society (Kiparsky *et al.*, 2013). Formal institutions are associated with the State and are present at central, provincial, regional and local levels. Informal institutions are based on social networks which emerge from the need for social safety nets triggered by the absence of State-driven initiatives (Lebel *et al.*, 2005; Hillmann, 2013). The intertwining or hybridization of these institutions can be observed here in the way formal (municipality) and informal (*samaj* – a self-help group, described below – and private water suppliers) work in cognizance of each other's abilities to monopolise, dominate or complement. This intertwining and overlapping is also evident in the way households are able to create their own 'water bundles' using a combination of water sources that they can access or afford and prefer. We use the term 'water bundle' in a similar context to the 'optimal consumption bundle' (Brunnermeier, 2004) of goods in micro-economics – where a consumer (a household in our case) is able to meet the net requirement of their goods' (here, the total domestic water requirement) needs through a particular combination of different goods (various water sources); where each of them has a known rule of preference over the other and has a differential cost (monetary or non-monetary) associated with acquiring them.

Mountains are often referred to as natural 'water towers' for humanity, highlighting their importance as a prominent water source for the arid and semi-arid lowlands but with little reference to water security in the mountains themselves. The Himalaya in the Indian context are perceived as an inexhaustible freshwater source but only a small fraction of the streamflow is stored and consumed by the communities living in these mountains. Water scarcity in these mountains is largely influenced by the unique local hydrogeology, climate and social characteristics. Harnessing and storing of water in large quantities is also impossible in this seismically active mountain range. Hence, springs have historically been an important and primary source of water for a significant proportion of the population, both for rural and urbanizing communities (Khawas, 2002; Sharma *et al.*, 2010; Boer *et al.*, 2011; Mahamuni & Kulkarni, 2012; Tambe *et al.*, 2012; Tiwari & Joshi, 2012; Mukherjee, 2013; Agarwal *et al.*, 2015; Ghatani, 2015; Basumajumdar, 2016; Drew & Rai, 2016).

This paper argues that the issue of water scarcity in the region has been wrongly equated to water shortage rather than searching for and finding solutions in the economic and institutional space – so, the problem, often misconstrued as a paradox, is actually more of a conundrum. The paper will primarily look at various drivers of water governance and social organization that translate to domestic water scarcity in Darjeeling (Darjeeling Municipal Town, in the district of the same name), including the central town itself which is catered for by a centralized formal water supply, as well as the surrounding sprawl whose water comes from multiple informal sources including water tankers and directly from springs. A description of methods and the selection of Darjeeling as a study site and its history are presented first. This is followed by a presentation of the history and current scenario of formal and informal water institutions in Darjeeling. Next, the paper draws insights into the coverage of the various institutions, their multiplicity in some cases and their intertwining in others, contributing to water scarcity at household level. Lastly, the associated risks, uncertainties and conflicts surrounding the creation of water bundles to handle the water scarcity so created are presented.

Methods

Topic-guided interviews were undertaken with key informants such as officials from the Darjeeling Municipality (DM), staff of the DM Water Works Department (WWD), and the DM's staff associated with the operation and maintenance of the water supply infrastructure. Focus group discussions and semi-structured interviews were carried out with households and targeted communities. Primary data gathering was through snowballing and purposive sampling. Secondary data were gathered from official municipal reports, archives and literature. Using the guidelines of the Asian Development Bank (ADB) (Ministry of Urban Development Government of India & Asian Development Bank, 2007), the status and performance of the DM as a water utility provider was assessed.

Geographic context

The Eastern Himalayan Region (EHR) is considered as one of the highest rainfall receiving regions in India and supposedly has the highest per capita and per hectare availability of water in the country (Sharma *et al.*, 2010). However, this region experiences acute temporal and spatial water scarcity in a paradoxically water-abundant region (Khawas, 2002; Chhetri & Tamang, 2013; Lepcha, 2013; Rasaily, 2014; Drew & Rai, 2016). The history of Darjeeling's water problems and its formal water institutions are documented in detail by Rasaily (2014; in Nepali language), a major source of the historical information cited in this paper. Darjeeling (located in the EHR) receives an average annual rainfall of around 3,500 mm, compared to the all India average of 750 mm. The EHR receives 90% of its annual rainfall during the south-west monsoon (summer), with the rest coming from the north-east monsoon (winter). A skewed temporal distribution of rainfall poses the challenge of creating sufficient surface water storage for the rest of the year from the rains of the south-west monsoon, in a terrain which is highly undulating, prone to landslides and seismically very active.

Residents of Darjeeling have faced a water crisis for decades, indicating the inefficacy of the water governance system and the apathy of formal institutions. Studies on water issues in Darjeeling have focused on these physical water systems or on the absence of adequate bulk water storage, and the inefficient and insufficient supply infrastructure (Boer *et al.*, 2011; Chhetri & Tamang, 2013; Basumajumdar, 2016; Drew & Rai, 2016; Bhutia, 2017; Tamang & Jana, 2017). Hence, the water paradox – the too much, too little syndrome – has been limited to how little water exists in storage and in aquifers for use post monsoon. In our assessment, the water scarcity and security of the region are manifested by the political economy and the development trajectory of the region which is similar to that of many mountain towns in the EHR. Studies which inaccurately describe the water crisis as a paradox imply a mismatch in low levels of water available to the communities in a region with a high amount of rainfall, missing the myriad of issues which also exist within the system of water governance. Hence, we argue that the crisis is rather a conundrum due to the intertwined problems across institutional capacity, political unwillingness, inadequate investments, and failure of cooperation between the state and regional institutions and governance.

The cultural and political history of Darjeeling

The development of Darjeeling (Figure 1) as an urban town was driven by various factors. The major junctures in recorded history which contributed to the increasing population of this region for the past *c.* 200 years were: the creation of Darjeeling as a sanatorium for British soldiers; colonial investments in tea and cinchona plantations; and its place as a permanent armed forces recruitment centre for the British Army.

The region's ethnic identity was less recognized by the State in the post-colonial period from both cultural and political perspectives, in part leading to the re-emergence of an ethnic and cultural struggle and to identity politics, with increased vigour and social impacts. The struggle for a separate state and the consequent related unrest has been going on for more than a century (Dasgupta, 1999; Khawas, 2002; Sharma, 2014), with a recent agitation in 2017 lasting for more than 100 days.

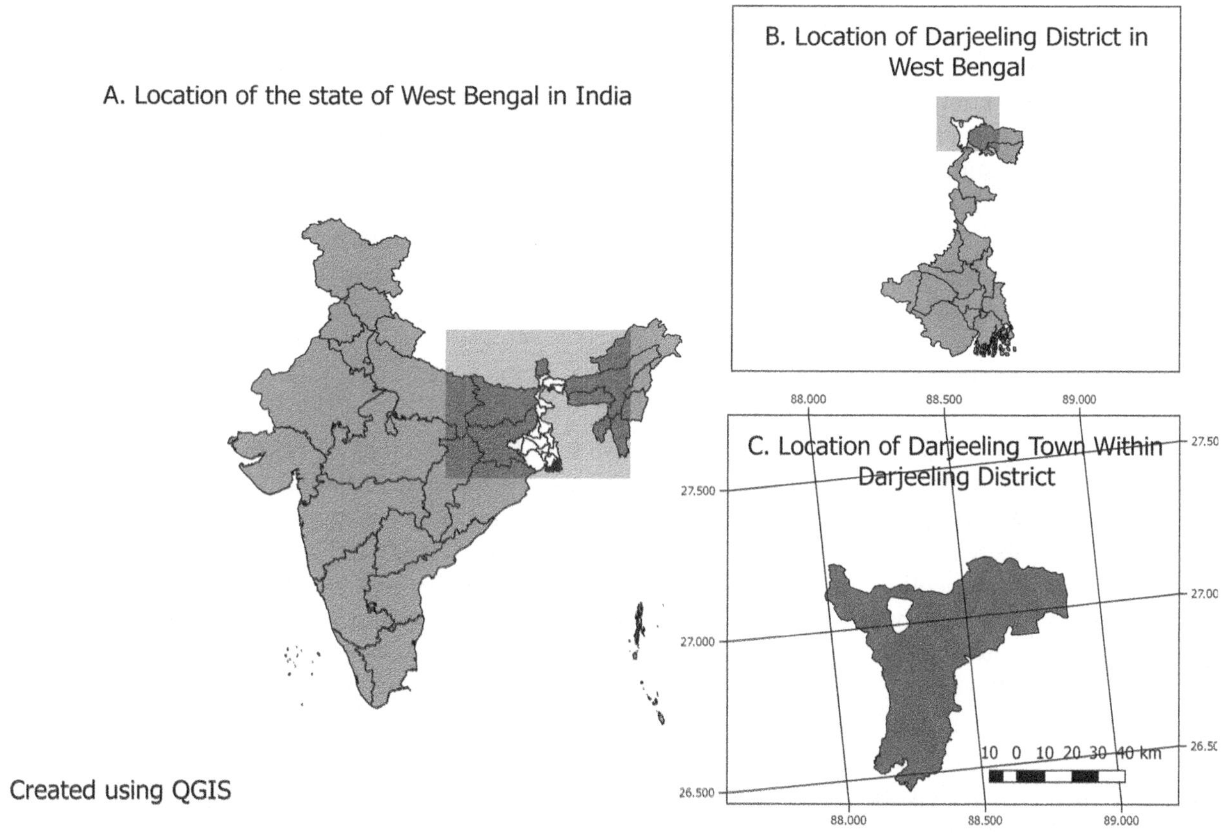

Fig. 1. Location of the DM town. (The Kalimpong district is included within the Darjeeling district boundaries, due to the unavailability of digitized new boundaries.)

Today, Darjeeling makes up roughly 11 km^2 of the entire geographical region and is the largest urban centre in the hilly regions of the states of West Bengal and Sikkim. Besides being a popular tourist attraction, it is often referred to as the 'financial, administrative and social confluence of the region' (Mell & Sturzaker, 2014). Darjeeling now falls under Gorkha Territorial Administration (GTA-2). GTA-2 is the autonomous hill council formed after the agitation that took place in 2017. It was preceded by GTA-1 (2012–2017) and the Darjeeling Gorkha Hill Council (DGHC 1988–2012) (Sharma, 2014; Wenner, 2015); the DGHC was the first autonomous hill council outside northeast India (Sonntag, 1999) and one outcome of the agitation for a separate state (1986 to 1988). Darjeeling has the dual character of being a growing urban town whilst also being located in the uplands which are quite rural in nature. The lands around Darjeeling belong to State-affiliated organizations such as the Indian military, the Forest Department, and Tea Board of India which each add to existing spatial constraints for urban expansion and development planning. With the rapid urbanization of smaller towns in the region and the increase in the urban population, Darjeeling presents an interesting case for understanding the challenges of the provisioning of basic amenities, including drinking water, given the region's distinctive geography, and social, cultural and political landscape dynamics.

Formal and informal water institutions

Due to the mixed urban–rural landscape of the towns in the EHR, numerous institutional arrangements exist alongside to meet domestic water needs. The town municipal supply (through private–individual, private–commercial or public connections) through the piped networks falls under a formal institution, managed by the DM, whilst spring-management and private water suppliers are part of informal institutions. In this section we untangle the genesis, structure and functioning of these formal (State-run) and informal (non-State) water institutions against the backdrop of the political economy of the region, throwing light on the status of water provisioning by both formal and informal institutions and how water scarcity is alleviated or reinforced through them. Table 1 lists the timeline of major projects and schemes initiated to augment domestic water supply within the Darjeeling town.

Darjeeling Municipality

Established in 1850, DM is the main civic administrative body covering 32 wards and represented by an elected board of councillors. Run by the DM, the town's water distribution system was originally designed for a stable population of around 10,000 during the 1920s. The primary water sources for this system are the twin North and South Lakes, constructed in 1910 and 1932, respectively, which lie 15 km south of Darjeeling inside the Senchal Wildlife Sanctuary and which are fed by 26 streams and springs (Government of West Bengal, 2012; Bhutia, 2017; Tamang & Jana, 2017). Water from these lakes then flows to Jorebungalow (which has five filtration units) and on to storage tanks at St Paul's and Rockville. With the exception of one household which lies above St Paul's and to which water is pumped by the municipality, the entire supply and distribution system operates under gravity through the mains and distribution network (Boer *et al.*, 2011; Rasaily, 2014; Bhutia, 2017) and all the way to individual tap connections in households, businesses, army camps and schools.

Several projects were proposed and initiated by the State to augment the town water supply. The Public Health and Engineering Department, Government of West Bengal (PHED), is responsible for the construction, operation and maintenance of both the rural water supply and bulk water supply for

Table 1. Chronology of major water supply projects initiated in Darjeeling.

Year	Event
1850	Darjeeling Municipality established
1910	North Lake constructed in Senchal Sanctuary
1920	Jorebungalow filter house constructed
1932	South Lake constructed in Senchal Sanctuary
1969–72	Rambi *Jhora* Phase-I
1993	Rambi *Jhora* Phase-II
1984	Sindhap Lake constructed in Senchal Sanctuary
1989–90	Water crisis worsened; the DM and DGHC supplied water using tankers
1991–93	Khong *Khola* proposal
1995–97	Balasun River Scheme proposal
2002	Rungdung River Scheme proposal
2004	Darjeeling residents stage protests against water crisis; PHE revives the Balasun scheme.
2005	Balasun River Scheme foundation laid
2010	Balasun River Scheme scheduled to be completed
2015	Balasun River Scheme is operational during summer when the flows to Senchal lakes diminish
2016	AMRUT (Atal Mission for Rejuvenation and Urban Transformation) scheme of Government of India sanctioned

urban areas (Government of West Bengal, 2018). The PHED worked on the Rambi *Jhora* Phase-I project from 1969 to 1972. In 1993 the Rambi *Jhora* Phase-II project proposal, under the Darjeeling Water Supply Improvement Scheme, was submitted to the then state Finance Minister, Asim Dasgupta (Ghatani, 2015). The commissioning of the project was undertaken by the PHED under the orders of the Government of West Bengal (GoWB) but was later abandoned due to a non-feasibility report prepared by the PHED (Bhutia, 2017). The *Bokshi Jhora* Water Project was another augmentation project taken up jointly by the PHED and the DM at the suggestion of the Citizens' Welfare Committee (Rasaily, 2014).

In 1984, a third reservoir in Senchal, Sindhap lake ('*dhap*' meaning 'lake' in Nepali), was constructed (Biswas, 2013; Chhetri & Tamang, 2013; Basumajumdar, 2016) at a lower altitude than the twin lakes. Due to flaws in its design and to avoid costly repairs, the reservoir is now only being used at up to 50% of its capacity. From January 1989 – May 1990 the town's water crisis worsened with several reports of violence, such as '*Khukhuri taana taan*' (the drawing of khukhuri – a kind of knife used in and around this region – by people involved in an argument). Rasaily (2014) noted that '*all records of scarcity were broken*' and the communities had to divert their spending from purchasing food to procuring water. This situation prompted the DM and DGHC to start distributing water within the town using private tankers. Between 1991 and 1993, the DGHC came up with a proposal for two new reservoirs at Khong *Khola* (Rasaily, 2014), a small perennial stream fed by a series of springs from the Senchal catchment. The proposal was glorified to such an extent that it was said that people would need to 'build swimming pools on rooftops' to store the large quantities of water it was believed would be transported to the town.

In 1995, the DM requested the DGHC to form a committee to prepare a proposal for source augmentation options and to improve the supply infrastructure. The result was a proposal to lift water from the Balasun river into the two Senchal reservoirs (Boer *et al.*, 2011) at a cost of Indian Rupees (INR) 400 million (approximately USD 6 million, as on 14 May 2018). The Balasun River Scheme, under the PHE, was revised to also upgrade the pipe infrastructure at an additional INR 80 million (USD 1.2 million). Due to an insufficiency of funds, the GoWB hesitated to move forward on this project. The

DM sought funds from the World Bank with the agreement of the GoWB, but the application made no progress due to irreconcilable differences on the project design, implementation and operation between the three parties (the GoWB, World Bank and DGHC) (Chhetri & Tamang, 2013).

Under the GoWB's instruction, the DGHC then prepared the Rungdung River Scheme (Chattorpadhyay, 2005; Rasaily, 2014) but the project was stalled even before it started due to: (1) disagreements between contractors to build the road between Darjeeling and Rungdung to facilitate construction, (2) villagers' refusal to part with their land, a claim refuted by the villagers, and (3) the emergence of Gorkha Janmukti Morcha (GJM; the Gorka People's Liberation Front) and its movement for a separate state in 2007. GJM is a political entity in the region which fell out in 2007 with the Gorkha National Liberation Front, which also campaigns for the creation of a separate state *Gorkhaland* within India, out of the northern districts in West Bengal (Chettri, 2013).

In the meantime, 2004 also witnessed ground movement concerning water issues in Darjeeling. The Balasun River Project, shelved in 1995–97 due to lack of funds, was taken up by the PHED after reaching a consensus that only this project could bring resolve Darjeeling's water woes (Rasaily, 2014). The cost had increased to INR 560 million (USD 8 million) by the time work began. The foundation stone was laid in 2005 and work was to be completed by April 2010 (Rasaily, 2014; Bhutia, 2017). The project although was completed late, started operating only in 2015. Water from the Balasun river is pumped to the Senchal lakes during summer when the dry season flows from within the Senchal sanctuary decline.

During the period from 1967 to 2001, seven medium-sized and major projects were planned, of which only one, the *Bokshi Jhora* Project, was fully implemented (Ghatani, 2015). Since the time of its construction in the early 1900s, there have been no major repairs made to the supply system, except cleaning and painting jobs (Rasaily, 2014). The current infrastructure for water storage and distribution is insufficient to store the stream flow during the peak monsoon season and, as a result, half of the streams that feed these reservoirs are left free flowing and un-stored during the monsoon.

The municipal water supply runs for 60–90 minutes once a week during the dry season and 2–3 times a week during the monsoon. All formal, municipal water connections – domestic, commercial, or public standpipes – have the same frequency and duration of supply since they are connected to the same supply tank and distributaries. People with private connections receive 1,000–1,500 litres of water during each supply cycle, depending on the location, distance from the distribution line and the storage capacity of each household. Households who primarily depend on public standpipes receive far less or sometimes nothing during each supply cycle. Additionally, most public standpipes have no taps which results in massive wastage of water if the timing of supply is not known to the dependent community in advance. A local *samaj* often has to approach the DM to provide a public standpipe in their locality if one does not exist. On receiving such a request, the DM carries out feasibility checks and accordingly gives permission for construction. The cost of installation of a public standpipe is either borne by the DM or by the *samaj*, depending on negotiations and the political reach of the *samaj*. The DM apparently takes time to approve connection requests and is also said to be facing a financial crunch in recent years, and consequently the *samaj* have been bearing connection costs.

In a recent development, Darjeeling has been listed as one of the beneficiary cities of the Atal Mission for Rejuvenation and Urban Transformation (AMRUT) scheme. AMRUT was launched by the Ministry of Housing and Urban Affairs (MoHUA), Government of India, in 2015 to provide basic amenities like water supply, sewerage facilities and urban transport for the benefit of town residents, especially for poorer sections of society (Ministry of Housing and Urban Affairs, 2017). Among the many objectives of this Government of India scheme, AMRUT aims to provide tap water supply and sewerage connection to every household, and the Darjeeling component was sanctioned in 2016 (Chettri, 2016) and is expected to cost INR 2050 million (USD 30 million). The project is yet to achieve major milestones but citizens perceive that this project may alleviate their water stress to some extent, if not completely. However, the success of the project is also dependent on identifying and linking the new infrastructure to newer sources of water in Senchal and surrounding areas.

Community self-help groups: the Samaj

Springs are primary sources of water for a large section of the population living within Darjeeling and at least 50–60% of the population of 132,000 within the town primarily depend on them (personal interview: Assistant Engineer, 2014 and 2016). These springs and their source catchments are managed by the communities who live around them through self-help groups called '*samaj*' (literal translation: 'society').

A *samaj* is an 'inclusive social institution that organizes support for social events' and provides a platform for settling conflicts in a village (Wenner, 2015) and is quite commonly found in the Darjeeling–Sikkim Himalaya region (Boer *et al.*, 2011). The first known *Gaon Samaj* (Village Society) was the *Gorkha Dukha Niwarak Sammelan* (Gorkha Community Problem Solving Convention) founded in 1932 (Boer *et al.*, 2011). Originally, the primary objective of these groups was to deal with immediate and urgent community needs, such as medical and funeral expenses or other emergencies. Over time, *samaj* have been set up by communities with a similar self-help intent for various issues, including accessibility and availability of water (Drew & Rai, 2016) and recently also for the protection and management of springs. A *samaj* is formed on the basis of the location of households and for a purpose; a single household can be a member of more than one such *samaj*, depending on the primary objectives of the *samaj*, for example being a member of one *samaj* with the primary objective of maintaining a springshed (the recharge area of a spring) and another with the objective of being a micro-credit facility, or of another formed for one ethnic group.

Members of a springshed *samaj* share the responsibilities of managing the area surrounding springs, constructing reserve tanks, and also setting up rules of access collectively, and are expected to pay a monthly contribution. Each s*amaj* has a local authority and members are likely to lose their rights and privileges or even face a social boycott if rules are violated. People who have recently migrated to Darjeeling area are allowed to join the *samaj* with a probation period of a year or two (Boer *et al.*, 2011). *Lal Dhiki Ekta Samaj*, *Mangalpuri Samaj*, and *Frymal Sudhar Samiti* are *samaj* within Darjeeling which carry out various self-help activities including protection of springs and their sources. S*amaj* boundaries are determined by the membership and do not overlap with the administrative boundaries of wards (the smallest formal administrative unit) within a town. In Darjeeling there are around 130 *samaj* across 32 wards (Boer *et al.*, 2011).

Private water suppliers

There are four types of private water suppliers in Darjeeling and its surrounding area: water porters, hand-drawn water carts, water tanker trucks, and households who share/sell 'their' spring water or their surplus water (Chakraborty, 2018). As for informal institutions around springs, documentation of the emergence and semi-formalization of private water markets is minimal. These informal private water supply systems play a critical role and often form a significant portion of the water sourced by marginal and vulnerable communities.

Since the 1980s, private water suppliers have played an important role in supplying domestic water at the doorstep in almost every part of Darjeeling. Among the private water suppliers, water porters and water carts supply water in smaller quantities (less than 500 litres) whilst tankers supply larger quantities (4,000–6,000 litres). Some households form a significant portion of these informal private water suppliers by charging money to other households to access 'their' springs (springs on their private land) or with whom they share their surplus storage of water (Chakraborty, 2018).

Water porters collect water from the springs, pipe leaks or breaks within the town and deliver water to people's homes on a regular basis, on demand. A 60-litre can is priced at INR 70–90 (USD 1–1.5) with a monthly supply costing between INR 1300–1500 (USD 19.5–22.5) depending on the location of the house. Hand-drawn carts, locally called '*Gorkhey jeeps*', also collect water from multiple sources including springs, and purchase from large water tankers at INR 50 (USD 0.74) for 250 litres (Chakraborty, 2018) and then supply it to households with small profit margins. A typical water tanker of 6,000 litres costs around INR 1,000–1,200 (USD 15–18) depending on the season, demand and availability. The drying up of water sources overlaps with incoming tourists during the dry season which also affects water pricing.

Water tanker operators started emerging as key players in domestic water supply in Darjeeling during the 1980s, when areas of town started experiencing acute water shortages even during the monsoon due to a lack of formal water supply infrastructure in the region (Rasaily, 2014). Water tankers were also used by DGHC and DM to alleviate the water crisis in the early 1990s (Rasaily, 2014; Chakraborty, 2018). Water tankers supply to almost everyone in town; as one supplier noted, '*… Mithai dokaan dekhi liyera paan ko patta dhuune samma …*' ('… from sweet shops to the washing of betel leaves by shopkeepers …'). Demand for water tankers is relatively lower in areas serviced by the formal municipal water supply and during monsoon; however, some parts of the town, are heavily dependent on one or more private water suppliers throughout the year. (Telephone interviews with tanker truck owners for this study were held in 2018.)

The last category of private water suppliers are households which allow access to water from 'their' springs and those who supply from their surplus storage of water. The first group own springs located on their private land. This ownership of springs allows them exclusive use as well as the ability to sell water from it for profit (Chakraborty, 2018). Some households also sell the surplus water collected and stored in their households, creating another category of private water supplier. This group of private water suppliers are a minority and not organized collectively with formal or informal institutions which guide other individual operators (private tank operators or a *samaj*) and set up rules or prices. Sharing or selling of water among this group is much more informal and their transactions are either in kind or the price is negotiated.

The status of municipal water supply in Darjeeling

Darjeeling Municipal WWD has been studied in terms of its function, staff, and support staff for operations and customers serviced (Table 2) (Ministry of Urban Development Government of India & Asian Development Bank, 2007). Here, the WWD is placed in comparison with the water utilities of 20 other Indian cities. The insufficiency of this framework to incorporate the uniqueness of utilities in mountain towns and cities is discussed below. The ADB has an exhaustive list which could not be acquired for the

Table 2. ADB indicator values for WWD.

Indicators	Indicator values considering the actual available supply	Minimum to Maximum
Water supply coverage (%)	15	15–100
Weekly water availability (hrs)	1	0.3–12
Per capita consumption (l/c/d)	29	29–203
Production/population (m^3/d/c)*	0.03	0.03–0.80
Unaccounted for water (%)**	25	13–60
Connections metered (%)	0	0–100
Revenue collection efficiency (%)	75	45–189
Staff/1,000 connections ratio	26	0.4–26
New Connection Fee (INR)	17,000 (USD 250)	100–17,000 (USD 1.5– 250)

*Considering resident static population, not including tourists and students.
**Refers to leakage and other losses in the distribution system.

DM; hence, in this study, computations have been made using the available data. (This study was part of the first author's unpublished master's thesis; see Shah, 2015).

Coverage supplied by the WWD (10–15%) is the lowest in comparison to other assessed cities (Samanta & Koner, 2016; Bhutia, 2017). The other 85% is covered by a combination of informal supplies, springs and water suppliers (personal interview: DM WWD Assistant Engineer, 2014 and 2016). Per capita water consumption is far less than the 220 LPCD (litres per capita per day) drinking water requirement for a Class I city, as according to the National Commission on Integrated Water Resources Development. Production/population is again the lowest compared to similar sized towns (production/population = [annual production volume (m^3)/365]/[number of people served]). Revenue collection efficiency is at the higher end with 75% water supplied being paid for, compared to other cities in India which have low levels of revenue efficiency. The 'staff to connections' ratio is the highest; a lower value of this ratio indicates a lack of staff to handle service provisioning and high implies the system is highly inefficient: whilst high does indicate inefficiency, difficult terrain cannot be ignored as a reason for the need for a higher staff count. The connection charge is INR 17,000 (USD 250) with additional labour and pipeline material costs. For an immediate connection, charges shoot up to INR 35,000 (USD 520) with extra labour and piping charges. An application form costs INR 100 (USD 1.50) (see: http://dm.gensoftindia.co.in/Application-Forms.aspx) and needs to be accompanied by three supporting documents: a Khatian (a record of rights for identifying land) from the Land Reforms Department; a land registration document from the court; and a mutation document from the municipality as a proof of residency (Shah, 2015). An annual water fee for a private connection is about INR 500 (USD 7.50) People have been gradually moving from public standpipes to individual private connections, despite the time, effort and money needed to obtain a private connection. Public standpipes have an increasing number of dependents who are mostly migrants without the ability to procure the required documentation; private connections are convenient and reliable. (Interviews were conducted at household level for this study in 2014).

A low ratio of production to population of the municipality affects weekly water availability which translates to low per capita municipal water availability. Another factor that affects per capita consumption or availability from the municipal source is the lack of storage at a household level. Low levels of frequency, duration, and amount of supply in addition to high financial costs for acquiring a formal water connection are some of the reasons that determine the low coverage for this water utility.

However, the ADB report (Ministry of Urban Development Government of India & Asian Development Bank, 2007) did not assess any water utility for a mountain town and thus factors specific to such regions may not have been considered. The dynamic nature of the population, consisting of a very high seasonal influx of tourists and students, is an important factor which is missing.

Governance: multiplicity and hybridity of institutions

Water, unlike a classic commodity, is a multifaceted resource which reflects on the nature of institutions that govern it and their intertwining in terms of governance and access by communities (Mehta, 2006; Kiparsky *et al.*, 2013). Institutions that govern a water resource vary and so do associated property regimes. The fluidity of water creates competitive and overlapping water rights, which leads to a mix and intertwining of informal and formal arrangements (Mehta, 2006) which we refer to here as hybridity, since no single institution on its own is capable of meeting communities' demands for domestic water. In the case of Darjeeling, water is a public, private, or common property resource depending on the location, presence or absence of an institution and the nature of these institutions.

Multiplicity and intertwining of institutions makes urban water governance complex because the institutions involved have overlapping and fragmented responsibilities, jurisdictions and decision-making method silos (Anand, 2001; Hillmann, 2013; Kiparsky *et al.*, 2013; Fuenfschilling & Truffer, 2016). Fragmentation of institutions can be political, issue-based, due to gaps in design and implementation of programmes, and as a result of a mismatch between biophysical and political boundaries. Darjeeling hills come under GTA-2, an autonomous hill council in addition to the national and provincial governments. Urban water in Darjeeling is managed by the DM which is an elected body and by the PHED which has appointed engineers. Political fragmentation is created here with multiple sources of funding and multiple institutions responsible for implementation of projects (Kiparsky *et al.*, 2013). Issue-based fragmentation occurs because the land and water resources of Darjeeling and the surrounding areas lie with different State-affiliated organizations such as the military, Forest Department and Tea Board, which increases a physical constraint on the town and restricts access to the resources under the other organizations. The town municipality only manages the formal supply infrastructure but the areas served by springs which a large section of the town is dependent on fall outside the jurisdiction of the municipality. The multiplicity of institutions leads to an interrupted transition from planning to execution to implementation with a different institution being involved at every other step. The Balasun River Scheme is an example, with the PHED building the river-water lifting system and drawing pipelines to the supply systems of the municipality, from where the municipality has to handle the supply.

Another way of locating the multiplicity of institutions is through the infrastructure of the municipality's formal water supply system (Kiparsky *et al.*, 2013), from source to taps. The water harnessing infrastructure falls under the GTA and PHE, whilst the water supply distribution system falls under the municipality (Samanta & Koner, 2016). The water supply lines pass through forested areas under the Forest Department or army cantonment areas under the Indian Army. This multiplicity of formal institutions denotes a political fragmentation through an overlap but at the same time a conflict of responsibilities among various levels of government and agencies, making complex governance structures for urban water (Kiparsky *et al.*, 2013).

Formal and informal institutions handling water resources in Darjeeling and the rules with which these institutions are governed influence how they understand and handle issues of resource access, in this case, communities' ability to access water. Emphasis on resource access is necessary because it helps individuals and communities shape their development but the institutions might enable or curb such development processes (Mell & Sturzaker, 2014).

The intertwining or hybridity occurs at the institutional level where the State-run municipality and non-State entities work together and, at times, against each other to cater for the water requirements of the communities. The low coverage, low frequency and quantities of supply make it imperative for the municipality to seek ways to fulfil the demands of households who are dependent on its distribution lines. The municipality as a formal institution interacts with the private water suppliers, especially the water tankers in times of supply deficit in their system. Vehicles used for private water supply, such as carts and tankers, have to work around the traffic police of the town, underlining the interaction of the formal and the informal. When the municipality seeks the tankers, traffic rules are relaxed accordingly. *Samaj* look after the springs which are ignored by the formal institutions, hinting at the emergence of informal institutions in the absence or negligence of a formal body. *Samaj* also approach the town municipality on behalf of their members when there is a need for public standpipes in their vicinity, and even collectively pay for the installation costs which in most cases are borne by the municipality – highlighting their importance as a conduit between communities and the State.

In the history of water investments for the DM region (Table 1), projects are primarily augmentation projects with only the Balasun River Scheme and AMRUT scheme in recent times looking to restructure an age-old distribution system. This shows the dominance of an engineering outlook with 'add more' as the only solution, without reining in an adequate knowledge of ecology or socio-economics. The stress has to be shifted towards distribution infrastructure and effective institutional functioning to enable better access, which needs support from both political and executive departments. Lack of incentives for positive performances in the public sector makes that sector more conservative and risk-averse to innovations. Absence of innovation restricts initiatives for mountain regions especially to overcome the constraints posed by the topography of the region to build sufficient, effective and less energy-intensive water distribution systems (Bandyopadhyay, 1989; Kiparsky *et al.*, 2013). As water users, communities have to work around these multiple institutions and interact with both formal and informal, to cushion the water scarcity they face. The cushion is generally in the form of a water bundle created using a combination of water sources, as noted above. The composition and proportions of a water bundle are primarily dependent on the financial resources available to a household, and its spatial proximity to the water sources. In the presence of multiple institutions, and the amount of time and resources they have to spend navigating around them, we observed that the higher the number of resources households had access to, the more secure they felt, i.e. the more varied a water bundle was, the more secure a household felt because they could have an array of resources to turn to if they had insufficient water. With the low frequency of municipal supply, if a household can afford a private supply, the frequency increases because they can ask for water to be delivered to their households. If a public spring is present in the household water bundle, a household feels more secure, regardless of the time and effort involved in fetching water from the spring (Shah, 2015).

Any external or non-local interventions tend to overlook the informal and social institutions which have emerged across the years to manage common water resources, such as *samaj*, which makes the interventions counterproductive (Bandyopadhyay & Gyawali, 1994). Problems occur at various levels pointing to the need to understand the interconnectedness of biophysical, socio-economic, political and institutional issues at each level and across various levels to enable integrated planning (Badiger *et al.*, 2014).

Conclusion

Despite being located in a volumetrically water-rich region, Darjeeling has faced water scarcity for a number of decades now. We argue that the crisis is result of a conundrum due to the interlinked problems across political unwillingness, insufficient investments, failure of cooperation between the state and regional institutions, and inadequacies in local governance including institutional capacity. Translation of physical water availability to access for communities is an important factor that alleviates or reinforces water accessibility. The multiplicity of water institutions and the way in which they are intertwined reveals the complexity of urban water governance. Creation of a household water bundle by communities through complex interactions with these institutions illustrates the difficulties they face in acquiring a basic amenity such as water.

DM is unable to meet the water requirements of the communities living in the town. Water scarcity here is cushioned to a certain extent by the presence of springs and private water suppliers, both of which fall under the ambit of informal institutions. As an exclusive dependence on municipal supply is not enough to meet basic water needs, communities create a water bundle from a combination of available sources which are in turn dependent on a variety of factors such as socio-economic status and the spatial location of households, among many others.

In a similar way to other global experiences, engineering solutions have been the order of the day, with supply augmentation the only route pursued in Darjeeling. Success rates of such augmentations have been low and have not been enough to close the gap between increasing water demands and augmented supply. Low success rates show the need to look at aspects of both harnessing and supplying water. An acknowledgement of natural, social and traditional knowledge is needed to provide better solutions and water rights and, in this case, existing systems such as springs and the array of private water suppliers should not be left out.

Various departments need to work together for a better solution to the increasing water crisis. There are an absence and reluctance on the part of the government to push for development schemes with a thorough study. Difficult terrain, lack of funds and an imbalance between urban growth and the provisioning of services might intercept any initiatives taken or render them futile after a certain period of time.

Despite getting a substantial amount of rainfall, households in Darjeeling suffer from acute water scarcity due to a myriad of governance issues. Developing buffer storage is one critical measure needed to compensate for the skewed seasonality of precipitation and to deal with high fluxes in the floating population. At present, there are quite a few initiatives aimed at alleviating the water crisis in Darjeeling. Of the many, one, under the AMRUT scheme (Chettri, 2016) and the other under the National Adaptation Fund (Department of Environment, 2016) seem to show promise in alleviating the crisis. The National Adaptation Fund also emphasizes the importance of rainwater harvesting systems. With these progressive developments, initiatives seem to be making positive headway, with the former looking at reforms in water governance (specifically at the formal water distribution system) and the latter at harnessing rainfall. The implementation of these projects might lead to the alleviation of water scarcity in Darjeeling and will additionally provide pointers for effective intervention in similar towns within the EHR.

Acknowledgements

This work was carried out by the Himalayan Adaptation, Water and Resilience (HI-AWARE) consortium under the Collaborative Adaptation Research Initiative in Africa and Asia (CARIAA) with financial support from the United Kingdom's Department for International Development (DFID) and the International Development Research Centre (IDRC), Ottawa, Canada. We are grateful for the institutional support provided by the Ashoka Trust for Research in Ecology and the Environment, and by the Manipal Academy of Higher Education (formerly, Manipal University), and for the financial support provided by the Government of India's National Mission on Himalayan Studies Grant to the first author. We are also grateful to the many respondents in the Darjeeling region who provided valuable time and insights. The first author thanks her faculty at Tata Institute of Social Sciences who guided her masters' dissertation which laid groundwork for this study.

References

Agarwal, A., Agrawal, N. & Rathi, V. (2015). Water stress on springs of lesser Himalayan region. *British Journal of Applied Science & Technology 9*(3), 243–255. http://doi.org/10.9734/BJAST/2015/17140.

Agrawal, A. & Yadama, G. (1997). How do local institutions mediate market and population pressures on resources? Forest panchayats in Kumaon, India. *Development and Change 28*(3), 435–465. http://doi.org/10.1111/1467-7660.00050.

Anand, P. B. (2001). *Water 'Scarcity' in Chennai, India: Institutions, Entitlements and Aspects of Inequality in Access*. (No. 2001/140). WIDER Discussion Papers, World Institute for Development Economics (UNU-WIDER), Helsinki.

Anand, P. B. (2004). The political economy of water scarcity and issues of inequality. Entitlements and identities: a tale of two cases from southern India. *International Journal of Technology Management and Sustainable Development 3*(2), 115–132. http://doi.org/10.1386/ijtm.3.2.115/0.

Anand, N. (2011). Ignoring power: knowing leakage in Mumbai's water supply. In: *Urban Navigations: Politics, Space & the City in South Asia*. Anjaria, J. & McFarlane, C. (eds). Routledge Publishers, New Delhi, pp. 191–212.

Badiger, S., Gopalakrishnan, S. & Patil, I. (2014). Contextualizing rural–urban water conflicts: bio-physical and socio-institutional issues of domestic water scarcity. In: *Globalization of Water Governance in South Asia*. Narain, V., Gurung Goodrich, C., Chourey, C. & Prakash, A. (eds). Routledge Publishers, New Delhi.

Bandyopadhyay, J. (1989). Riskful confusion of drought and man-induced water scarcity. *Ambio 18*(5), 284–292.

Bandyopadhyay, J. & Gyawali, D. (1994). Himalayan water resources: ecological and political aspects of management. *Mountain Research and Development 14*(1), 1–24.

Basumajumdar, A. (2016). Impact of global warming on climate change regarding water supply in the Darjeeling Hills of the eastern Himalaya and change in mountain ecology. In: *Globalization and Marginalization in Mountain Regions*. Chand, R. & Leimgruber, W. (eds). Springer International Publishing, Switzerland, pp. 161–171.

Bhutia, S. (2017). A situational analysis of water resources in Darjeeling Municipal Town: issues and challenges. *International Journal of Research in Geography 3*(4), 52–60.

Biswas, S. (2013). Urban growth centers in Darjeeling Hills of West Bengal: a geographical analysis. *The Journal of Bengal Geographer 2013*(January), 30–48.

Boer, L., Joshi, D., Rai, S. & Thomas, S. (2011). *The Perennial Springs of Darjeeling, A Survey to Community Based Conservation*. Internship Report. Wageningen University, The Netherlands.

Brunnermeier, M. K. (2004). Learning to reoptimize consumption at new income levels: a rationale for prospect theory. *Journal of the European Economic Association 2*(1), 98–114.

Chakraborty, A. S. (2018). 'Hamro Jhora, Hamro Pani' (Our spring, Our water): water and the politics of appropriation of 'Commons' in Darjeeling Town, India. *Hydro Nepal 22*, 16–24.

Chattorpadhyay, S. S. (2005). Emphasis on Infrastructure. Retrieved April 11, 2018, from http://www.frontline.in/static/html/fl2201/stories/20050114005712000.htm

Chettri, M. (2013). Choosing the Gorkha: at the crossroads of class and ethnicity in the Darjeeling Hills. *Asian Ethnicity 14*(3), 293–308. http://doi.org/10.1080/14631369.2013.764763.

Chettri, V. (2016). *Centre Funds Darjeeling Water Supply Rejig*. Telegraph India, Siliguri, 26 April 2016. See: https://www.telegraphindia.com/1160426/jsp/siliguri/story_82251.jsp.

Chhetri, B. & Tamang, L. (2013). Population growth and associated problems: a case study of Darjeeling Town. *International Journal of Humanities and Social Science Invention 2*(5), 63–67.

Dasgupta, A. (1999). Ethnic problems and movements for autonomy in Darjeeling. *Social Scientist 27*(11), 47–68. See: http://www.jstor.org/stable/3518047.

Department of Environment. (2016). *Detailed Project Report for National Adaptation Fund: Rain Water Harvesting and Sustainable Water Supply to the Hilly Areas in Darjeeling as an Adaptive Measure to Potential Climate Change Impacts.*

Drew, G. & Rai, R. P. (2016). Water management in post-colonial Darjeeling: the promise and limits of decentralised resource provision. *Asian Studies Review 40*(3), 321–339. http://doi.org/10.1080/10357823.2016.1192580.

Falkenmark, M. & Lundqvist, J. (1998). Towards water security: political determination and human adaptation crucial. *Natural Resources Forum 22*(1), 37–51. http://doi.org/10.1111/j.1477-8947.1998.tb00708.x.

Fuenfschilling, L. & Truffer, B. (2016). The interplay of institutions, actors and technologies in socio-technical systems – an analysis of transformations in the Australian Urban Water Sector. *Technological Forecasting and Social Change 103*, 298–312.

Ghatani, S. (2015). *Sustainable Urban Water Management in Darjeeling.* Master of Philosophy Dissertation. Sikkim University, Gangtok, India. See: http://14.139.206.50:8080/jspui/bitstream/1/3112/1/Suvechha Ghatani(Geography).pdf

Government of West Bengal. (2012). *West Bengal State Action Plan on Climate Change.*

Government of West Bengal (2018). *Public Health Engineering.* Retrieved July 12, 2018, from https://wb.gov.in/portal/web/guest/public-health-engineering

Hillmann, H. (2013). Economic institutions and the state: insights from economic history. *Annual Review of Sociology 39*(1), 251–273. http://doi.org/10.1146/annurev-soc-071811-145436.

Hoekstra, A. Y. & Mekonnen, M. (2012). The water footprint of food. *Proceedings of the National Academy of Sciences 109*(9), 3232–3237. http://doi.org/10.1016/B978-0-12-799968-5.00007-5.

Khawas, V. (2002). *Environment and Rural Development in Darjeeling Himalaya: Issues and Concern.* Centre for Environment Planning and Technology, Ahmedabad, India.

Kiparsky, M., Sedlak, D. L., Thompson, B. H. & Truffer, B. (2013). The innovation deficit in urban water: the need for an integrated perspective on institutions, organizations, and technology. *Environmental Engineering Science 30*(8), 395–408. http://doi.org/10.1089/ees.2012.0427.

Lebel, L., Garden, P. & Imamura, M. (2005). The politics of scale, position, and place in the governance of water resources in the Mekong region. *Ecology and Society 10*(2), 18. http://www.jstor.org/stable/26267731.

Lepcha, N. K. C. (2013). Problems and prospects of water resource of kurseong municipality, Darjeeling District, West Bengal. *Science Park Research Journal 1*(19), 1–11.

Mahamuni, K. & Kulkarni, H. (2012). Groundwater resources and spring hydrogeology in South Sikkim, with Special Reference to Climate Change. In: *Climate Change in Sikkim – Patterns, Impacts and Initiatives* (M. L. Arrawatia & S. Tambe, eds). Gangtok: Information and Public Resources Department, Government of Sikkim, pp. 261–274. Retrieved from http://www.sikenvis.nic.in/writereaddata/16-Chapter_Groundwater_Resources_and_Spring_Hydrogeology_in_South_Sikkim.pdf.

Mehta, L. (2006). *Water and Human Development: Capabilities, Entitlements and Power.* United Nations Development Programme.

Mehta, L., Xenos, N., Hartmann, B., Fine, B., Luks, F., Samuel, S., Robert, J., Thompson, M., Hildyard, N., Scoones, I., Millstone, E., Lankford, B., Jairath, J., Gyawali, D. & Dixit, A. (2011). Introduction. In: *The Limits to Scarcity: Contesting the Politics of Allocation.* (L. Mehta, ed.). NOIDA: Orient Blackswan Private Limited, 1–8.

Mell, I. C. & Sturzaker, J. (2014). Sustainable urban development in tightly constrained areas: a case study of Darjeeling, India. *International Journal of Urban Sustainable Development 6*(1), 65–88. http://doi.org/10.1080/19463138.2014.883994.

Ministry of Housing and Urban Affairs (2017). *Atal Mission for Rejuvenation and Urban Transformation – AMRUT.* Retrieved July 11, 2018, from http://mohua.gov.in/cms/amrut.php

Ministry of Urban Development Government of India, & Asian Development Bank. (2007). *2007 Benchmarking and Data Book of Water Utilities in India, 104.* See: http://www.adb.org/Documents/Reports/Benchmarking-Data-Book-Utilities-in-India/2007-Indian-Water-Utilities-Data-Book.pdf.

Molden, D. J., Vaidya, R. A., Shrestha, A. B., Rasul, G. & Shrestha, M. S. (2014). Water infrastructure for The Hindu Kush Himalayas. *International Journal of Water Resources Development 30*(1), 60–77. http://doi.org/10.1080/07900627.2013.859044.

Mukherjee, D. (2013). Critical analysis of challenges of Darjeeling Himalaya: water, natural resources, hazards, and the implications of climate change. *International Journal of Agriculture Innovation and Research 2*(1), 13–22.

Ohlsson, L. & Turton, A. R. (1998). *The Turning of A Screw: Social Resource Scarcity as a Bottle-Neck in Adaptation to Water Scarcity. SOAS Occasional Paper.* SOAS-KCL Water Issues Group, London.

Rasaily, D. S. (2014). *Darjeeling Pahadka Nagarpalika Kshetra ko Vikas ra Khane Paani Ko Itihaas Sanchipta ma, San. 1835–2012.* New Delhi.

Samanta, G. & Koner, K. (2016). Urban political ecology of water in Darjeeling, India. *SAWAS Journal 5*(3), 42–57.

Seckler, D., Molden, D. & Barker, R. (1999). *Water Scarcity in the Twenty-First Century.*

Shah, R. (2015). *Understanding Water Availability and Accessibility Issues From a Perspective of Climate Variability and Climate Change: A Study in the Darjeeling Municipality.* Master of Science dissertation. Tata Institute of Social Sciences, Mumbai, India.

Sharma, G. (2014). GORKHALAND – Darjeeling Gorkha Hill Council (DGHC) to gorkhaland territorial administration (GTA) – what next? *Asian Journal of Multidisciplinary Studies 2*(6), 44–50.

Sharma, B., Riaz, M. V., Pant, D., Adhikary, D. L., Bhatt, B. P. & Rahman, H. (2010). *Water Poverty in the Northeastern Hill Region (India): Potential Alleviation Through Multiple-Use Water Systems – Cross Learnings From Nepal Hills.* New Delhi.

Sonntag, S. K. (1999). Autonomous councils in India: contesting the liberal nation-state. *Alternatives: Global, Local, Political 24*(4), 415–434.

Srinivasan, V., Seto, K. C., Emerson, R. & Gorelick, S. M. (2013). The impact of urbanization on water vulnerability: a coupled human-environment system approach for Chennai, India. *Global Environmental Change 23*(1), 229–239. http://doi.org/10.1016/j.gloenvcha.2012.10.002.

Tamang, P. & Jana, S. K. (2017). Willingness to Pay for improved water services: a case of Darjeeling, India. *Asian Journal of Water, Environment and Pollution 14*(2), 51–59. http://doi.org/10.3233/AJW-170015.

Tambe, S., Kharel, G., Arrawatia, M. L., Kulkarni, H., Mahamuni, K. & Ganeriwala, A. K. (2012). Reviving dying springs: climate change adaptation experiments from the Sikkim Himalaya. *Mountain Research and Development 32*(1), 62–72. http://doi.org/10.1659/MRD-JOURNAL-D-11-00079.1.

Tiwari, P. C. & Joshi, B. (2012). *Urban Growth in Himalaya.* Mountain Research Initiative Newsletter, Berne, Switzerland.

Vaidya, R. A. (2015). Governance and management of local water storage in The Hindu Kush Himalayas. *International Journal of Water Resources Development 31*(2), 253–268. http://doi.org/10.1080/07900627.2015.1020998.

Wenner, M. (2015). Legitimization through patronage? Strategies for political control beyond ethno-regional claims in Darjeeling, India. *Geoforum 66*, 234–243. http://doi.org/10.1016/j.geoforum.2014.11.002.

Wolfe, S. & Brooks, D. B. (2003). Water scarcity: an alternative view and its implications for policy and capacity building. *Natural Resources Forum 27*, 99–107.

Received 1 June 2018; accepted in revised form 21 August 2018. Available online 8 October 2018

Adapting to urban flooding: a case of two cities in South Asia

Ismat Ara Pervin[a], Sheikh Mohammad Mahbubur Rahman[a], Mani Nepal[b,*], Abdul Kalam Enamul Haque[c], Humayun Karim[a] and Ganesh Dhakal[d]

[a]*Institute of Water Modelling, Dhaka, Bangladesh*

[b]*South Asian Network for Development and Environmental Economics at the International Centre for Integrated Mountain Development, Kathmandu, Nepal*

**Corresponding author. E-mail: mani.nepal@icimod.org*

[c]*Asian Centre for Development, Dhaka, Bangladesh*

[d]*Freelance researcher, Kathmandu, Nepal*

Abstract

Cities in South Asia are experiencing storm water drainage problems due to a combination of urban sprawl, structural, hydrological, socioeconomic and climatic factors. The frequency of short duration, high-intensity rainfall is expected to increase in the future due to climate change. Given the limited capacity of drainage systems in South Asian cities, urban flooding and waterlogging is expected to intensify. The problem gets worse when low-lying areas are filled up for infrastructure development due to unplanned urban growth, reducing permeable areas. Additionally, solid waste, when dumped in canals and open spaces, blocks urban drainage systems and worsens urban flooding and waterlogging. Using hydraulic models for two South Asian cities, Sylhet (in Bangladesh) and Bharatpur (in Nepal), we find that 22.3% of the land area in Sylhet and 12.7% in Bharatpur is under flooding risk, under the current scenario. The flood risk area can be reduced to 3.6% in Sylhet and 5.5% in Bharatpur with structural interventions in the drainage system. However, the area under flood risk could increase to 18.5% in Sylhet and 7.6% in Bharatpur in five years if the cities' solid waste is not managed properly, suggesting that the structural solution alone, without proper solid waste management, is almost ineffective in reducing the long-term flooding risk in these cities.

Keywords: Climate change; Hydraulic modelling; South Asia; Urban flooding; Waste management; Waterlogging

doi: 10.2166/wp.2019.174

1. Introduction

Unplanned urban growth and development interferes with the natural flow of water and hence usually increases the risk of urban flooding (Booth, 1991; Douglas *et al.*, 2008). For the purposes of this paper, there are two consequences of urban development in South Asia: (i) it augments run-off and hence increases the risk of the breaching of embankments if a city abuts a river; and (ii) it reduces the flow of drainage by gravity due to drainage congestion, as urban drainage systems are unable to keep pace with urban development. In general, urban development, if unplanned, often reduces the effectiveness of urban drainage systems and increases the risk of flooding and waterlogging (Clemens & Veldhuis, 2010). Developing countries are particularly susceptible to urban flooding because their drainage systems are limited, mismanaged and often congested for a variety of reasons, including because of the dumping of solid waste in drains and canals (Zurbrugg, 2003; Haque, 2013).

Solid waste dumping in drainage routes reduces the carrying capacity of the drainage system, resulting in unwanted waterlogging in the event of heavy rainfall. Waterlogging due to drainage congestion contaminates ground water, which in turn creates a public health problem if cities use ground water or polluted surface water for domestic purposes (Phanuwan *et al.*, 2006; ten Veldhuis *et al.*, 2010). Most of the fast-growing cities in South Asia are located on riversides and developed without proper urban planning and appropriate drainage routes. Urban flooding is exacerbated as a result of rapid but unplanned urbanization and the resulting increased impervious surface (Huong & Pathirana, 2013).

Evidence suggests that poor households are disproportionally affected by seasonal flooding, as they mostly settle in flood-prone, low-lying marginal land, where land prices are relatively cheaper and the areas are unrestricted, allowing for unplanned settlement (IPCC, 2007; Nchito, 2007; Douglas *et al.*, 2008). With high-intensity and high-frequency climate and weather events such as excessive rainfall in short duration, waterlogging and flooding pose a serious risk for such unplanned and unprepared cities in terms of property damage, disruption of the daily lives of the residents, and water contamination and the resulting health problems (Patz *et al.*, 2005). Alderman *et al.* (2012) reported an increased risk of post-flood disease outbreaks in areas with poor hygiene and displaced populations. Potential diseases include hepatitis E, gastrointestinal diseases and leptospirosis. The rapid rural to urban migration, land use change and poor urban infrastructures in developing countries exacerbate flooding and waterlogging. Cities need to prepare climate-resilient development pathways to adapt to extreme climate events and disaster risk reduction, as planned adaptation helps to reduce disaster risk significantly (IPCC, 2018).

Lele *et al.* (2018) elaborated on a framework by Srinivasan *et al.* (2013) which illustrates how multiple stressors, multiple concerns and links between the basin, water supply and sewerage utility and household scales of analysis are explicitly incorporated in the climate change adaptation debate in an urban context. The authors put emphasis on pre-existing conditions, the urban water infrastructure, equity and institutions as the major normative issues that need to be considered when analysing water-related outcomes (drinking as well as wastewater) in the urban environment. In order to maintain good environmental conditions in cities, some of the main adaptation options include constructing an adequate drainage system, and a comprehensive solid waste collection and disposal system, amongst others (SANDEE & ACD, 2017). There is, however, very little literature examining the effectiveness of soft interventions such as proper management of solid waste by changing human behaviour to reduce the risk of urban flooding and waterlogging in the context of extreme climatic events in South Asia.

In this paper, we examine the existing designs of the drainage systems in Sylhet (in Bangladesh) and Bharatpur (in Nepal) to understand how the inadequate capacity of a drainage system and the unplanned dumping of solid waste result in higher risks of waterlogging and flooding in the low-lying areas of two South Asian cities. We develop drainage network models and use different scenarios to analyse how improved drainage infrastructure and management combined with the proper management of solid waste, as adaptation strategies, could reduce the risk of urban flooding and waterlogging under different climatic and non-climatic scenarios.

2. Methodology

2.1. Description of study area

For research purposes, we selected two medium-sized South Asian cities (with populations close to half a million or less) which are situated by rivers and foothills. The selection of these cities was guided by the idea that the research could influence the adaptation policies of these cities, as they are growing fast but are still manageable given their size, and there is a willingness amongst the city officials to be guided by research outcomes while preparing adaptation plans.

There are a number of areas in both Bharatpur and Sylhet that suffer from waterlogging and standing water after heavy raining during monsoon. Due to socioeconomic development and rural to urban migration, land use in these cities is changing continuously from agriculture to a built-up environment, leading to an increased volume and speed of run-off. The existing drainage systems fail to carry the huge volume of the increased amount of storm water, increasing the risk of flooding and waterlogging.

2.1.1. Sylhet. Sylhet city was declared a metropolitan area in 2002 with an area of 508.7 km^2. The Surma River divides the Sylhet City Corporation (SCC) area into two parts: the northern part, with an area of 18.18 km^2, and the southern part, covering 6.52 km^2. Only the northern core city area has been considered in this study, as it is a relatively high flood-risk area (Figure 1).

2.1.1.1. Climate. Sylhet has a tropical climate. In summer (March–May), the mean temperature reaches around 27°C, while in winter, the mean temperature stays at around 20°C. The mean annual rainfall is 4,000 mm. Most of the rainfall occurs during the summer monsoon season (June–August).

2.1.1.2. Flooding and drainage system. In Sylhet, two types of flood risks occur: flash floods in the Surma River, and localized floods caused by heavy monsoon rainfall. As the main drainage canals flowing through the city originate from sloped hills, the run-off from upstream areas travels fast towards the urban area, which is comparatively less sloped. Therefore, during heavy rainfall events, the rainfall–run-off causes flash floods.

Every year, the riverbed gets higher due to heavy siltation. Consequently, during events of moderate to high rainfall, the run-off cannot be evacuated through the canals (that are ultimately linked to the Surma River) due to the backwater effect. The run-off spills over into surrounding areas, causing widespread flooding and waterlogging. The waterlogging remains for 5–8 hours when the water level of the Surma River is low. However, when its water level is more than the danger level, which is 10.05 metres above mean sea level (mMSL), the backwater effect becomes very pronounced and the waterlogged situation continues for about 7–10 days until the water level of the river declines.

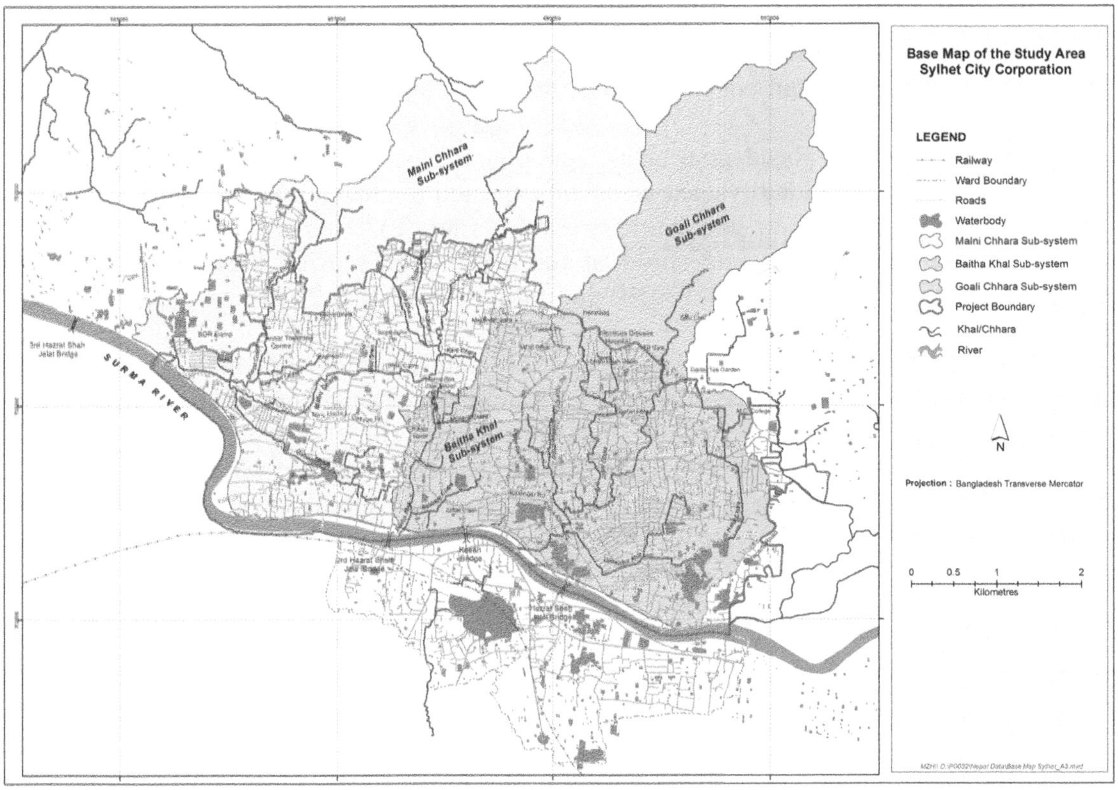

Fig. 1. Study area, Sylhet. (*Source*: the authors).

Subsidiary canals are locally called chhara or khal, and there are three main drainage sub-systems in the study area of Sylhet: Malni chhara, Bhaita khal and Goali chhara (Figure 1). These subsidiary canals drain from north to south and fall into the Surma River. Malni chhara and Goali chhara originate in hilly areas. Brief details of the drainage sub-systems are given in Table 1.

The capacity of these drainage canals is not sufficient in many areas to carry even moderate rainfall–run-off mainly due to unauthorized land use and solid waste dumping. Sylhet City Corporation sometimes cleans the existing chhara/drain when the situation becomes very bad but the city does not have a regular cleaning programme. After heavy rainfall, the roads and outside areas around housing become inundated in many places. Sometimes people build temporary brick walls around their houses as flood

Table 1. Sylhet city drainage sub-systems.

Canal	Number of tributaries	Total length, including tributaries (m)	Watershed area (km^2)
Malni chhara	17	23,600	16.26
Baitha khal	5	4,245	2.79
Goali chhara	12	18,013	12.98

Note: The number and length of tributaries are based on a survey and watershed areas are estimated after analysis of a Digital Elevation Model (DEM).

protection. Recently, Sylhet City Corporation adopted a 5-year Capital Investment Plan to improve environmental conditions, tourism facilities and infrastructure development, etc. Reducing waterlogging by constructing reinforced cement concrete drains, regulators and pumps is a major component of the project.

2.1.1.3. Land use patterns. Residential areas, agricultural plots and tea gardens dominate land use in Sylhet city. The agricultural and vacant areas are expected to change into residential and commercial land use in the future due to population pressure (Table 2). This will add more impermeable surfaces and increase the volume and speed of run-off, thus increasing the risk of flooding. The land use maps for the present situation (2010) and future conditions (2030) for Sylhet are presented in Figure 2.

2.1.1.4. Solid waste management. The Sylhet City Corporation's solid waste management programme requires improvement. Households pay for waste collection services that are managed by private clubs. The payment per household ranges between Bangladeshi taka (BDT) 50–150 per month, and about 80% of households participate in the waste collection service. The clubs pick up waste from households six times a week. However, they only take it to transfer stations; final disposal is done by the city. Currently, households and institutions that generate waste do not segregate it at source, and hence there has been no practice of composting, few materials are recovered and there is a low rate of recycling. The city provides large street bins but the collection is not regular so the bins are over-filled and the waste gets scattered, ending up in the streets and ultimately in the drains, which reduces the conveyance capacity of the drainage system and often blocks the flow of water. Households that do not subscribe to the collection services also dispose of their solid waste in open spaces, which ultimately lead to the drainage canals.

Moreover, the utility tunnels at canal crossings also decrease the conveyance capacity of the drainage network below the culverts and obstruct the free flow of water, which results in local flooding. The city does not have enough capacity (both technical and financial) to manage solid waste; it does not fully cover the residential areas to pick up waste from residents, and it does not have a sanitary landfill site for safe disposal. To improve the capacity of Sylhet City Corporation, additional resources (human, material and financial) and proper training of human resources are required. In the case of segregating waste at source, the resources needed for collection and disposal of the waste will be less, since proper segregation at source means more material recovery (for reuse and recycling) and less waste for disposal when organic wastes are composted at the community or household level (suitable for those who have kitchen gardens). The research team set up a community level demonstration of composting organic waste in a few areas on an experimental basis. After the demonstration, residents expressed their willingness to adopt the technology if provided with suitable training and technological support.

Table 2. Existing and proposed Sylhet city land use. (*Source*: Government of the People's Republic of Bangladesh, 2010.)

	% of area			% of area	
Land Type	Existing	Proposed	Land Type	Existing	Proposed
Residential	39.4	54.3	Vacant Land	13.3	–
Agriculture	7.2	3.4	Education & Research	4.4	4.5
Tea Garden	4.9	4.7	Transport & Communication	5.0	7.5
Others	26.0	25.7			

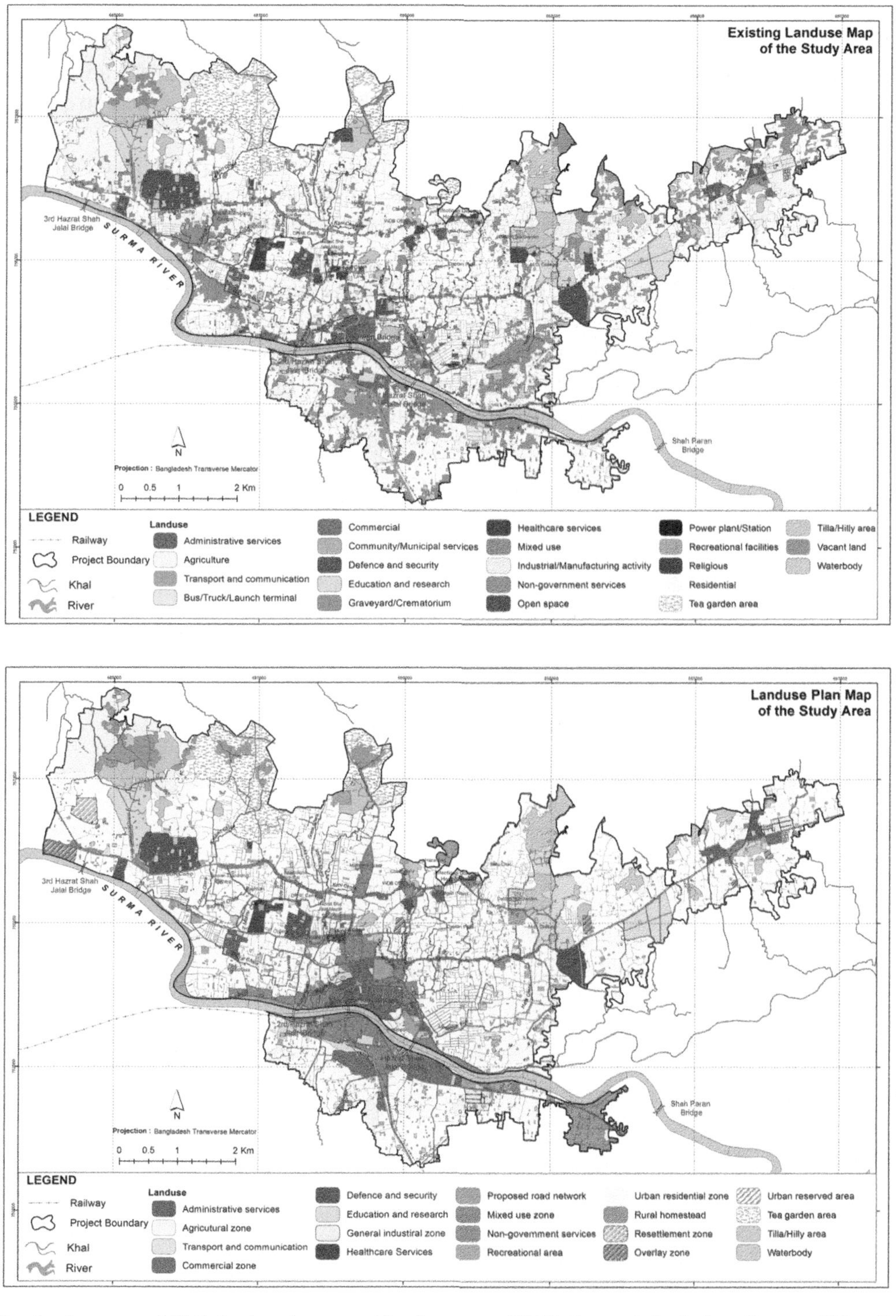

Fig. 2. Existing land use map (2010) (top) and proposed land use map (2030) (bottom) for Sylhet. (*Source*: Urban Development Directorate (UDD), 2010.)

2.1.2. Bharatpur. Bharatpur city is situated in south-central Nepal, in Chitwan district, which lies on the left bank of the Narayani River. Bharatpur municipality was declared a sub-metropolitan city in 2014 (with a total area of 77.91 km^2) and then a metropolitan city in 2016 (with a total area of 418.8 km^2). The study area is located in the then municipality area, which is the core area of the current metropolitan city (Figure 3).

2.1.2.1. Climate. Bharatpur has a humid, subtropical climate. The temperature peaks in June–August, when the mean temperature is approximately 28°C. During the winter season (December–February), the mean temperature is around 19°C. The mean annual rainfall is around 2,100 mm. The heaviest rainfall occurs during June–September.

2.1.2.2. Flooding and drainage system. There are two main drainage canals flowing through Bharatpur city: the Pungi canal and Kerunga canal. Both canals originate in the northern hilly areas and travel in a southwesterly and southerly direction, and finally meet the Narayani and Rapti rivers, respectively. There is no backwater problem in Bharatpur. However, even after moderate rainfall, run-off cannot find a proper drainage route to reach the existing canals, which causes flooding. The waterlogged situation

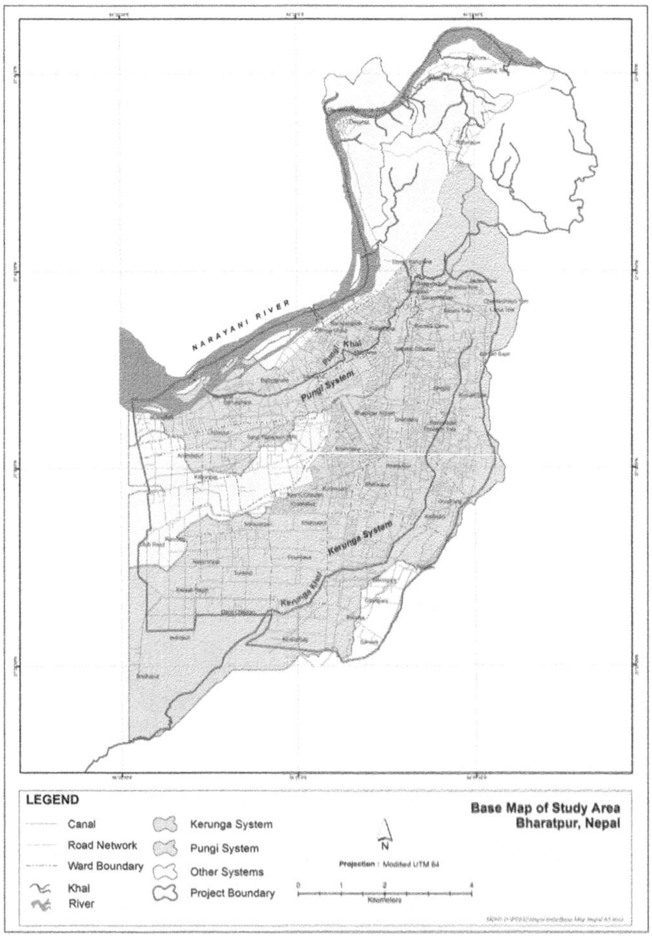

Fig. 3. Study area, Bharatpur. (*Source*: Sayenju & Talchabhandari, 2008.)

persists for a period ranging from a few hours to 1–2 days in some parts of the city. After the rain stops, the run-off flows slowly through natural drainage routes and reaches the canals.

The Pungi canal is 6.6 km long, extending from Aptari Bishal Marg to the outfall at Ramghat on the Narayani River. The length of the Kerunga canal is 28.4 km, extending from Shahid Dharma Bhakta Marg to the outfall on the Rapti River. About 14 km of the Kerunga canal lies within the study area. There are some underdeveloped areas on the western side of the study area, where no defined drainage system exists. The storm water from these parts flows to the western low-lying areas and ultimately into the Narayani River. The watershed of the Pungi canal covers 13.03 km^2, while that of the Kerunga canal covers 41.75 km^2.

2.1.2.3. Land use patterns. Most of Bharatpur city's land area is characterized by agricultural (52%) and forest land (21%). The economy of Bharatpur is mainly dependent on agriculture. Nevertheless, some agricultural lands are converting into residential areas due to rapid population growth in the city and the resulting change in economic activities. The extent of land use change is quite substantial, as the population growth rate is over 6% annually, but there is no definite plan in Bharatpur city for urban development and thus there is a risk of erratic urbanization. A land use map from 2008 is shown in Figure 4.

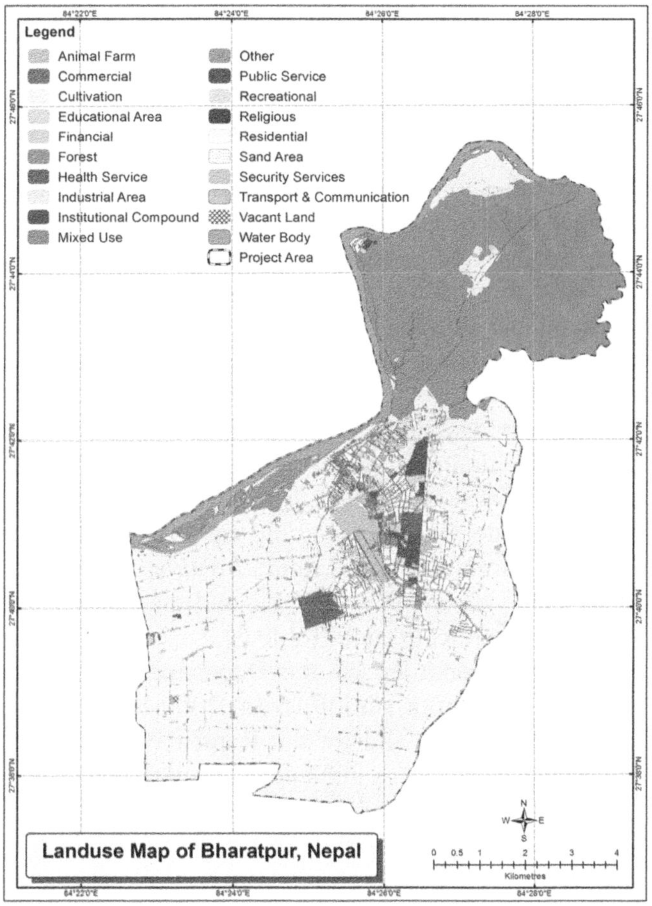

Fig. 4. Land use map for Bharatpur city. (*Source*: Sayenju & Talchabhandari, 2008.)

2.1.2.4. Solid waste management. In Bharatpur, solid waste is managed based on a public-private partnership approach. Two private firms are responsible for collecting and disposing of solid waste from the study area. They provide a door-to-door collection service and charge a certain amount based on the frequency of service provided (to the households) and the amount of waste generated (in the case of hotels, business and institutions). For households, the fee ranges between Nepalese rupee (NPR) 30–100 (which is close to US Dollar (USD) 0.30–1.00) per month. The money collected is not sufficient to manage the municipal solid waste and the city provides an additional budget to the private firms. As the collection service is based on a service charge payment and is voluntary, not all households pay the fee and the participation rate is reported to be around 70%–80%. The rest of the households do not pay the service fee and mainly dispose of their waste in open spaces and drains.

Households who pay the service fee reported that the waste pick-up service is often irregular and they do not know the exact time. The contractors reported that the resources they have generated from the users, on top of the transfer from the city, are not enough to properly manage the municipal solid waste and, therefore, they cannot provide the service at a pre-determined time. Due to the mismatch between the time households put their waste outside for collection and the exact collection time by the collectors, rag pickers often scatter the waste in search of reusable and recyclable materials and the waste ends up in open spaces, on streets and in drainage canals.

2.2. Data collection

The research team visited both cities to get first-hand information on existing drainage networks, episodes of flooding and potential causes, and solid waste management practices. Meetings with city officials, community leaders and private companies who collect and dispose of solid waste were used to get necessary background information. Several focused group discussions with residents were organized in order to understand the flooding and waterlogging issue and their perceived causes in both cities. After rounds of such meetings, the research team determined the study area within each city in consultation with city officials.

Since there was a lack of necessary data, the research team collected field level data in the summer of 2017 from both cities to develop the drainage model. Data pertaining to the primary water level, water discharge, and existing drainage networks and their cross-sections were collected from the field. The primary data were used for both model development and calibration. Moreover, historical water-level data (from 1938 to 2015) for the Surma River and rainfall data at Sylhet recorded by the Bangladesh Water Development Board (BWDB) (from 1957 to 2011) and the Bangladesh Meteorological Department (BMD) (from 1957 to 2015) were collected (Table 3).

Our model included rainfall data recorded by the Department of Hydrology and Meteorology (DHM) for Bharatpur city at the Bharatpur station (from 2000 to 2016; Table 3), water-level data for the Narayani River at Devghat station, and for the Rapti River at Rajayia station. The bankfull levels of the Narayani and Rapti rivers were considered as the design water level for Bharatpur. The surveyed cross-sections of the Narayani and Rapti were interpolated to generate bankfull water-level data at the outfalls of the Pungi and Kerunga canals. For Sylhet city, the research team had access to water and rainfall data. Therefore, a once in 50 year return period was considered as the design flood level for the Surma River, for the design of the regulator and pumps (see section 2.3.3).

Table 3. Historical data analysis of rainfall and water level.

	Sylhet station			Bharatpur station			Surma River
Return period	1-day max rainfall	2-day max rainfall	5-day max rainfall	1-day max rainfall	2-day max rainfall	5-day max rainfall	Maximum water level (mMSL)
2	191	278	804	141	208	282	10.03
2.33	201	291	835	152	221	297	10.07
5	249	345	987	202	275	361	10.36
10	289	**389**	1,114	246	**313**	407	10.53
25	343	445	1,278	304	355	462	10.72
50	385	486	1,404	349	383	500	**10.84**
100	428	528	1,531	395	410	536	10.95

Note: Rainfall values are in mm; the 2-day and 5-day max rainfalls indicate cumulative maximum rainfalls for that duration.

2.3. Data analysis

The study developed mathematical models for urban drainage systems and calibrated the models to represent the present situation (as of 2017). Once the model was tested for robustness, climate change and other scenarios were superimposed on them. The following steps were taken while developing and applying the models: (i) understanding the existing drainage system and the causes for drainage congestion; (ii) cross-verification of the collected data; (iii) developing a one-dimensional drainage model and calibrating/validating it; and (iv) simulating different scenarios using the calibrated/validated model to analyse the impact in the cities, and suggest measures for drainage improvement.

MIKE11 modelling tools from the Danish Hydraulic Institution (DHI) were used to develop the drainage models for Sylhet and Bharatpur. MIKE11 is a one-dimensional modelling system suitable for rivers and channels. The drainage model contains two main systems: a hydrological and a hydrodynamic system. The hydrological model was developed using Mike Urban A, a lumped hydrological model. It uses the 'Time–Area' method, by which the run-off amount is controlled by the initial loss, size of the catchment area, percentage of the impervious area in the catchments, and losses due to evapotranspiration and depression storage. The output of the hydrological model was used as an input for the hydrodynamic model.

The delineation of the catchments for Sylhet and Bharatpur were made based on Digital Elevation Models (DEMs) (Figure 5). Seventy-seven sub-catchments were delineated for the three Sylhet drainage systems. There are no defined drainage routes in Bharatpur other than the Pungi and Kernga canals. There are two main catchments (watersheds) for Bharatpur city and some minor catchments, from where water flows into the Narayani River directly, or into low-lying areas following the natural gradient.

The catchment parameters were chosen to reflect their characteristics. Using a spatial analysis based on aerial images and land use data, impervious areas were determined for each of the catchments.

Three individual sub-models (the Malni chhara, Baitha khal and Goali chhara sub-models) were developed for Sylhet. The sub-models consist of major canals/drainage channels, along with their tributaries, on the right bank of the Surma River. The drainage network for each individual canal was developed using the survey data of the canal's cross-sections and its topography in 2016 by the Institute of Water Modelling, Dhaka. The boundary data consist of hydrometeorological data, including rainfall

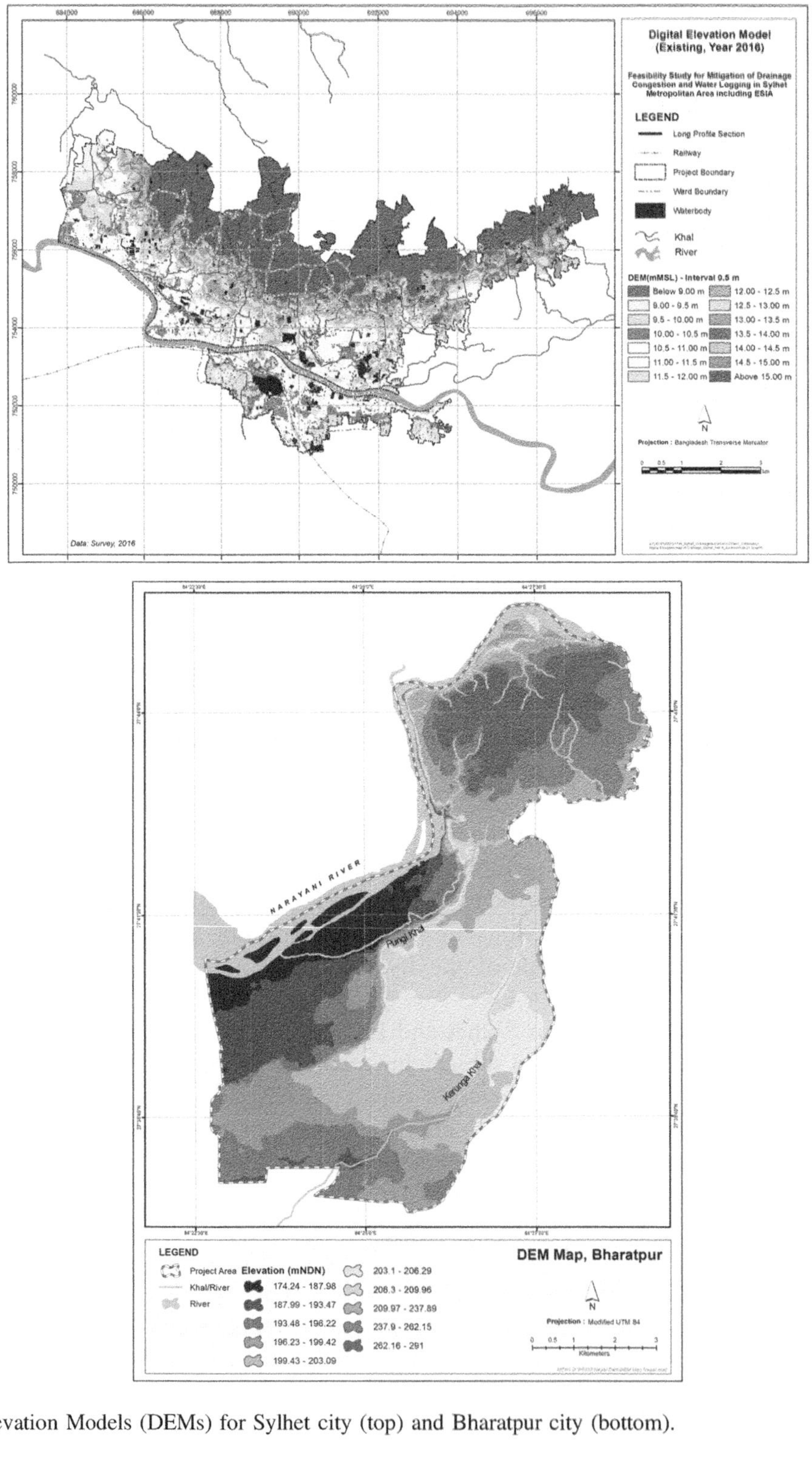

Fig. 5. Digital Elevation Models (DEMs) for Sylhet city (top) and Bharatpur city (bottom).

and evapotranspiration data from the Sylhet station, and downstream water-level data for the Surma River.

For Bharatpur, two sub-models (the Pungi canal and the Kerunga canal sub-models) were developed using surveyed cross-sections of these canals. The boundary data consist of hydrometeorological data, including rainfall data from the Bharatpur station and evapotranspiration data from the Kaligandaki station, and downstream bankfull water-level data for the Narayani and Rapti rivers as water levels.

2.3.1. Model calibration and validation. The Sylhet model was calibrated and further validated against water-level data from 2016 and 2017, respectively. The Bharatpur model was calibrated against collected water-level and discharge data for 2017. Calibration and validation plots are shown in Figures 6–10. In both cities, the catchments are rapid response catchments, due to their steep gradients from hilly areas. For calibration/validation of steep slope catchments, 10 min or 30 min interval rainfall data are necessary to match the peak flows. However, such shorter interval rainfall data are not available. Therefore, the calibration/validation of the models are not as accurate as expected since we used three-hour interval data for Sylhet and daily data for Bharatpur.

2.3.2. Climate change and socioeconomic scenarios. According to the synthesis of the IPCC's (Intergovernmental Panel on Climate Change) *Fifth Assessment Report* (IPCC, 2014), climate change will lead to three major changes in Asia: (i) changes in glaciers, snow, ice and permafrost; (ii) flooding of rivers/lakes and/or droughts; and (iii) coastal erosion and/or sea level rise and other effects. However, Sylhet and Bharatpur are not coastal cities: Sylhet city is about 350 km north of the Bay of Bengal and sea level rise will not affect water levels in the Surma River; a similar situation applies to Bharatpur city which is away from any coastal area. Therefore, we developed climate change scenarios focussing solely

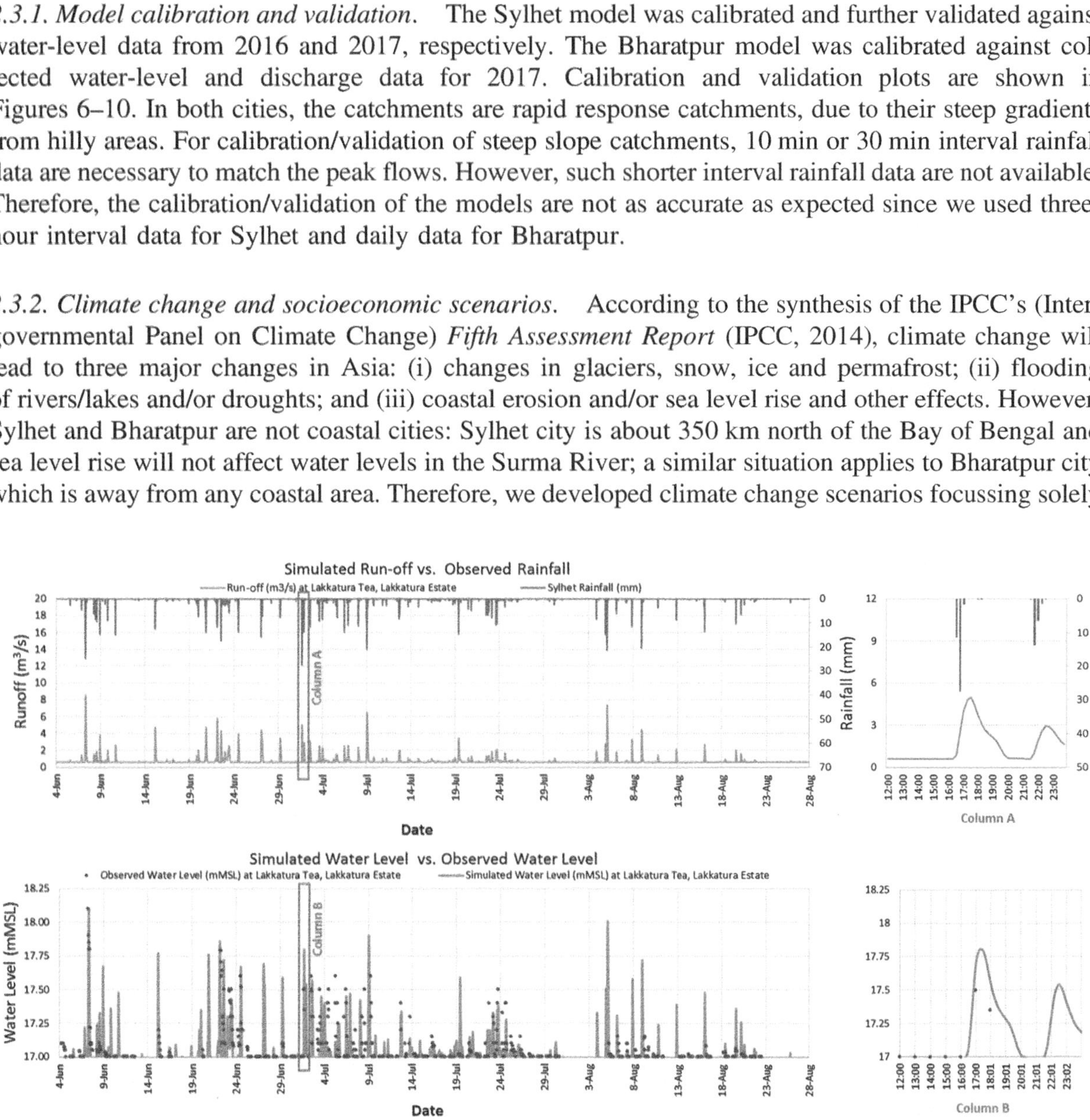

Fig. 6. Model calibration plot for the Lakkatura Tea Estate on Lakkatura Chhara, Malni Chhara sub-system.

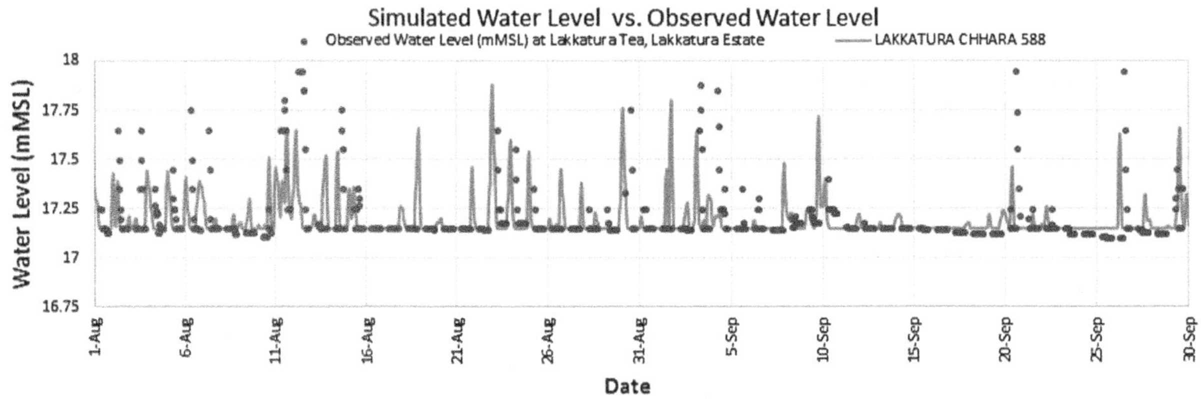

Fig. 7. Model validation plot for the Lakkatura Tea Estate on Lakkatura Chhara.

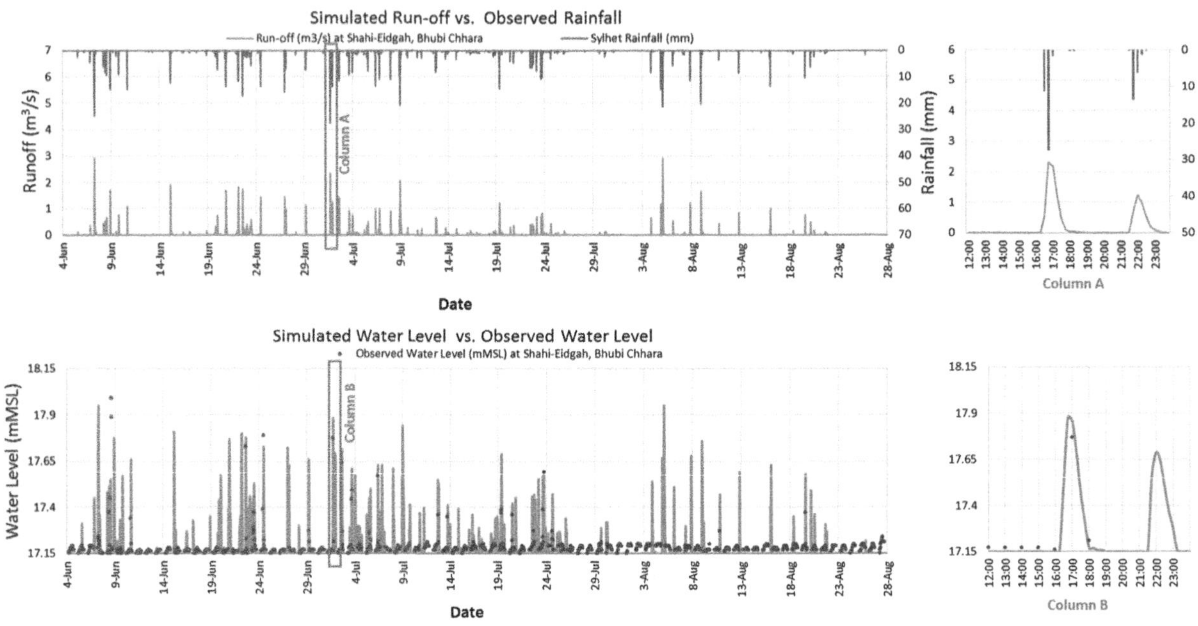

Fig. 8. Model calibration plot for Arambag on Goali Chhara, Goali Chhara sub-system.

on changing rainfall patterns. Climate change is happening and many climate models show that, due to climate change, the volume of rainfall will increase during the summer monsoon (May–October). The models also indicate reduced rainfall during the dry period (December–March).

Moreover, the frequency of short duration, high-intensity rainfall is expected to increase during the monsoon in both cities. This will result in more run-off in short periods, as both cities lie at the foothills which would aggravate drainage congestion. The rainfall pattern for the high emission Representative Concentration Pathway (RCP) 8.5 scenario was used to access the worst conditions using the CNRM ARPEGE climate model, which is suitable for the Ganges basin (in which Bharatpur lies) and the

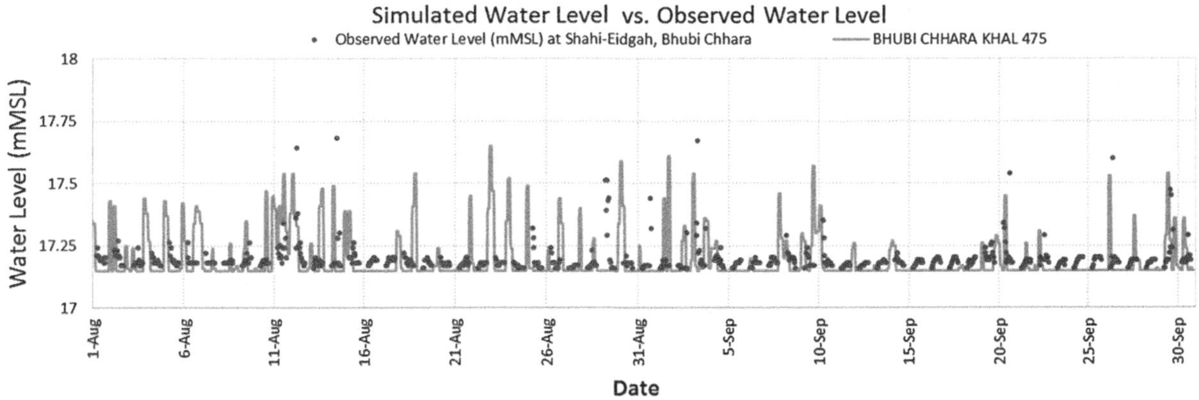

Fig. 9. Model validation plot for Shahi Eid-gah on Bhubi Chhara, Goali Chhara sub-system.

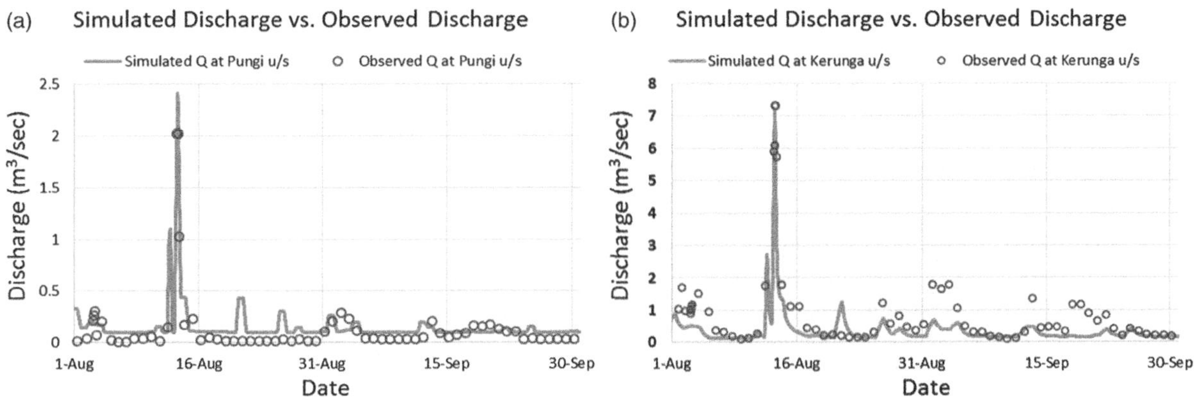

Fig. 10. Model calibration plots for locations (a) near Tinkune Chawk on the Pungi canal and (b) near Sree Naya Kiran Secondary School on the Kerunga canal.

Meghna basin (for Sylhet) (IWM, 2016). The CNRM model suggests 11% and 20% increases in rainfall intensity during the summer monsoon for Sylhet, in 2030 and 2050, respectively. The model indicates a 7% and 13% increase in rainfall intensity for Bharatpur, in 2030 and 2050, respectively, during the summer monsoon (IWM, 2016).

This study also considers socioeconomic scenarios. In terms of economic trends, both of the cities are growing fast with a population growth rate of more than 6%. High urban population growth leads to rapid changes in land use, from agriculture to built-up areas. With economic growth, the land prices in these cities have been increasing rapidly. As a result, land filling is taking place in both cities, converting low-lying agricultural land into housing and commercial spaces. The study, therefore, used urban population growth in these cities and land filling to predict land use changes in the future (in 2030 and 2050).

2.3.3. Identifying different scenarios. Calibrated/validated drainage models for Sylhet and Bharatpur were used to study different structural and non-structural adaptation options/scenarios to understand flooding/waterlogging risks. The scenarios are presented in Table 4. The simulated results are presented in Tables 5 and 6 and, based on the results, suitable adaptation option(s) for reducing the future flooding/waterlogging risks in these two cities were discussed.

These adaptation scenarios or lack thereof were developed after having several focus group discussions and in-depth interviews with the municipal authorities, local residents and key informants in both cities. Scenario Sc-0 is the baseline scenario and simulates the current conditions of the drainage

Table 4. Scenarios for improving drainage.

Scenario ID	1. Scenarios for Sylhet	2. Scenarios for Bharatpur
Sc-0 (Baseline)	*Existing situation with design event* ○ Rainfall: 2-day cumulative, once in 10 year rainfall ○ Water level (WL): once in 50 year water level of the Surma River ○ Socioeconomic conditions: existing land use	*Existing situation with design event* ○ Rainfall: 2-day cumulative, once in 10 year rainfall ○ Water level (WL): bankfull level of the Narayani and Rapti rivers ○ Socioeconomic conditions: existing land use
Sc-1	*Existing situation with design event and climate change* ○ Sc-0 + climate change impacts (in 2030 and 2050)	*Existing situation with design event and climate change* ○ Sc-0 + climate change impacts (in 2030 and 2050)
Sc-2	*Rehabilitation, expansion, and re-sectioning of drainage channels* ○ Rehabilitation/expansion/re-sectioning of the existing inadequate channels ○ Construction of new drains ○ Two silt traps on the upstream side of Malni and Goali chharas ○ Reconstruction of the inadequate culverts ○ Solid waste disposal facilities	*Rehabilitation, expansion, and re-sectioning of drainage channels* ○ Rehabilitation/expansion/re-sectioning of the existing inadequate channels ○ Construction of new drains ○ Solid waste disposal facilities
Sc-3	*Structural interventions using regulators and pumps* ○ Sc2 + regulators and pumping stations at outfalls	(Not relevant)
Sc-4	*Physical interventions and climate change* ○ Sc-3 + climate change impacts (in 2030 and 2050)	*Physical interventions and climate change* ○ Sc-2 + climate change impacts (in 2030 and 2050)
Sc-5	*Trends in economic growth* ○ Sc-3 + socioeconomic conditions: Future land use (in 2030 and 2050)	*Trends in economic growth* ○ Sc-2 + Socioeconomic conditions: predicted land use change based on population for 2030 and 2050
Sc-6	*Trends in economic growth and climate change* ○ Sc-5 + climate change impacts (in 2030 and 2050)	*Trends in economic growth and climate change* ○ Sc-5 + climate change impacts (in 2030 and 2050)
Sc-7	*Solid waste dumping in the drainage system* ○ Sc-2 + no solid waste disposal and cleaning facilities	*Solid waste dumping in the drainage system* ○ Sc-2 + no solid waste disposal and cleaning facilities

Source: Developed by the authors based on information from stakeholders, including officials from the respective municipal offices.

Table 5. Summary of drainage model simulations by scenario (Sylhet).

Scenarios >>>>>>>	Sc-0 Baseline	Sc-1 Climatic change in 2050 with existing situation	Sc-2 Rehabilitation/ expansion/re-sectioning of drainage canals	Sc-3 Structural interventions	Sc-4 Structural interventions and climate change in 2050	Sc-5 Changes in socioeconomic conditions in 2050	Sc-6 Changes in socioeconomic conditions and climate change in 2050	Sc-7 No waste management
Flooded area (ha)	486	568	374	164	193	109	106	421
Percentage area of flooding in northern city (%)	22.3	27.1	15.8	3.6	5.2	2.8	2.6	18.5
Total area that benefits (ha)	–	–82	112	322	293	377	380	47
Additional area that benefits (ha)	–		–	210	–	55	3	–
Additional investment (USD million)	–		31.62	43.46	11.25	0	11.25	–
Investment per ha (USD million)	–		0.28	0.21	–	0	3.75	–

Note: 1 $km^2 = 100$ ha; 1 ha $= 0.01\ km^2$.

Table 6. Summary of drainage model simulation using scenarios (Bharatpur).

Scenarios >>>>>>>	Sc-0 Baseline	Sc-1 Climate change in 2050 with baseline	Sc-2 Rehabilitation/ expansion/re-sectioning of drainage channels	Sc-4 Structural interventions and climate change in 2050	Sc-5 Changes in socioeconomic conditions in 2050	Sc-6 Changes in socioeconomic conditions and climate change in 2050	Sc-7 No waste management
Flooded area (ha)	166	175	72	77	86	94	99
Flooded area (%) (Pungi catchment)	12.7	13.5	5.5	5.9	6.6	7.2	7.6
Total area that benefits (ha)	–	–	93	89	80	71	66
Additional inundated area (ha)	–	–	–	4	14	8	27
Additional investment (USD million)	–	–	0.38	–	–	–	–
Investment per ha (USD million)	–	–	0.004	–	–	–	–

Note: 1 $km^2 = 100$ ha; 1 ha $= 0.01$ km^2.

system. Scenario Sc-1 introduces climate change over Sc-0, to understand the impacts of climate change if no drainage interventions were applied. Scenario Sc-2 includes an improved drainage system that is superimposed on the baseline scenario Sc-0. Scenario Sc-3 is a further superimposition with control on backwater flows from the Surma River. This scenario applies to Sylhet but not for Bharatpur as there is no backwater effect in Bharatpur. In scenario Sc-4, climate change impacts over the improvement scenarios – Sc-3 (Sylhet) and Sc-2 (Bharatpur) – are simulated. Scenario Sc-5 considers changes in land use patterns in the future, based on current trends. Scenario Sc-6 includes changes in socioeconomic conditions (income and living standards, measured as changes in housing patterns) and climate change scenarios in terms of variability and changes in rainfall patterns. Scenario Sc-7 describes the effects of poor waste management and dumping solid waste in the chhara/khal and canals, which is superimposed on scenario Sc-2.

3. Results

3.1. Sylhet city

3.1.1. Existing situation with design event. In scenario Sc-0, due to the backwater effect of the Surma River in Sylhet, the catchment run-off cannot drain out quickly, which causes flooding and the situation can exist for about seven days (observed from historical records). The backwater effect of the Surma River progresses up to 3,000 metres inside the Malni chhara, 900 m inside Boitha khal, and 2,100 m inside the Goali chhara. The simulated area of inundation is around 486 ha, including approximately 103 ha of water bodies (ponds, lakes etc., other than canal areas) (Figure 11).

3.1.2. Existing situation with design event and climate change. Due to climate change (as described in section 2.3.2), run-off in the monsoon could increase in 2030 and 2050, which will increase the inundation area from the Sc-0 scenario. The simulated inundation area would be around 548 ha in 2030 and 568 ha in 2050, including water bodies covering 103 ha. Therefore, due to climate change, an additional 62 hectares (ha) in 2030 and 82 ha in 2050 would be inundated if no adaptation measures are taken.

3.1.3. Rehabilitation, expansion and re-sectioning of drainage channels. To improve the drainage situation from the baseline scenario (Sc-0), rehabilitation/expansion/re-sectioning of the existing inadequate channels, with the provision of some new drains and culverts, was incorporated in scenario Sc-2. The simulated inundation area would be 374 ha, which includes the 103 ha of water bodies. With these interventions, an area of 112 ha would benefit, in terms of avoiding flooding.

3.1.4. Structural interventions using regulators and pumps. For further improvement in the drainage system in Sylhet, structural solutions like regulators and pumps were considered at the outfall of Malni chhara and Goali chhara under scenario Sc-3. These regulators would prevent the backwater effect of the Surma River and would permit drainage by gravity when the outfall water level is higher than the water level in the Surma River. When the water level in the Surma River rises to 10 mMSL, the gate would close and the pump would start expelling the excess run-off generated by heavy rain. The pump capacity was estimated in such a way as to maintain the water level in the canals and channels close to this danger level (10.05 mMSL). After several trials, the optimum pump capacity for both canals was estimated at

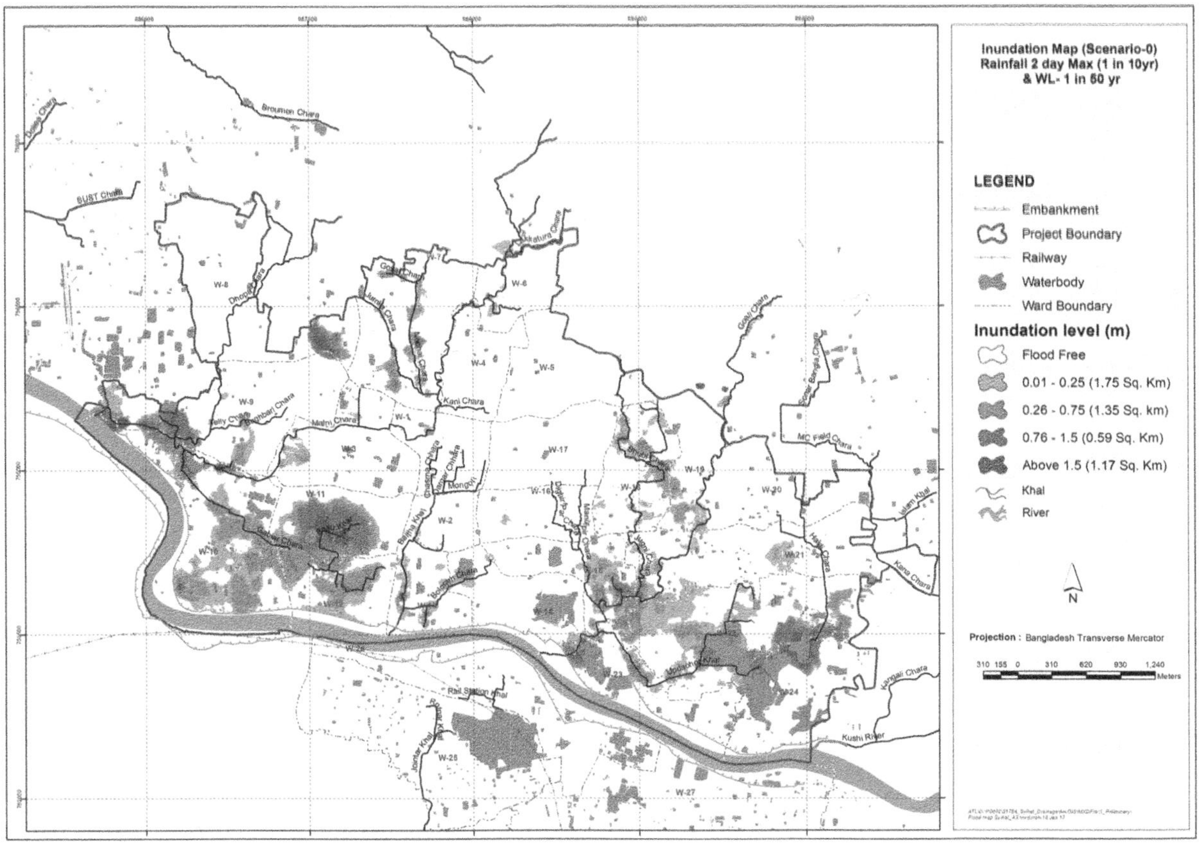

Fig. 11. Flood map for Sc-0 (baseline), Sylhet.

22 m^3/sec, whereby the water level remains above the danger level for only 1.2 hours in Goali chhara and 6 hours in Malni chhara. Due to these structural interventions, 322 ha of land benefit by avoiding waterlogging. The simulated inundation area would be 164 ha, which includes the 103 ha of water bodies (Figure 12).

3.1.5. Physical interventions and climate change. Climate change impacts in 2030 and 2050 over the improved scenario (Sc-3) were simulated in scenario Sc-4. It was observed that the extent of flooding and duration of inundation both increased over Sc-3, due to additional run-off generated by climatic conditions. In order to adapt this, one additional pump of 7.33 m^3/s would be required in 2030 for Malni chhara, and another pump of 7.33 m^3/s in 2050. However, no additional pump is required for Goali chhara in 2030 or 2050. Due to these structural interventions, 294 ha of land would benefit in 2030, and 293 ha in 2050 in terms of avoided waterlogging. The simulated inundated area would be 192 ha and 193 ha, in 2030 and 2050 respectively, which includes water bodies covering 103 ha.

3.1.6. Trends in economic growth. Socioeconomic development would alter land use. The land use plan for 2030 proposed by the city was considered for Sylhet city. The land use plan will increase

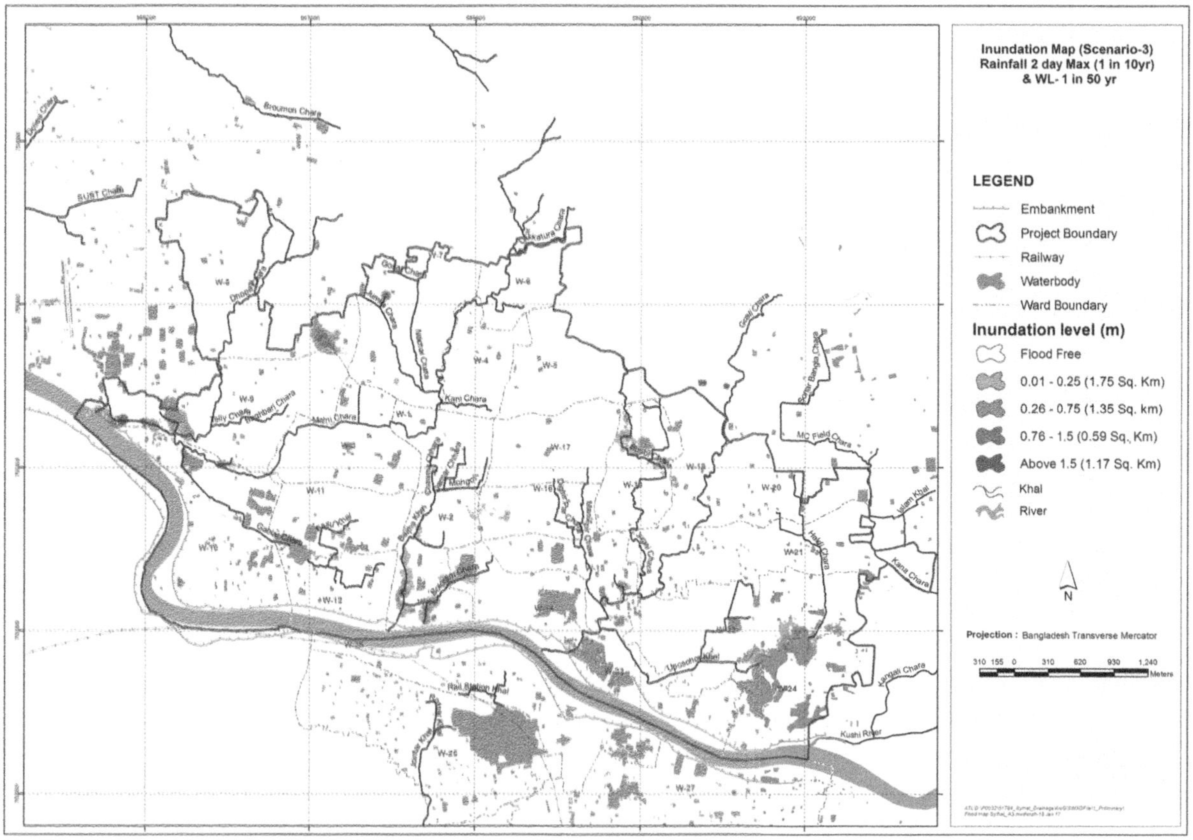

Fig. 12. Flood map for Sc-3 scenarios, Sylhet.

the extent of the impervious area in each sub-catchment, and will increase run-off patterns, which were simulated in scenario Sc-5. In 2030, the elevation of the land would increase to a level of 12 mMSL due to land development and land use changes. The land development level is considered 1.15 m above the once in 50 year water level of the Surma River. Due to land development, which is one form of adaptation, the inundated area would be reduced in 2030, and no additional pump (as estimated in Sc-3) would be required for both Malni chhara and Goali chhara. About 377 ha of additional land area would be free from waterlogging as a result of land development, but the duration of the water above the danger level would increase to 7 hours in Malni chhara and 3 hours in Goali chhara. The simulated area inundated would be 109 ha, which includes water bodies covering 60 ha (the area covered by water bodies declines due to land use changes by 2030).

3.1.7. Trends in economic growth and climate change. Future climate change, superimposed over an altered socioeconomic condition, was simulated in scenario Sc-6. It was considered that the proposed land use plan would be implemented in 2030, even as the city would simultaneously experience the effects of climate change with increased economic activities. To cope with the excess storm water that would ensue due to climate change and to ensure adequate drainage facilities, an additional pump (7.33 m^3/s) would be required for both Malni chhara and Goali chhara in 2030. Yet another pump would be required in 2050. The land area that would benefit by being free of inundation would be 380 ha. The simulated inundated area would cover 1.06 km^2, including water bodies covering 60 ha.

3.1.8. Solid waste dumping in the drainage system. Solid waste dumping in the drainage canals reduces conveyance capacity (Figure 13). It was observed from the field survey that due to the dumping of waste, the depth of the canal decreased by 11 cm, 10 cm, and 20 cm in Malni, Baitha and Goali canals, respectively, in a one-year period. If no cleaning or dredging operation is carried out with the rehabilitation/expansion/re-sectioning of existing canals, the depth will decrease continuously. The possible effects after five years of solid waste dumping in the canals were simulated under scenario Sc-7. It was observed that the inundated area would be 421 ha, including water bodies spread over 103 ha. Therefore, the area that would benefit after rehabilitation/expansion/re-sectioning would decrease by approximately 47 ha.

Since the indiscriminate dumping of solid waste critically impairs the conveyance capacity of the drains, improving the drainage infrastructure alone without considering solid waste management would be a waste of resources. Therefore, improving waste management systems by adding segregation at source and placing waste collection bins in the streets are important adaptation activities. Waste collection bins were considered in improvement scenarios Sc-2 to Sc-6, at a cost of USD 0.51 million.

Therefore, the important adaptation options include both infrastructural change and behavioural change. Infrastructural changes in Sylhet include: (i) the rehabilitation/expansion/re-sectioning of existing drainage canals; (ii) the construction of new drains; (iii) reconstruction of culverts; (iv) the construction of silt traps; (v) the placement of solid waste collection bins; (vi) regulators; and (vii) a drainage pumping station. The behavioural change includes improved solid waste management with segregation at source and separate collection of different types of waste for proper disposal. The extent of flooding and the cost of investment for each scenario are given in Table 5.

3.2. Bharatpur

3.2.1. Existing baseline situation with design event. In the baseline scenario for Bharatpur, Sc-0, the existing situation of the canal, bankfull water levels for the Narayani and Rapti rivers, and cumulative

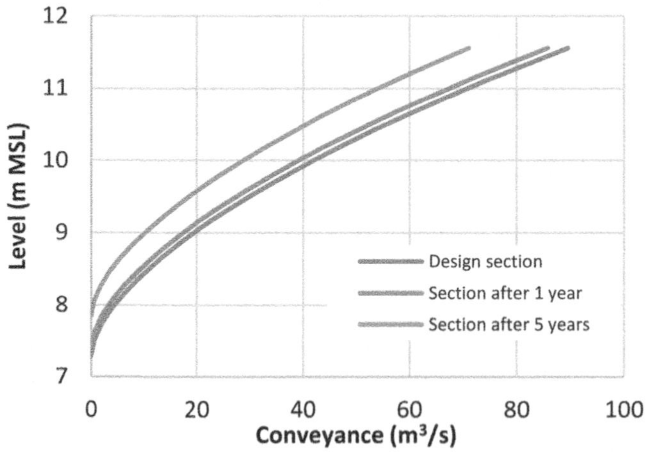

Fig. 13. Change of conveyance capacity due to solid waste dumping.

two-day, once in 10 year rainfall data were used. As there is no backwater effect, water flows through the canals by gravity. The drainage capacity of the Pungi canal is insufficient in some places, which contributes to flooding in a number of areas. The drainage capacity of the Kerunga canal is sufficient to carry the storm water it receives and hence no flooding occurs in the Kerunga canal catchment. The simulated inundated area is around 166 ha, which falls mainly in the Pungi canal catchment under this scenario (Figure 14).

3.2.2. Existing situation with design event and climate change. Due to climate change, the run-off during the monsoon season would increase in 2030 and even more in 2050, which would increase inundation over the Sc-0 scenario. The simulated inundated area is almost the same in 2030 and in 2050, being roughly 175 ha, meaning that an additional 9 ha area, in addition to the 166 ha flooded area under Sc-0, would be inundated in 2030 and 2050 due to climate change, if no adaptation measures are taken.

3.2.3. Rehabilitation, expansion and re-sectioning of drainage channels. To improve the drainage situation from the baseline scenario, rehabilitation/expansion/re-sectioning of the existing inadequate

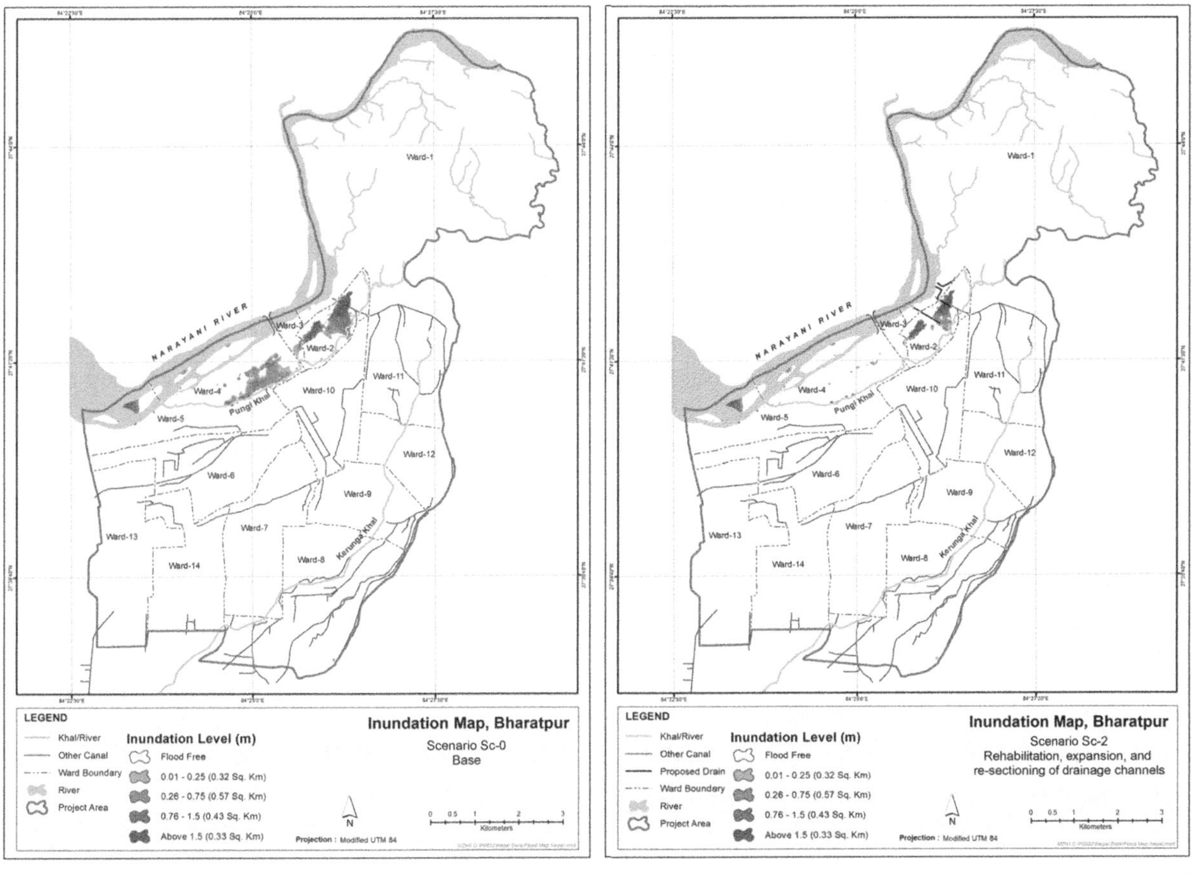

Fig. 14. Flood maps for scenarios Sc-0 and Sc-2, Bharatpur.

canals and the construction of some new drains in the Pungi canal system were incorporated into scenario Sc-2. The bottom level of the canal becomes silted up in some places, or the channel sections become narrow due to encroachment. While designing the section, a gentle slope is maintained along the canal and widened at places where necessary. These improvement works would increase the canal's water-carrying capacity and would make 94 ha land free of waterlogging in the Pungi catchment area, and the simulated inundated area would be about 72 ha (Figure 14). The areas that are flooded are mainly agricultural lands at a lower elevation.

3.2.4. Physical interventions and climate change. Climate change impacts in 2030 and 2050 over the improved scenario (Sc-2) were simulated in scenario Sc-4. The extent of flooding and the duration of inundation both increase slightly over scenario Sc-2 due to the additional run-off under the climate change scenario (change in rainfall pattern). The simulated inundated area was found to be 77 ha in the Pungi catchment in both 2030 and 2050. Thus, the area to benefit would be decreased (over scenario Sc-2) by just 5 ha in 2030 and 2050. As the additional inundated area is not significant, no additional structural improvement is suggested for the future.

3.2.5. Trends in economic growth in Bharatpur. Due to socioeconomic developments, land use patterns are changing continuously in Bharatpur. However, there is no land use plan for the near future in Bharatpur, even though the population has been growing at about 6% per year during the last decade. Agricultural land and vacant land is being rapidly converted into residential, commercial and service areas as development and population growth continues. The major reasons that Bharatpur attracts growing numbers of people are its proximity to Kathmandu, the capital city, and moderate year-round weather, with almost all facilities needed for urban living (health service facilities, education institutions, businesses, employment opportunities, industries and transportation, including quick air links to the major cities, Kathmandu and Pokhara). Due to rapid expansion of the built-up areas, the impervious surface area would increase by 40% in 2030 and by 70% in 2050 above the present situation, as simulated under scenario Sc-5. Considering these factors, simulated results show an increased run-off and accordingly the drainage situation was simulated under this scenario. The simulated inundated area covers 77 ha in 2030 and 86 ha in 2050, slightly higher than scenario Sc-2. Due to the increased run-off, the inundation-free area would decrease by 5 ha in 2030 and 14 ha in 2050. These areas are mainly agricultural areas. With land use changes, these low-lying areas would eventually be urbanized and developed after landfill. Stormwater drains of proper design and connected to existing canals need to be implemented to avoid waterlogging in newly developed areas.

3.2.6. Trends in economic growth and climate change. Climatic change with future socioeconomic development (as described above) is simulated in scenario Sc-6. Economic development in Bharatpur would create larger impervious areas in future and climate change would aggravate the situation by increasing the rainfall amount. The simulated inundated area would be 79 ha in 2030 and 94 ha in 2050 – higher than scenario Sc-5 due to the increase in rainfall–run-off. The inundation-free area would decrease further by 2 ha in 2030 and 8 ha in 2050 from Sc-5. However, the increase in the flooded area is not significant, so no improvement is recommended under scenario Sc-6. If the land is developed as discussed above, the extent of flooded areas would reduce in due course. An additional risk faced by Bharatpur is riverbank erosion of the Narayani River, which requires a proper embankment. However, this issue has not been considered in this study.

3.2.7. Solid waste dumping in the drainage system. There is no field level or survey data for Bharatpur regarding reduced depth of drainage due to the dumping of solid waste in the existing canals. Considering the rapid population growth in Bharatpur, it is assumed that the decrease in depth due to the dumping of waste in the canals would be 10 cm in the Pungi and Kerunga system per year (in Sylhet it is between 10 and 20 cm per year and we assumed the lowest effect of 10 cm in the absence of historical records data for Bharatpur). Solid waste dumping without any cleaning or dredging operations in the expanded canals and new drains (Sc-2) after five years was simulated under scenario Sc-7. It is estimated that the simulated inundated area is 99 ha. Therefore, the additional area that would get waterlogged is approximately 27 ha more than in scenario Sc-2.

A proper solid waste management system was considered under scenarios Sc-2 to Sc-6. The important infrastructural changes in Bharatpur for drainage improvement include: (i) the rehabilitation/expansion/re-sectioning of the existing drainage system in the Pungi canal; (ii) the construction of new drains in the areas where new development is occurring; and (iii) soft measures for improving solid waste management. The extent of flooding and the cost of investment for each adaptation scenario are given in Table 6.

4. Discussion and conclusions

The existing drainage networks, waterlogging and flooding risks were studied under different scenarios in two medium-sized Asian cities. Results from the simulation exercises show that improving the existing drainage canals through rehabilitation/expansion/re-sectioning and the construction of new canals reduces the flooding risk in the short term, but without proper management of solid waste, the flooding risks increase significantly in five years. With engineering solutions, soft measures such as changing human behaviour to properly manage solid waste and not dispose of it in canal and drainage systems would help the cities reduce flooding and waterlogging risks. This cheaper but more effective solution could potentially save city resources that otherwise would be needed to reduce urban flooding and waterlogging.

In order to properly manage a city's solid waste, an improvement in the collection and disposal system is needed. City governments should improve the current practice of non-segregation at source in order to properly managing solid waste. During our interactions, city residents suggested that they would be willing to segregate waste (organic, plastics, metals, bottles and the rest), recycle and dispose properly if the waste collection service also provided systematic collection of the waste from households (putting organic waste separately from recyclable and other waste). A social campaign regarding solid waste management would be an important accompaniment, along with awarding waste collection and management contracts to private firms that are willing to compost the organic waste, and to recover and recycle useful materials.

Sylhet city faces the issue of backwater flow from the Surma River which requires structural interventions in addition to improving the existing drainage canals and proper solid waste management. Such interventions help prevent about 322 ha of land from flooding. The structural interventions include the construction of new drains, reconstruction of culverts, construction of silt traps and regulators. The cost of such improvements is estimated to be at least USD 75.08 million (based on the project estimates that Sylhet city has been working on). This is about USD 233,000 per hectare in terms of public investment. Under the climate change scenario but without socioeconomic development (Sc-4), an additional public investment of USD 11.25 million is required to prevent flooding and waterlogging, although the area that would benefit would reduce by 29 ha from the 'no climate change' scenario.

Under scenario Sc-5, no additional public investment is required because this scenario assumes that, with improvements in economic conditions, city residents will make investments (using landfill measures) to make their land flood-free. This means that the city will gain an additional 55 ha of land without any additional burden on the public exchequer. Private owners will recover their costs, however, through the increased value of their land. For landowners, it is an investment to increase the value of their land.

Under the climate change scenario, there is a need for additional investment in Sylhet to make more land flood-free because climate change will bring in additional rainfall (Sc-6). The estimated cost of adaptation under this scenario is around USD 11.25 million. However, such measures will only make another 3 ha of land flood free, resulting in a large average cost of USD 3.75 million per hectare.

In Bharatpur, the estimated cost of improving the existing drainage channels and constructing three new drains is USD 0.38 million, which would reduce flooding over 94 ha land in the Pungi canal catchment. With urbanization, more new drains need to be constructed to avoid local flooding – which is not planned and has not been analysed in this paper.

The simulation results of scenario Sc-7 show that the carrying capacity of the drainage channels would reduce due to the poor management of solid waste even after the engineering solution. The mismanagement of solid waste would possibly result in an additional 47 ha of flooding in Sylhet and an additional 27 ha flooding in Bharatpur within five years, despite the additional investment in drainage expansion. In order to prevent such a situation, awareness-building of city dwellers, the continuous monitoring of waste management and maintaining the drainage conveyance capacity are all needed in both cities. If households cannot be motivated to practice proper solid waste management, and if they continue disposing of solid wastes in open spaces and in drainage canals, the carrying capacity of the drainage canals will be reduced and the benefit of additional investment in the drainage infrastructure will be neutralized. Our results suggest that, without the proper management of solid waste, the cities cannot maintain a congestion-free drainage environment under the climate change risk.

To be more precise, under the current scenario, about 22.3% of the land area in Sylhet and 12.7% of the land area in Bharatpur is under flooding risk. The flood risk area can be as low as 3.6% in Sylhet and 5.5% in Baratpur with structural interventions combined with proper solid waste management. However, the area under flood risk goes back up to 18.5% in Sylhet and 7.6% in Bharatpur in five years if solid waste is not managed properly, suggesting that the structural solutions without properly integrating effective solid waste management become almost ineffective in reducing flooding risk in these cities.

Under the proper solid waste management scenarios (Sc-2 to Sc-6), we assumed that the capacities of the city councils in both cities are improved, and that 100% of the solid waste is either placed in a suitable dumpsite or composted properly as organic fertilizer and the remaining non-recyclable waste is properly managed. The strategies proposed in the scenarios include: (i) at-source separation of kitchen waste by residents; (ii) managing plastic properly; and (iii) local composting of organic waste or disposing of it properly in landfill sites by the waste collectors. These options were discussed with the concerned officials in both cities and are feasible to implement within reasonable and short time periods.

Acknowledgments

This work was carried out by the Himalayan Adaptation, Water and Resilience (HI-AWARE) consortium under the Collaborative Adaptation Research Initiative in Africa and Asia (CARIAA) with financial support from the United Kingdom's Department for International Development (DFID) and the International Development Research Centre (IDRC), Ottawa, Canada. This research was jointly carried out by the South Asian Network for Development and Environmental Economics (SANDEE), Asian Center for Development (ACD) and Institute of Water Modelling (IWM) with the financial support from the International Development Research Center (IDRC), Ottawa, Canada (Grant #08283-001), which the authors gratefully acknowledge. The authors would like to acknowledge David Molden and Anjal Prakash, the editors of the special issue, for their support at different states of manuscript development and insightful comments; two anonymous reviewers for their critical comments that improved the quality of this research; and the Himalayan Adaptation, Water and Resilience (HI-AWARE) initiative of the International Center for Integrated Mountain Development (ICIMOD) for making it an open access. The authors received support in the field from Birat Ghimire and city officials from Bharatpur and Sylhet. Rajesh Rai, Madan Khadayat and several researchers/enumerators provided support during the inception meeting, focus group discussions and the fieldwork.

References

Alderman, K., Turner, L. R. & Tong, S. (2012). Floods and human health: a systematic review. *Environment international 47*, 37–47.

Booth, D. B. (1991). *Urbanization and the Natural Drainage System: Impacts, Solutions, and Prognoses*. See: https://digital.lib.washington.edu:443/researchworks/handle/1773/17032

Clemens, F. H. L. R. & Veldhuis, J. A. E. T. (2010). *Quantitative Risk Analysis of Urban Flooding in Lowland Areas* (unpublished doctoral thesis). Retrieved from http://resolver.tudelft.nl/uuid:ef311869-db7b-408c-95ec-69d8fb7b68d2

Douglas, I., Alam, K., Maghenda, M., Mcdonnell, Y., Mclean, L. & Campbell, J. (2008). Unjust waters: climate change, flooding and the urban poor in Africa. *Environment and Urbanization 20*(1), 187–205. https://doi.org/10.1177/0956247808089156.

Government of the People's Republic of Bangladesh (2010). *Master Plan for Sylhet Divisional Town (2010–2030). Urban Development Directorate*, Government of the People's Republic of Bangladesh, Dhaka, Bangladesh.

Haque, A. K. E. (2013). *Reducing Adaptation Costs to Climate Change Through Stakeholder-Focused Project Design: The Case of Khulna City in Bangladesh (No. G03520)*. International Institution for Environment and Development, London. Retrieved from http://pubs.iied.org/G03520/?k=enamul.

Huong, H. T. L. & Pathirana, A. (2013). Urbanization and climate change impacts on future urban flooding in Can Tho city, Vietnam. *Hydrology and Earth System Sciences 17*(1), 379–394.

Institute of Water Modelling (2016). *Assessment of State of Water Resources*. Water Resource Planning Organization, Ministry of Water Resources, Government of the People's Republic of Bangladesh.

Intergovernmental Panel on Climate Change (IPCC) (2007). *Climate Change 2007: Synthesis Report. Contributions of Working Groups I, II, and III to the Fourth Assessment Report of the Intergovernmental Panel on Climate Change*. IPCC, Geneva.

Intergovernmental Panel on Climate Change (IPCC) (2014). *Climate Change 2014: Synthesis Report. Contributions of Working Groups I, II, and III to the Fifth Assessment Report of the Intergovernmental Panel on Climate Change*. IPCC, Geneva.

Intergovernmental Panel on Climate Change (IPCC) (2018). *Global Warming of 1.5°C: Summary for Policy Makers*. IPCC, Incheon, Republic of Korea.

Lele, S., Srinivasan, V., Thomas, B. & Jamwal, P. (2018). Adapting to climate change in rapidly urbanizing river basins: insights from a multipleconcerns, multiple-stressors, and multi-level approach. *Water International 43*(2), 281–304. DOI: 10.1080/02508060.2017.1416442.

Nchito, W. S. (2007). Flood risk in unplanned settlements in Lusaka. *Environment and Urbanization 19*(2), 539–551. https://doi.org/10.1177/0956247807082835.

Patz, J. A., Campbell-Lendrum, D., Holloway, T. & Foley, J. A. (2005). Impact of regional climate change on human health. *Nature 438*(7066), 310.

Phanuwan, C., Takizawa, S., Oguma, K., Katayama, H., Yunika, A. & Ohgaki, S. (2006). Monitoring of human enteric viruses and coliform bacteria in waters after urban flood in Jakarta, Indonesia. *Water Science and Technology 54*(3), 203–210.

SANDEE and ACD (2017). *Climate Smart Cities: Research Into Action*. Retrieved on 13 April 2018, from http://acdonline.org/climate-smart-cities-research-action/

Sayenju, N. & Talchabhandari, L. (2008). *Resource Maps and Spatial Profile of Bharatpur Municipality*, Vol. 5. GENESIS/Department of Urban Development and Building Construction, Government of Nepal, Kathmandu.

Srinivasan, V., Thomas, B. K., Jamwal, P. & Lélé, S. (2013). Climate vulnerability and adaptation of water provisioning in developing countries: approaches to disciplinary and researchpractice integration. *Current Opinion in Environmental Sustainability 5*(3–4), 378–383. doi:10.1016/j.cosust.2013.07.011.

ten Veldhuis, J. A. E., Clemens, F. H. L. R., Sterk, G. & Berends, B. R. (2010). Microbial risks associated with exposure to pathogens in contaminated urban flood water. *Water Research 44*(9), 2910–2918.

Urban Development Directorate (UDD) (2010). *Master Plan for Sylhet Divisional Town (2010–2030), Structural Plan, Urban Area Plan and Detailed Area Plan*. June 2010. UDD, Ministry of Housing and Public Works, Government of Bangladesh, Dhaka.

Zurbrugg, C. (2003). *Solid Waste Management in Developing Countries*. SWM Introductory Text on www.Sanicon.Net, 5.

Received 20 August 2018; accepted in revised form 6 November 2018. Available online 8 February 2019

Equity impacts of informal private water markets: case of Kathmandu Valley

Aditi Raina[a], Yogendra Gurung[b] and Bhim Suwal[c]

[a]*Corresponding author. Infrastructure, PPPs and Guarantees (IPG) Group, World Bank, 10 Marina Boulevard, #12-01, Singapore 018983. Email: rainaditi@gmail.com*

[b]*Central Department of Population Studies, Tribhuvan University, Kirtipur 44618, Nepal*

[c]*Faculty of Humanities and Social Sciences, Tribhuvan University, Kirtipur 44618, Nepal*

Abstract

South Asia is a hotspot for populations and economies adversely impacted by poor water security. This is evident in the case of Nepal where it has been estimated that 20% of households have no access to a domestic water source and two-thirds of the urban households live with inadequate water supply. Therefore, many depend on private solutions, such as private wells and purchasing water from informal water markets, to meet household water needs. Within this context, this paper examines whether private water vendors provide equitable access to both poor and wealthy households, whether they practice discriminatory pricing and whether poor households face a greater financial burden in meeting their household water needs. The analysis uses primary data collected from a 1,500-household survey conducted in 2014 and uses regression analyses to derive the results. The results reveal patterns of inequity in the private water market, but seemingly not purely due to an inherent bias against poorer households. Regardless, the market does not serve the poor adequately and given the lack of alternatives that poor households have, it points to an urgent need for the government to step in to counterbalance the deficiencies of the market.

Keywords: Equity; Informal water market; Kathmandu Valley; Nepal; Water vendor

1. Introduction

Water supply is an essential service which, at the most basic level, requires that every resident receives safe, adequate and affordable water. It is imperative that the disadvantaged are not excluded from these services (Bird & Busse, 2006). Yet, in South Asia, no country has universal coverage for its urban population and inequity is estimated to be the highest within the Asia and Pacific region

doi: 10.2166/wp.2018.138

(AWDO, 2013). The gap between urban services that are delivered and those that are needed has for a long time been met by the informal sector (Portes & Haller, 2010). Empirical studies have shown that private water vendors serve a significant percentage of urban residents in Africa, Asia and South America (Conan, 2004). However, the question has remained whether these informal service providers ameliorate the situation by filling the void or whether they instead exacerbate the inequity already existing in the public delivery system by taking advantage of the desperate situation, especially for those who lack alternative ways of obtaining the essential resource of water.

In the urban water supply sector, the formal providers are typically the government utilities or private companies that are officially contracted by the government to distribute networked water (Allen *et al.*, 2006). Informal providers include the diverse range of water vendors who sell water to households, outside the formal piped network system, through various means, such as tanker trucks, push carts, donkey carts, or their own piped network (Snell, 1998; Sansom, 2006; Bakker *et al.*, 2008; Kjellén & McGranahan, 2009). Most of these private players operate in the informal system because they do not fall under government purview or are not subject to any regulatory frameworks or environmental/quality standards (UN Habitat, 2010; Allouche, 2011).

Private water vendors present a critical policy challenge. On the one hand, they are greatly beneficial to society as they reach customers in some of the world's most difficult to reach areas. They also provide them access to an essential service that the government is unable to provide. Even though not all customers who utilize the services of water vendors are poor (Komives *et al.*, 2001), in many cities, it is the low-income households that are worst served by the utility and most dependent on the private water vendors (Pangare & Pangare, 2008; Banerjee & Morella, 2011). On the other hand, anecdotal evidence suggests that these vendors exploit the urban poor and make large profits (Whittington *et al.*, 1989; Cairncross & Kinnear, 1991). In fact, households without piped water often pay more per unit price of water than those connected to the piped network, because alternative sources of water tend to be costlier than the subsidized government water supply (Whittington, *et al.*, 1991; Cairncross & Kinnear, 1992; McPhail, 1994). Further, being an unregulated market, the potential risk of over-exploitation of the resource is larger since the volume of water extracted is not monitored. Gerlach & Franceys (2012), in their analysis of regulatory experience of the water service sector in 11 metropolitan cities, found that any form of oversight over the operations of water vendors was limited and not enforced, especially in terms of price monitoring. In addition, specifically in India, Jordan and Indonesia, they found that these informal businesses were associated with over-extraction of groundwater and intrusion of seawater into freshwater aquifers.

While reports and studies recognizing the significant role played by small private water vendors in urban water supply have grown over the last few years, they remain largely based on anecdotal evidence and case studies, rather than on rigorously collected large-scale primary data (Kariuki & Schwartz, 2005; Opryszko *et al.*, 2009). Further, hardly any aim to empirically assess equity impacts on the market, such as whether vendors provide equal access or charge the same price to households irrespective of their socio-economic status. It is essential to examine whether the private water vendors perform better than the public utility from an equity perspective because, most often, the urban poor are unable to afford their own private solutions to water scarcity. Therefore, their dependence on vendors should not translate into lower-income households being further exploited.

Nepal is one of the fastest urbanizing nations, with an average urban population growth rate per year of 6% since the 1970s (UNDESA, 2012). However, this growth has been taking place in the context of a decade-long Maoist insurgency (1996–2006) which caused an influx of migrants from rural areas to the capital city of Kathmandu, a trend that the government has been unable to cope with. Extreme unplanned development has seen Kathmandu grow to an urban sprawl of 2.7 million people (Central Bureau of Statistics, 2012) with high levels of air and water pollution, a looming energy crisis, increasing traffic congestion, inadequate water supply, solid waste disposal and sewerage systems. The rapid increase in the valley's population during the last decade and a lack of investment in water supply provision and system maintenance have resulted in less than 20% of the population receiving a reliable supply of piped drinking water (ADB, 2010). The daily demand for water in the valley is around 360 million litres, but the supply is much less – approximately 76 million litres a day in the dry season and 123 million litres a day in the rainy season (KUKL, 2014). In light of this, unregulated private vendors have stepped in to fill the water demand–supply gap and therefore, an extensive and vibrant private water market exists.

Using primary data based on a 1,500-household survey conducted in Kathmandu, Nepal, in 2014, the broad purpose of this paper is to examine the equity impacts of the informal water markets, i.e. the manner in which water is made available, allocated and priced in the private market. Many cosmopolitan cities increasingly show a similar reliance on water vending activities to meet the water demand of their residents. Hence, lessons learned from this analysis could help further policy planning in other developing cities.

Specifically, this paper looks at three key issues:

1. Does the private water market provide equitable access to households across different socio-economic indicators?
2. Do the private water vendors practice discriminatory pricing and charge the poor a higher price for the same water?
3. Do poorer households bear a greater financial burden in meeting their household water needs?

The paper is structured as follows. The second section reviews the literature and explains the theoretical perspective on equity that is used in this paper and the way it is applied in the private water markets. The third section describes the data collection methods as well as the methodology employed in the study. The fourth section explains the results of the regression analysis, and the last section consists of the discussion of the results as well as the overall conclusion about the equity impacts of the private water market.

2. Operationalizing equity in the context of water markets

From a review of literature, it is evident that the poor, who are not connected to the public water systems, have to purchase water at prices higher than the utility from private players (Brookshire & Whittington, 1993; Crane, 1994; Banerjee *et al.*, 2010). In fact, in one of the earliest studies to examine water vending, Zaroff & Okun (1984) found that households spent a median of over 20% of their income on buying vended water, and that the prices charged by vendors were typically more than 10 times (up to 50 times) the tariff charged by the public utility. Fass (1988) estimated that in Port-au-Prince, the poor were paying some of the highest water prices in the world, ranging from US$1.1 to US$5.5 per cubic metre. Crane (1994) found that in Jakarta, vendors' prices were approximately 30 times more than the public utility price of water per cubic metre. The high price of privately delivered water was assumed to be the result of vendors' abusive pricing practices (Kjellén & McGranahan, 2009). Later studies, which closely examined vendors' initial investments and operating costs, concluded that many vendors made only a relatively modest profit (Whittington *et al.*, 1991; Solo, 1999; Collignon & Vezina, 2001; Gulyani *et al.*, 2005; Kariuki & Schwartz, 2005; Opryszko *et al.*, 2009).

Private water vendors do not possess the same economies of scale as the government water supply network, and, also, do not benefit from any subsidies. Therefore, the mere fact that vended water is more expensive than public utility water does not imply that the water vendors in the private water market follow inequitable pricing strategies. While the studies reported earlier find that the unit price of water charged by water vendors is generally higher than other options (Brookshire & Whittington, 1993; Banerjee *et al.*, 2010; UNDP, 2011), they did not go further to see whether there is any discrimination (on factors such as income, race, gender) in the way vendors price water for different households.

Only three studies have empirically looked at equity issues in the context of informal urban water markets. First, a study by Ayalew *et al.* (2014), based in Kenya and Ethiopia, reported that even though low-income households spent a significantly greater amount of income on water, it was not due to profiteering amongst water vendors, even though they earned large profits in their business. Second, a study by Sima *et al.* (2013) found that vendors in Kisumu, Kenya, do not try to extract the highest price from high demand areas. Instead the prices reflected the ability of consumers to pay in each neighbourhood, where the highest price was charged to businesses in the central market area and in the wealthier neighbourhoods. Lastly, a recent study by Wutich *et al.* (2016) examining the role of vendors in squatter settlements of Cochabamba, Bolivia, found that vendors are concerned about distributive justice and perform better when unionized. Therefore, this study aims to contribute to knowledge on the equity impacts of the private water vending market, which is currently an under-studied area, because it has important implications for managing urban water resources and ensuring that every citizen receives their basic right to safe, affordable and adequate water.

An empirical estimation of equity can only take place if the term 'equity' itself is clearly defined. In the literature, there is no standard way of operationalizing equity. In the context of water distribution by the private water market, this paper uses the concept of distributional justice that refers to the socially just allocation of goods in a society. Fair allocation takes into account the total amounts of goods that are distributed, the procedure of distribution as well as the pattern of distribution that results.

The principles of distributive justice used here are based on the works of the Greek philosopher Aristotle, who distinguished between horizontal and vertical equity (Roemer, 1998). *Horizontal equity* refers to the idea that people who are equal with respect to certain relevant characteristics should be treated equally. Therefore, horizontal equity occurs when all similar individuals are treated the same, implying an absence of discrimination (Wagstaff & van Doorslaer, 2000) on the grounds of gender, race, profession, etc. When applied to the case of private water markets, horizontal equity in water service provision by private vendors would mean equal physical access to the service across different communities and varying income levels. For horizontal equity to exist all households should have an equal chance to purchase water, should they need to, and there should be no difference in access based on household characteristics. This is crucial in developing countries, where service coverage by public utilities is typically low, usually ranging from 15% to 75% of the population (Fauconnier, 1999).

According to the idea of *vertical equity*, people who are unequal in relation to specific and relevant characteristics should be treated unequally, and in proportion to those inequalities. This principle recognizes that not all households are the same and that their starting points relative to other households should be considered in an analysis of equity. In this context, it relates particularly to cost and affordability. This is an essential equity impact to explore in the context of private water markets, where prices can differ across households and, therefore, place financial burdens in differing ways.

Vertical equity typically requires that as an individual's capability to pay increases, so should the net fiscal burden (which can be negative) (Wagstaff & van Doorslaer, 2000). Since many people are unequal with respect to income and wealth (ability to pay), for a water market system to be vertically equitable, those with greater income and wealth should pay a greater proportion of their income to meet their household needs, since water is an essential and non-substitutable good.

In general, the most economically efficient or socially optimal way to price is said to be marginal cost pricing. This refers to setting the price of a product at or slightly above the variable cost to produce it. However, this does not always happen as it depends on the market power of the sellers. It is possible that vendors base their price according to how dependent the consumers are on the water, i.e. which lack access to alternative sources of water. Therefore, if the market is vertically inequitable, then low-income households will pay more per cubic metre because the tanker drivers will respond to the differential willingness-to-pay (and elasticity of demand) by charging the low-income households more to enhance their profits.

An overview of the equity principles and how they are applied in this paper is displayed in Table 1. The two equity principles and their associated guiding questions, indicators and measures are included.

In large metropolitan cities, a significant number of people continue to lack access to the public water utility system (UNDP, 2006). In fact, many slums and other informal settlements are excluded from municipal water infrastructure and are likely to remain so in the foreseeable future (Njiru, 2004; Fox, 2014; Mehta *et al.*, 2014). Therefore, assessing whether the main alternative system for them, i.e. the private water market, is equitable or not is of crucial importance.

3. Data collection and methodology

The data used in this paper were drawn from a household survey conducted in the Kathmandu Valley. The household survey was conducted for 1,500 households across the five municipalities of Kathmandu Valley – Kathmandu, Lalitpur, Madhyapur, Kirtipur and Bhaktapur. It was a part of a re-survey conducted of 1,500 households selected in 2001 based on a multi-stage clustered random sampling

Table 1. Equity principles and their application in the study.

Type of equity	Guiding question	Outcome variable	Equity measure
Horizontal equity	A. Do factors such as income, ethnicity or location of residence mediate household access to tanker water?	Whether the household can be accessed by tanker trucks	Equal access to vended water for those in equal need of vended water
Vertical equity	A. Do the tanker water operators practice discriminatory pricing?	Unit cost per metre cube of water purchased by household	Equal price of water for those in equal situations
	B. Do poorer households spend a greater proportion of their income on purchasing water than wealthier homes?	The proportion of household income spent on purchase of water	Equal (or less) proportion of income spent on purchasing water by low-income households when compared to higher income households

procedure (Pattanayak *et al.*, 2005). Clusters were located using aerial maps provided by the Central Bureau of Statistics for the 1996/97 World Bank Living Standard Measurement Survey for Kathmandu. In three of the five municipalities in the Kathmandu Valley (Kathmandu, Lalitpur and Bhaktapur), a previously conducted complete enumeration of all households was used as the sample frame (SILT Consultants and Development Research & Training Centre, 1999). In Kirtipur and Madhyapur, the 1991 population census was used as the sampling frame. Further details on the sampling strategy, data collection and replacement process can be found in Supplementary Material Appendix 1.

From the equity framework developed above, three outcome variables are relevant: (i) a dummy for whether tanker trucks can access the household's water storage reservoir/overground tank (i.e. the truck can come close enough for water to be delivered by pipes which are usually 30–60 metres long), (ii) the unit cost of water (cost per cubic metre) households pay to purchase water from tankers and (iii) proportion of income spent on meeting household water needs, including both public and private sources. In the study, all these variables were analysed using regression models.

The independent variable consists of the wealth group and the control variables include ethnicity (six major categories as reported in the UNDP Nepal Human Development Report, 2014), location of residence (ward level) and the type of road that is closest to the household (dirt road, narrow brick road or paved road). The household wealth was estimated using a wealth index based on key asset ownership and housing characteristics, following the principal component analysis method of Filmer & Pritchett (1999). It included aspects such as house characteristics (e.g. lot size of the house, number of rooms and ownership of home), construction materials used (e.g. materials for roofs, walls and floors), access to utilities and infrastructure (e.g. type of sanitation facility, electricity and sources of water), and ownership of selected durable assets (e.g. TVs, radios, computers, Internet access, telephones, mobile phones, VCD/DVD players, refrigerators, washing machines, motorbikes and cars). Using the wealth index values, three wealth groups were created – richest 20%, middle 40% and the poorest 40%. An asset-based indicator was chosen for the analysis because, firstly, asset ownership rather than consumption expenditure is considered to be a better estimate of long-run economic status of the poor (Filmer & Pritchett, 2001). Secondly, information collected on assets is believed to be more accurate than that on consumption, thereby reducing the likelihood of measurement error (Labonne *et al.*, 2007). Appendix 2 details the analytical and empirical strategy and explains the choice of regression models as well as other control variables used in the study. Further, Appendix 3 provides the descriptive statistics for all the variables used in the regression analyses. For robustness check, all three regressions were also run with a log of household income (except for the third because income forms a part of the dependent variable) and wealth as a continuous variable.

4. Results

This section is divided into three sub-sections, each looking at the results related to the three key research questions.

4.1. Does the private water market provide equitable access to households across different socio-economic indicators?

Table 2 presents the marginal effects for the regression results of the probit model that looks at whether tanker trucks can access the household's yard. Column 1 includes only a measure of wealth and indicates

Table 2. Factors mediating access to tanker truck water.

Dependent variable – access to tanker trucks (access = 1, no access = 0)							
Variables	(1)	(2)	(3)	(4)	(5)	(6)	(7)
Wealth group: middle	0.095***	0.084***	0.067**	0.065*	0.063*		
	(0.029)	(0.03)	(0.032)	(0.035)	(0.035)		
Wealth group: rich	0.128***	0.107***	0.101**	0.085**	0.082**		
	(0.035)	(0.036)	(0.039)	(0.041)	(0.042)		
Wealth index						0.003	
						(0.003)	
Log of household income							0.020
							(0.022)
Road 2: pitch				0.244***	0.243***	0.242***	0.259***
				(0.045)	(0.045)	(0.045)	(0.045)
Road 3: stone/brick				−0.331***	−0.331***	−0.329***	−0.331***
				(0.043)	(0.043)	(0.043)	(0.043)
Storage					0.010	0.026	0.022
					(0.033)	−0.032	(0.032)
Ethnicity FE		Yes	Yes	Yes	Yes	Yes	Yes
Ward FE			Yes	Yes	Yes	Yes	Yes
N	1,394	1,380	1,335	1,335	1,335	1,335	1,355
Pseudo R-sq	0.009	0.013	0.119	0.236	0.236	0.234	0.243

Notes: Base group for wealth is poor and for road is dirt road. Standard errors in parentheses $^*P < 0.1$, $^{**}P < 0.05$, $^{***}P < 0.01$. FE: fixed effect.

that being in the middle or rich wealth bracket significantly enhances the household's access as compared to poor households. A middle wealth household is likely to have 9.5% greater likelihood of being accessible to tanker trucks, while rich households are 12.8% more likely. Columns 2 and 3 show that this general pattern is repeated even when ethnicity of the household and location of residence are controlled for. Column 4 shows that the type of road closest to the household plays a significant role in access, with living near narrow brick roads leading to a 33.1% decrease in access. The explanatory power of the model with the inclusion of type of road near the household increases drastically, while at the same time, dramatically lowering the predicted effect of wealth on access. Despite this, it is important to note that, being rich enhances access by 6.5% and being of middle wealth enhances access by 8.5%. In Column 5, the dummy on storage is included and does not have a significant effect on access. The complete table for this estimation can be found in Appendix 4, Table A3.

The Hosmer-Lemeshow goodness-of-fit test (with the null being that the model is correctly specified) on the complete model (with a result of chi-square value being 7.44 and an associated P-value of 0.9 with nine groups) reveals that the model has been specified correctly. Additionally, when the Pregibon's link test is conducted, the prediction squared does not have explanatory power and indicates that the independent variables have been correctly specified[1]. Lastly, the Wald test was also

[1] The Pregibon link test is a model specification test that consists of using the predicted values and the predicted values squared as the only predictors for a second-round model. The expectation is that if the model is specified correctly, then the prediction squared would have no explanatory power. The predicted value would, of course, be significantly different from zero because it is the predicted value from the model. If the predicted squared is significant, then it implies the presence of omitted variables.

conducted (chi-square value of 332.85 with an associated P-value of less than 0.00001) which implies that the selected independent variables collectively affect the outcome variable in a statistically significant manner.

4.2. Do the private water vendors practice discriminatory pricing and charge the poor a higher price for the same water?

The regression results in Table 3 reveal that even while controlling for factors such as ethnicity, location of residence and infrastructure like roads, poorer households pay a higher unit price for tanker water. The number of observations drops to approximately 450 households because this analysis takes into account only those households that participate in the private water market. Column 1 shows that wealth does affect the price of tanker water where the rich households pay significantly less for the water they purchase from tanker trucks. This amounts to 155.2 NPR (1.45 USD) per cubic metre less than poor households. Columns 2 and 3 show that there is a significant difference in the amounts charged to different ethnic groups (Table A4 in Appendix 4). Ethnic groups lower on the human development index (as per the UNDP Nepal Human Development Report, 2014) tend to pay less than the higher caste Newars and Brahmins, which shows that there is price discrimination, although in a positive way when it comes to ethnic differences. Column 4 shows that living near narrow brick roads, which in Kathmandu are usually found in the older sections of the city, leads to a higher price charged to households (Table A4 in Appendix 4). This is likely because longer pipes are needed to access the homes in such areas. It further reveals that having a storage tank (overground or underground) reduces the price

Table 3. Factors affecting the unit price per metre cube that households have to pay for tanker water.

Dependent variable: price/m^3 for private tanker water per household (in Nepali Rupees)						
Variables	(1)	(2)	(3)	(4)	(5)	(6)
Wealth group: middle	−73.10	−71.17	−67.29	−35.12		
	(61.44)	(61.91)	(62.54)	(63.7)		
Wealth group: rich	−155.2**	−163.6**	−173.8**	−131.1**		
	(59.64)	(60.00)	(63.04)	(62.8)		
Wealth index					1.01	
					(3.8)	
Log of household income						−35.38
						(36.16)
Storage				−151.48***	−178.29***	−197.13***
				(69.15)	(67.28)	(69.17)
Ethnicity FE		Yes	Yes	Yes	Yes	Yes
Ward FE			Yes	Yes	Yes	Yes
Road FE				Yes	Yes	Yes
Constant	522.0***	543.7***	392.9**	272.52**	283.56***	691.4*
	(51.00)	(54.42)	(52.09)	(77.23)	(73.8)	(398.0)
N	454	454	454	454	454	453
R-squared	0.012	0.022	0.105	0.142	0.134	0.129

Notes: Base group for wealth is poor. Robust standard errors in parentheses $*P < 0.1$, $**P < 0.05$, $***P < 0.01$. FE: fixed effect.

faced by the household, presumably because they are able to purchase in greater quantities. The R-squared of the complete model in Column 4 of Table 5 is also substantially greater than before. However, as before, despite accounting for the effects of other meditating factors like ethnicity and road and storage, wealth is still a significant determinant of the price charged to the household for tanker water. Households from the rich wealth bracket pay 131 NPR (1.2 USD) less per cubic metre of water than poor households.

In order to detect whether there was any misspecification in the model, the Ramsey RESET (Regression specification-error test for omitted variables) was conducted (Ramsey, 1969). The result of the test where the null hypothesis states that the model is specified correctly (f-statistic of 1.25 with an associated P-value of 0.3) confirmed that there were no omitted variables in the model. Additionally, Pregibon's link test was also conducted, the results of which also confirmed that the model is specified correctly.

4.3. Do poorer households bear a greater financial burden in meeting their household water needs?

The results in Table 4 show that poor households spend a greater proportion of their income on meeting their household water needs (accounting for expenditure on both public and private water sources),

Table 4. Factors affecting percentage of income spent on purchasing water to meet household needs.

Dependent variable: percentage of income spent on all purchase costs related to water, including both private and public water sources					
Variables	(1)	(2)	(3)	(4)	(5)
Wealth group: middle	−0.225*	−0.267*	−0.256*	−0.338**	
	(0.134)	(0.148)	(0.154)	(0.152)	
Wealth group: rich	−0.214	−0.279*	−0.272*	−0.382**	
	(0.151)	(0.167)	(0.163)	(0.162)	
Wealth index					−0.0223**
					(0.0112)
Household size	−0.0741***	−0.0682***	−0.0711***	−0.0692***	−0.0994***
	(0.0271)	(0.0256)	(0.0253)	(0.0233)	(0.0258)
Dummy for whether household buys *only* tanker water				0.715***	0.699***
				(0.127)	(0.125)
Dummy for whether household buys *both* tanker and bottled water				1.685***	1.660***
				(0.306)	(0.302)
Dummy for whether household buys *only* bottled water				1.281***	1.262***
				(0.152)	(0.154)
Ethnicity FE		Yes	Yes	Yes	Yes
Ward FE			Yes	Yes	Yes
Road FE			Yes	Yes	Yes
Constant	1.562***	1.473***	0.952***	0.894***	0.862***
	(0.206)	(0.184)	(0.284)	(0.277)	(0.261)
N	1,353	1,353	1,353	1,353	1,353
R-squared	0.010	0.016	0.056	0.130	0.126

Notes: Base group for wealth is poor. Standard errors in parentheses $*P < 0.1$, $**P < 0.05$, $***P < 0.01$. FE: fixed effect.

Table 5. Average unit price of water (NPR per cubic metre) in the dry season by household wealth group and access to government piped connection.

		Connected to government piped network		
Wealth group		*No*	*Yes*	Total
Rich 20%	NPR/m^3	249.82	371.58	366.80
	N	4	98	102
Middle 40%	NPR/m^3	357.20	454.28	448.95
	N	10	172	182
Poor 40%	NPR/m^3	373.04	545.36	522.04
	N	23	147	170

while controlling for household size. Column 1 and Column 2 include only wealth and ethnicity predictor variables and test results show that they are not well-specified models. The Wald chi-square test showed that the independent variables together do not significantly contribute to the determination of the outcome variable. Column 3 shows that when ethnic differences between households and location of residence and roads (which affected price of water as seen in the previous regression result) are controlled for, rich households spend 27.2% less on water when compared to the poor households at 0.1 level of significance. Column 4 includes variables that look at the household's involvement in the private market in terms of what type of water they buy. Here it can be seen that all households that purchase water in the private market, regardless of type, spend significantly more. Again, while controlling for all these factors, households from higher income brackets are likely to spend between 34 and 38% less on meeting their household water needs than poor households. The level of significance at which the coefficient is statistically significant also increases to 0.5 for the wealthier households in the complete model.

The Wald test was conducted to identify the suitability of explanatory variables included in the model. Since the associated *P*-value of the chi-square statistic is zero, it implies that the independent variables contribute significantly as a group to the explanation of the determinants of percentage of income spent of meeting household water needs. The Ramsey RESET also confirmed that there were no omitted variables in the model (f-statistic of 30.39 with an associated *P*-value of 0). Additionally, Pregibon's link test was also conducted, the results which also confirmed that the model is specified correctly.

As a further robustness check, all the three estimations used above were re-run 50 times with randomized wealth groups. The mean of the distribution of the coefficients (results available in Supplementary Material Appendix 5) were found to not be statistically different from zero, illustrating the fact that the wealth effects seen are not due to a random effect and that the findings are robust.

5. Discussion and conclusions

Equity is an important concept that can be used to evaluate the fairness and effectiveness of a distribution system (Rawls, 1972). It attempts to identify what is socially just and aims to address unequal outcomes (Walzer, 1984; Sayed, 2000). The results of this study reveal there is evidence of both horizontal and vertical inequity in the Kathmandu informal water market. This implies that poor households have poorer access to and pay more for the water they purchase compared to rich households. Being among the richest 20% of the households enhances the access to vended water by 8.2% and reduces per cubic metre price of vended water by 1.22 USD[2]. While infrastructure, specifically roads, impedes access to households if they are not paved, and more so if they are narrow brick roads, wealth still has a significant effect on access. In addition, the poor also spend a greater proportion of their income on meeting their household water needs, implying a greater financial burden. Rich households spend 38.2% less on water than poor households. All these results point to a need to bring the informal water economy into a more formal fold in order to counterbalance the deficiencies of the market.

Looking at the price differences between the poor and non-poor households more deeply, it is evident that (as seen in Table 5), in general, households connected to the government water network pay more per cubic metre of tanker water than those not connected. Therefore, this would prima facie imply that vendors do not charge a higher price to those who are more dependent on them.

However, consistently, poor households (whether connected or not to the government network) pay more than the middle and rich households, i.e. between 53–155 NPR more per cubic metre. One explanation for this could be that vendors give a discount for those who purchase in larger quantities. From Table 6, it can be seen that poorer households do purchase a lower quantity of water each time they buy water. This would seem to be a plausible reason for the difference in pricing because the cost of providing a small quantity of water is higher than delivering in bulk quantities.

This explanation is partly supported by the regression results, where households with a storage facility tend to pay less (approximately 151 NPR or 1.42 USD) than those who do not. Purchasing in larger quantities is only possible when a household has sufficient storage space in the form of an overground tank or an underground storage, which is more feasible for wealthier homes, in terms of both physical space as well as associated costs. This resulting inequity is not necessarily due to the profiteering mindset of the vendor, but related to the marginal costs of delivering different quantities of water. This kind of practice is known as second degree price discrimination (Pigou, 1932) where price varies by the quantity demanded and often, larger quantities are sold at a lower unit price. By setting different prices for different groups of people, vendors would be able to capture a greater portion of the total market surplus. While this in itself is not discriminatory towards poor households, the pricing scheme leads to a greater financial burden on the poor that could lead to a trade-off with other important essential expenditures, because they are forced to spend a high proportion of their income on obtaining water. However, it is essential to note that, despite accounting for having storage, the results show that

Table 6. Quantity of water purchased from tanker truck drivers each time in the dry season by different wealth groups.

Wealth group	Quantity of water purchased each time (mean)	Median	Std. Dev.	Freq.
Rich 20%	4,660.8	5,000	3,086.5	102
Middle 40%	3,727.3	3,000	3,165.2	182
Poor 40%	2,559.2	1,000	2,868.0	170
Total	3,499.6		3,138.8	454

[2] Being among the middle wealth group enhances the access to vended water by 6.3%, reduces per cubic metre of vended water price by 0.32 USD and reduces proportion of income spent on meeting household water needs by 33.8%.

the richest 20% of the sample still pay less than the poorest 40%. This implies that there is still some differentiation in the price charged to households based on wealth, which could be due to the lower bargaining power of the poor because of their greater dependence on the vendors. It has already been seen that half of the poorer households consume water in quantities deemed insufficient to meet basic hygiene and consumptions needs. Since water as a good is not substitutable, affordability and access are key criteria of equity.

These results demonstrate that the market does not serve poor households adequately. Poor households tend to have less access to government connections and ability to afford their own private wells, and so exhibit inelastic demand for water and, therefore, have to accept the price that the vendors set. The limitation faced by the poor in the quantity of water they can afford to buy and store exacerbates the inequities since it further raises the price of the water they buy. Thus, taking these results together, it is evident that an inequitable situation exists in Kathmandu whereby the poor households have poorer access to affordable public water and have to rely on the informal private water market, where they pay more than the rich households to buy water, leading to low levels of consumption that can have negative socio-economic and health impacts[3]. Therefore, the results highlight the need for the government to step in and address these inequities because, at the same time, one cannot deny that the vendors are still indispensable providers of water across the different types of households.

The water usage data from this study's survey showed that the vendors are an important source of water for poor households without a piped water connection, along with providing supplementary water to middle and rich income households, because of the gross inadequacy of the public water utility system (Raina, 2017). Richer households tend to use private wells to meet most of their household needs (56% of their water in the dry season comes from private wells). Therefore, without a doubt, these water vendors in the private water market, who are filling the demand and supply gap, are providing an essential service. However, at the same time, the results of this study show that the private water market in Kathmandu is not ideal in the way it currently functions. It seems to favour wealthier households in access and, possibly due to purchasing power differences, charge poor households more per unit of water than richer households. This leads to a vicious cycle, where the consequent prices reduce the ability of poor households to purchase more water and so, they eventually consume less and pay more.

Given the choice, households would always prefer a piped water supply if it is reliable and supplies adequate water, because of ease of access and affordability. This ideal is also not unachievable as has been witnessed in Phnom Penh, Cambodia, where service coverage was enhanced from 38% to 90% of the population and from 10 to 24 hours per day between 1993 and 2008 (Chan, 2009). The Phnom Penh Water Supply Authority was also able to reduce the supply of non-revenue water to 6%, provide subsidies to the poor and achieve successful billing collection of over 99% (Chan, 2009). This requires high governmental commitment as well as engineering solutions, which given the current turbulent political and social issues in Nepal are unlikely in the near future. Therefore, in the short to medium term, optimizing the services of the water vendors is extremely important and would require regulation in order to correct the market inadequacies and enhance consumer welfare across different wealth groups. Water vendors should be incentivized through market instruments such as taxes or subsidies to provide water to more under-served areas. In order to ensure affordability, there are primarily two ways a

[3] The health impacts here refer to the health risks associated with inadequate amount of water for drinking, cleaning or sanitation, not due to poor quality (an aspect that is not measured in this research).

government can regulate prices. One is by enforcing prices that are set by the regulating body. This requires a deep knowledge of the cost structure of the different types of firms, so as to ensure that the imposed tariffs do not drive the vendors out of business. Making water vending financially unsustainable could have an unintended negative consequence of enhancing inequity and making the urban poor significantly worse off without their access to the privately vended water. Therefore, this path requires high government capability. The second method is to use market mechanisms to encourage efficiency gains and reduction in prices by inducing competition. This can be done by making vendors bid for operating licences and permits, providing financial support for new market entrants and possibly setting up a publicized benchmarking regime that would incentivize vendors to improve their performance as well as compete on their pricing structure. This method reduces monitoring and implementation costs, even though it does require technical capability in drafting the contracts and setting up competitive tenders.

Thus, there are several policy implications of this research with regard to improving equity in urban water service delivery and incorporating the informal water vendors into the formal water supply system. However, at the same time, it is important to note that given the numbers of and diversity in the type of water vendors, their small-scale operations and current informal nature, formulating and implementing regulations (even drafting of contracts and holding of tenders) would involve non-trivial costs which would need to be mitigated. Above all, the ecological risk of such regulation should also be analysed by assessing the permissible limits of groundwater abstraction. There are also issues of corruption, weak contract enforcement and rent-seeking behaviour that need to be addressed. Therefore, identifying the exact regulation that would be effective will require further research and strong political capacity to effectively implement the policy instruments.

There are certain limitations in the analysis conducted in this study. First, while roads play a significant role in the access to and price of water, and the type of road captures a lot of the different elements that could affect delivery, the analysis could have been stronger if it had been possible to include the width (along with the type) of road. Secondly, knowledge of exactly what size of storage tank is used specifically for the purpose of storing vended water would have allowed for a deeper analysis of the link between the price and storage, since water storage tanks are used for storing water from all major sources including the government piped connection and private wells. Thirdly, the 2014 sample is wealthier than the average population of Kathmandu. This implies that it is likely the levels of inequity in the current context of Kathmandu are greater than those seen in this study because there has been a surge of migrants living in peri-urban and slum neighbourhoods in the last decade that are not accounted for in our sample. Lastly, the presence of price discrimination in the market could be reflective of market power, which would require anti-trust issues to be addressed by the government. Developing effective anti-trust policies requires understanding of firm behaviour, which is where further research is suggested. Understanding the nature of the market and the profitability levels of vendors is essential to gauge the scope of reducing water prices and enhancing consumer welfare (without forcing them out of business) through various different governmental policies and regulations.

Acknowledgements

This work was carried out by the Himalayan Adaptation, Water and Resilience (HI-AWARE) consortium under the Collaborative Adaptation Research Initiative in Africa and Asia (CARIAA) with financial support from the United Kingdom's Department for International Development (DFID) and the International Development Research Centre (IDRC), Ottawa, Canada. We gratefully acknowledge the funding for the project, provided by the Institute of Water Policy at the Lee Kuan Yew School of Public Policy, National University of Singapore (NUS). We thank Bal Kumar, Dale Whittington, Wu Xun, Jane Zhao, Namrata Chindarkar and Yvonne Chen for their contributions to survey design, management of field work, and comments.

Supplementary material

The Supplementary Material for this paper is available online at https://dx.doi.org/10.2166/wp.2018.138.

References

ADB (2010). *Project Preparatory Technical Assistance. Consultants' Report for Kathmandu Valley Water Supply and Wastewater System Improvement Study*. Asian Development Bank, Manila, Philippines.

Allen, A., Davila, J. D. & Hofmann, P. (2006). *Governance of Water and Sanitation Services for the Peri-Urban Poor: A Framework for Understand and Action in Metropolitan Regions*. Development Planning Unit, London, UK. http://discovery.ucl.ac.uk/52345/1/Allen_Davila_Hofmann_2006_WSS_PUI_book.pdf

Allouche, J. (2011). The role of informal service providers in post-conflict reconstruction and state building. In: *Water and Post Conflict Peacebuilding*. Troell, J., Weinthal, E. & Nakayama, M. (eds). Earthscan, London and New York, pp. 34–42.

AWDO. (2013). *Asian Water Development Outlook 2013: Measuring Water Security in Asia and the Pacific*. Asian Development Bank, Mandaluyong City, Philippines.

Ayalew, M., Chenoweth, J., Malcolm, R., Mulugetta, Y., Okotto, L. G. & Pedley, S. (2014). Small independent water providers: their position in the regulatory framework for the supply of water in Kenya and Ethiopia. *Journal of Environmental Law 26*, 105–128. doi:10.1093/jel/eqt028.

Bakker, K., Kooy, M., Shofiani, N. E. & Martijn, E. -J. (2008). Governance failure: rethinking the institutional dimensions of urban water supply to poor households. *World Development 36*(10), 1891–1915. doi:10.1016/j.worlddev.2007.09.015.

Banerjee, S. G. & Morella, E. (2011). *Africa's Water and Sanitation Infrastructure: Access, Affordability, and Alternatives*. The World Bank, Washington, D.C. http://elibrary.worldbank.org/doi/book/10.1596/978-0-8213-8457-2.

Banerjee, S., Luengo, M. & Keener, S. (2010). *Provision of Water to the Poor in Africa : Experience with Water Standposts and the Informal Water Sector. WPS5387*. The World Bank, Washington, D.C. http://documents.worldbank.org/curated/en/(2010)/07/12593378/provision-water-poor-africa-experience-water-standposts-informal-water-sector.

Bird, K. & Busse, S. (2006). *Pro-Poor Policy: An Overview*. Overseas Development Institute, United Kingdom.

Brookshire, D. S. & Whittington, D. (1993). Water resources issues in the developing countries. *Water Resources Research 29*(7), 1883–1888. doi:10.1029/92WR02988.

Cairncross, S. & Kinnear, J. (1991). Water vending in urban Sudan. *International Journal of Water Resources Development 7*(4), 267–273. doi:10.1080/07900629108722522.

Cairncross, S. & Kinnear, J. (1992). Elasticity of demand for water in Khartoum, Sudan. *Social Science & Medicine (1982) 34*(2), 183–189.

Central Bureau of Statistics (2012). *National Population and Housing Census 2011 (National Report)*. http://unstats.un.org/unsd/demographic/sources/census/wphc/Nepal/Nepal-Census-2011-Vol1.pdf.

Chan, E. S. (2009). Bringing safe water to Phnom Penh's city. *International Journal of Water Resources Development 25*(4), 597–609. doi:10.1080/07900620903306323.

Collignon, B. & Vezina, M. (2001). *Independent Water and Sanitation Providers in African Cities – Full Report of A ten-Country Study. 22270*. The World Bank. http://documents.worldbank.org/curated/en/2001/04/1490088/independent-water-sanitation-providers-african-cities-full-report-ten-country-study.

Conan, H. (2004). *Small Piped Water Networks: Helping Local Entrepreneurs to Invest*. Asian Development Bank. http://www.adb.org/publications/small-piped-water-networks-helping-local-entrepreneurs-invest.

Crane, R. (1994). Water markets, market reform and the urban poor: results from Jakarta, Indonesia. *World Development 22*(1), 71–83. doi:10.1016/0305-750X(94)90169-4.

Fass, S. M. (1988). *Political Economy in Haiti: The Drama of Survival*. Transaction Publishers, New Brunswick, Canada.

Fauconnier, I. (1999). The privatization of residential water supply and sanitation services: social equity issues in the California and international contexts. *Berkeley Planning Journal 13*(1), 37–73. http://escholarship.org/uc/item/44s7s44v.

Filmer, D. & Pritchett, H. L. (1999). The effect of household wealth on educational attainment: evidence from 35 countries. *Population and Development Review 25*(1), 85–120.

Filmer, D. & Pritchett, H. L. (2001). Estimating wealth effects without expenditure data-or tears: an application to educational enrollments in states of India. *Demography 38*(1), 115–132. doi:10.1353/dem.2001.0003.

Fox, S. (2014). The political economy of slums: theory and evidence from Sub-Saharan Africa. *World Development 54*(February), 191–203. doi:10.1016/j.worlddev.2013.08.005.

Gerlach, E. & Franceys, R. (2012). *Regulating Water and Sanitation for the Poor: Economic Regulation for Public and Private Partnerships*. Earthscan, Abingdon, UK.

Gulyani, S., Talukdar, D. & Kariuki, R. M. (2005). Universal (non)service? water markets, household demand and the poor in urban Kenya. *Urban Studies 42*(8), 1247–1274. doi:10.1080/00420980500150557.

Kariuki, R. M. & Schwartz, J. (2005). *Small-Scale Private Service Providers of Water Supply and Electricity: A Review of Incidence, Structure, Pricing, and Operating Characteristics*. Policy Research Working Papers. The World Bank, Washington, D.C. http://elibrary.worldbank.org/doi/abs/10.1596/1813-9450-3727.

Kjellén, M. & McGranahan, G., International Institute for Environment and Development, and Human Settlements Programme (2009). *Informal Water Vendors and the Urban Poor*. International Institute for Environment and Development, London.

Komives, K. & Cowen, P. J. B. (1998). *Expanding Water and Sanitation Services to Low-Income Households: the Case of the La Paz-El Alto Concession. 178*. Public Policy for the Private Sector. The World Bank, Washington, D.C.

Komives, K., Whittington, D. & Xun, W. (2001). Infrastructure coverage and the poor : the global perspective (No. WPS2551). The World Bank, Washington, D.C. http://documents.worldbank.org/curated/en/191111468739486750/Infrastructure-coverage-and-the-poor-the-global-perspective.

KUKL (2014). *Kathmandu Upatyaka Khanepani Limited, Sixth Anniversary Report*. Kathmandu Upatyaka Khanepani Limited, Kathmandu, Nepal.

Labonne, J., Biller, D. & Chase, R. (2007). "Inequality and relative wealth: do they matter for trust? evidence from poor communities in the Philippines. The World Bank, Washington, D.C. http://documents.worldbank.org/curated/en/2007/03/7536422/inequality-relative-wealth-matter-trust-evidence-poor-communities-philippines.

McPhail, A. A. (1994). Why don't households connect to the piped water system? observations from Tunis, Tunisia. *Land Economics 70*(2), 189–196.

Mehta, L., Allouche, J., Nicol, A. & Walnycki, A. (2014). Global environmental justice and the right to water: the case of peri-urban Cochabamba and Delhi. *Geoforum 54*(July), 158–166. doi:10.1016/j.geoforum.2013.05.014.

Njiru, C. (2004). Utility-small water enterprise partnerships: serving informal urban settlements in Africa. *Water Policy 6*(5), 443–452.

Opryszko, M. C., Huang, H., Soderlund, K. & Schwab, K. J. (2009). Data gaps in evidence-based research on small water enterprises in developing countries. *Journal of Water and Health 7*(4), 609–622. doi:10.2166/wh.2009.213.

Pangare, G. & Pangare, V. (2008). Informal water vendors and service providers in Uganda: the ground reality. *The Water Dialogues: Informal Water Vendors, Uganda*. http://waterdialogues.org/documents/InformalWaterVendorsandServiceProvidersinUganda.pdf.

Pattanayak, S. K., Yang, J.-C., Whittington, D. & Bal Kumar, K. C. (2005). Coping with unreliable public water supplies: averting expenditures by households in Kathmandu, Nepal. *Water Resources Research* 41 (2): W02012–WO2023. doi:10.1029/2003WR002443.

Pigou, A. C. (1932). *The Economics of Welfare*, 4th edn. Macmillan and Co., London.

Portes, A. & Haller, W. (2010). The informal economy. In: *The Handbook of Economic Sociology*. Smelser, N. J. & Swedberg, R. (eds). Princeton University Press, Princeton, NJ, USA, pp. 403–426.

Raina, A. (2017). Inequities in household water consumption in Kathmandu, Nepal. Global Water Forum. http://www.globalwaterforum.org/2017/12/04/inequities-in-household-water-consumption-in-kathmandu-nepal/

Ramsey, J. B. (1969). Tests for specification errors in classical linear least-squares regression analysis. *Journal of the Royal Statistical Society. Series B (Methodological) 31*(2), 350–371.

Rawls, J. (1972). *A Theory of Justice*. Clarendon Press, Oxford.

Roemer, J. E. (1998). *Theories of Distributive Justice*. Harvard University Press, Cambridge, MA, USA.

Sansom, K. (2006). Government engagement with non-state providers of water and sanitation services. *Public Administration and Development 26*(3), 207–217. doi:10.1002/pad.419.

Sayed, Y. (2000). *Post-Apartheid Educational Transformation: Policy Concerns and Approaches*. April. http://eric.ed.gov/?id=ED442177.

SILT Consultants Ltd and Development Research and Training Center (SILT-DRTC) (1999). *Consumer Survey for Project on Urban Water Supply and Sanitation Rehabilitation. Final Report*. Submitted to the Nepal Water Supply Corporation, Kathmandu.

Sima, L. C., Kelner-Levine, E., Eckelman, M. J., McCarty, K. M. & Elimelech, M. (2013). Water flows, energy demand, and market analysis of the informal water sector in Kisumu, Kenya. *Ecological Economics 87*(March), 137–144. doi:10.1016/j.ecolecon.2012.12.011.

Snell, S. (1998). *Water and Sanitation Services for the Urban Poor: Small-Scale Providers – Typology and Profiles. 33106*. The World Bank. http://documents.worldbank.org/curated/en/1998/12/6093452/water-sanitation-services-urban-poor-small-scale-providers-typology-profiles.

Solo, T. M. (1999). Competition in water and sanitation and the role of small-scale entrepreneurs. In: *The Private Sector in Water: Competition and Regulation*. The World Bank, Washington, DC, pp. 33–36. http://www-wds.worldbank.org/external/default/WDSContentServer/WDSP/IB/(2009)/08/18/000333037_(2009)0818013559/Rendered/PDF/499590WP0Box341ition0and0-Regulation.pdf#page=35.

UNDESA (2012). *World Urbanization Prospects, The 2011 Revision: Highlights*. United Nations Development of Economic and Social Affairs, New York, N.Y https://esa.un.org/unup/pdf/WUP2011_Highlights.pdf.

UNDP (2006). *Human Development Report 2006 – Beyond Scarcity: Power, Poverty and the Global Water Crisis*. UNDP, New York. http://www.undp.org/content/undp/en/home/librarypage/hdr/human-development-report-2006.html.

UNDP (2011). *Small-Scale Water Providers in Kenya: Pioneers or Predators?* UNDP. http://www.undp.org/content/undp/en/home/librarypage/poverty-reduction/inclusive_development/small-scale_waterprovidersinkenyapioneersorpredators.html.

UNDP (2014). *Nepal Human Development Report 2014*. UNDP, Kathmandu, Nepal. http://www.np.undp.org/content/nepal/en/home/library/human_development/human-development-report-(2014).html.

UN Habitat (2010). *State of the World's Cities*. Earthscan, London.

Wagstaff, A. & van Doorslaer, E. (2000). Income inequality and health: what does the literature tell us? *Annual Review of Public Health 21*, 543–567. doi:10.1146/annurev.publhealth.21.1.543.

Walzer, M. (1984). *Spheres of Justice: A Defense of Pluralism and Equality*, Reprint edition. Basic Books, New York.

Whittington, D., Lauria, D. T. & Xinming, M. (1989). *Paying for Urban Services : A Study of Water Vending and Willingness to Pay for Water in Onitsha, Nigeria*. WPS363. The World Bank, Washington, D.C. http://documents.worldbank.org/curated/en/1989/03/700561/paying-urban-services-study-water-vending-willingness-pay-water-onitsha-nigeria.

Whittington, D., Lauria, D. T. & Xinming, M. (1991). A study of water vending and willingness to pay for water in Onitsha, Nigeria. *World Development 19*(2), 179–198. doi:10.1016/0305-750X(91)90254-F.

Wutich, A., Beresford, M. & Carvajal, C. (2016). Can informal water vendors deliver on the promise of a human right to water? results from Cochabamba, Bolivia. *World Development 79*, 14–24.

Zaroff, B. & Okun, D. A. (1984). Water vending in developing countries. *Aqua 5*, 289–295.

Received 19 June 2018; accepted in revised form 16 July 2018. Available online 3 September 2018

IWA Publishing's authorised EU representative for General Product Safety Regulations is Diane D'Arras, 15 rue Duret, 75116 Paris, France, e-mail: safety@iwap.co.uk.

Printed and bound by CPI Group (UK) Ltd, Croydon, CR0 4YY
12/05/2026
02108880-0018